Kjell Nilsson • Stephan Pauleit • Simon Bell •
Carmen Aalbers • Thomas Sick Nielsen
Editors

# Peri-urban futures: Scenarios and models for land use change in Europe

Springer

*Editors*
Kjell Nilsson
Center for Forest & Landscape
University of Copenhagen
Frederiksberg
Denmark

Stephan Pauleit
Chair for Strategic Landscape Planning and Management
Technical University of Munich
Freising
Germany

Simon Bell
OPENspace Research Centre
Edinburgh College of Art
University of Edinburgh
Edinburgh
United Kingdom

Carmen Aalbers
Team Stadsregionale Ontwikkeling
Alterra Green World Research
Wageningen
Netherlands

Thomas Sick Nielsen
Department of Transport
Technical University of Denmark
Lyngby
Denmark

ISBN 978-3-642-30528-3 ISBN 978-3-642-30529-0 (eBook)
DOI 10.1007/978-3-642-30529-0
Springer Heidelberg New York Dordrecht London

Library of Congress Control Number: 2012950869



Printed on acid-free paper

Springer is part of Springer Science+Business Media (www.springer.com)

# Preface

The PLUREL project (2007–2011) envisaged research on urban, peri-urban and rural interaction and strategies focusing on sustainable land use changes such as the preservation of green open areas, agricultural lands and other unbuilt areas under urbanization pressures. The research was done through close cooperation with the stakeholders from the case-study regions. For the purpose of better cooperation, a Board of Stakeholders was formed—to follow the research closely and recommend improvements in the way of communication between research and policy.

The research addressed important issues for all the regions with urban cores, peri-urban edges and rural hinterlands. The stakeholders, representing planners, local politicians, NGOs, nature protection organizations or networks of relevant actors, took an active role in the discussions, analysis and choices on the way to the end results.

The expectations were high on both sides. The stakeholders expected user-friendly recommendations, models, scenarios and interactive tools. The amount of work delivered in the end was impressive. The experience, through the interaction between researchers and policy makers, was highly valuable. The two worlds with different languages and perceptions came closer to each other. Ideas were freely exchanged and common strategies worked out.

PLUREL offers a wide range of products which will indisputably have impact on policies. These products include analysis and recommendation on governance and sustainable peri-urban land use, planning and financial instruments in peri-urban areas that will enhance sustainable land use, strategies for growth management, urban containment by conservation and densification, development of consistent blue and green infrastructure, promotion of local production and short transport circles, development of eco-services, and ways of assessing impacts on quality of life. All these are numerous examples of successfully worked out recommendations for preserving and developing the peri-urban areas as open, green and vital areas.

Three European regions from Western Europe, three from the former Eastern Europe and one from China—all with different traditions in planning and land use—learned a lot from each other in the process. The regions which are traditionally strong in planning—Montpellier, The Hague and Manchester could offer their planning schemes and governance strategies as guides in the search for better balance in land use. The regions of Warsaw, Koper and Leipzig could offer their tradition in more social cohesive strategies. China offered decisive measures for vital green spaces and for containment of urban sprawl.

The whole process of attaining the research goals together—researchers and practitioners—seems to offer a good model for success.

The representative of the political post holders on behalf of the case-study regions in the Board of Stakeholders, Marcel Houtzager, portfolio holder 'Green, recreation and tourism' in The Hague Region, actively participated in the whole process. He is satisfied with the good communication between research and practice and with the overall results of the project. 'I was confident when we started the process in PLUREL and I was confident when we finished it. Now we have excellent end-products, ready to use. PLUREL was a very complex project about a complex problem—the interaction between urban, urban-peri and rural areas and the preservation of landscape identities, balancing environmental and social values with economic interests. The results as we see them presented in this book show how successful researchers and practitioners have been in analyzing the problems, communicating well to all the parties involved, and recommending ways to solve those problems.

I strongly hope that this book will help in the search of effective methods for preserving open green space and improving the balance between urban and peri urban areas. This is needed for our quality of life, the climate change challenges and the economic viability of our regions. We live in demanding times and this book offers some of the recipes to find a way between economic prosperity and environmental healthy attitudes.'

The Hague
March 2012

Marcel Houtzager

# Acknowledgements

The book would not have been possible without the commitment of all of its authors and more widely the entire team of researchers involved in the Plurel project. It was not possible to include the large number of researchers that participated in the project in the writing of this book, nor can we name them all, but they all deserve to be acknowledged and thanked for their valuable contribution to the project.

The Board of Stakeholders should be particularly highlighted for its enthusiastic support of the project, especially in providing feedback to the project's output from the practitioners' perspectives. It was appointed from policy makers at the EU level, professional organizations, NGO's and/or representatives from the case-study regions. Members of the Board were Tomasz Sławiński, Head of Office for Regional Planning, Mazovian Province; Marcel Houtzager, Regional Portfolio Holder on Finance, The Hague Region; Pam Warhurst, CBE, Chair of Forestry Commission; Jean-Paul Gambier, Chef du Service Foncier, Montpellier Agglomeration; Andrej Medved, Director of Birdlife Slovenia. Equally, the members of the project's Scientific Advisory Board played an important role by their constructive advice throughout the project and in particular at critical times. The Board members were: Prof. Hilda Blanco, University of Washington; Prof. Christer Bengs, Swedish University of Agricultural Sciences; Dr. Kai Böhme, SWECO EUROFUTURES AB.

We also would like to thank the internal and external reviewers that helped to greatly improve the chapters of this book, these were: Prof. Hilda Blanco, University of Washington; Prof. Adrienne Grêt-Regamey, ETH Zurich; Dr. Erik Koolman, University of Delft; Prof. Andrzej Lisowski, University of Warsaw; Prof. Steve Littlewood, Leeds Metropolitan University; Prof. Iris Reuther, University of Kassel; Dr. Emmanuel Negrier, Research Director, CNRS; Prof. Hojca Golobic, University of Ljubljana; Prof. Zhiyong Li, Chinese Academy of Forestry; Prof. Bob Evans, University of Newcastle; and Prof. Christer Bengs, Swedish University of Agricultural Sciences.

Finally, we wish to express our gratitude to the programme officers at the European Commission Karen Fabbri, Julia Acevedo-Bueno and Astrid Kaemena; the reviewers Ole Damsgaard, Michèle Sanglier, Bob Evans, Peter Nowicki and Marion Potschin; and to the publisher for their support in the production process.

# Acknowledgment

# Contents

# Chapter 1
# Introduction

**Kjell Nilsson, Stephan Pauleit, Simon Bell, Carmen Aalbers, and Thomas S. Nielsen**

## 1.1 A Dream that Became a Nightmare

A spectre is haunting Europe. The name of the spectre is not communism (Engels and Marx 1848) but urban sprawl, a phenomenon which has nothing to do with communism, but the metaphor of describing urban sprawl as a spectre is not new, especially in the United States where sprawl is endemic and there is a widespread concern over its environmental and public impact (Schmidt 1998). Urban sprawl can be defined as the low-density expansion or leapfrog development of large urban areas into the surrounding rural land. Several definitions also include the fact that the development takes place without systematic regional land use planning and control (e.g. EEA 2006; Reckien and Karecha 2007).

The origin of urban sprawl is the so-called American Dream. The term was first used by James Truslow Adams in his book The Epic of America (1931). Although

K. Nilsson (✉)
Center for Forest & Landscape, University of Copenhagen, Rolighedsvej 23, 1958 Frederiksberg, Denmark
e-mail: kjni@life.ku.dk

S. Pauleit
Chair for Strategic Landscape Planning and Management, Technical University of Munich, Emil-Ramann-Str. 6, 85354 Freising, Germany

S. Bell
OPENspace Research Centre, Edinburgh College of Art, University of Edinburgh, Lauriston Place, Edinburgh, EH3 9DF, United Kingdom

C. Aalbers
Team Stadsregionale Ontwikkeling, Alterra Green World Research, Droevendaalsesteeg 3, 6708 PB Wageningen, Netherlands

T.S. Nielsen
Department of Transport, Technical University of Denmark, Bygningstorvet 116B, 2800 Kgs. Lyngby, Denmark

K. Nilsson et al. (eds.), *Peri-urban futures: Scenarios and models for land use change in Europe*, DOI 10.1007/978-3-642-30529-0_1,

Adams himself argued that the American Dream "... is not a dream of motor cars and high wages merely, but a dream of a social order in which each man and each woman shall be able to attain to the fullest stature of which they are innately capable, and be recognized by others for what they are, regardless of the fortuitous circumstances of birth or position." (Adams 1931, p. 404), no American Dream has broader appeal than home ownership (Cullen 2003).

Today we see the consequences of a dream that might turn into a nightmare. Under the Clinton Administration, US former vice president and Nobel prize winner Al Gore led the campaign against urban sprawl. He even predicted that if steps were not taken to curb sprawl, urbanization will consume so much farmland that the United States may run out of enough agricultural land to feed itself in the twenty-first century and, for the first time in the nation's history, become a net importer of food (Carlisle 1999).

During the 10 year period 1990–2000 the growth of urban areas and associated infrastructure throughout Europe consumed more than 8,000 $km^2$, equivalent to 0.25 % of the combined area of agriculture, forest and nature land (EEA 2006). A quarter of 1 % may not seem to be worth worrying about. However, we are talking about an almost irreversible process. Less than 10 % goes the opposite way, i.e. is transferred from urban land into brownfields and only a minor part of these are reclaimed to arable land or nature.

More than 20 years ago, Newman and Kenworthy (1989) studied the relationship between energy consumption and the population density of cities around the world. They showed that resource consumption increased the less densely the city was populated. Low density American and Australian cities with high per-capita energy consumption were at one extreme end of the spectrum while dense Asian cities, with low per-capita energy consumption were at the other end. One main reason for this is that the demand for individual, car-based transport increases in low-density cities. Urban sprawl in combination with increasing levels of per-capita energy consumption leads to an ever increasing urban ecological footprint. For London, it has been estimated that a hypothetical area of more than the size of the entire United Kingdom is needed to keep this megacity alive (Girardet 2004). Clearly, this consumption pattern is unsustainable.

An important driving force behind urban expansion is, of course, the growth of the urban population. However, when talking about Europe, this is not a sufficient explanation. Since the mid-1950s European cities have expanded in size on average by 78 %, whereas the population has grown only by 33 % (EEA 2006). This is not a surprise in densely populated regions like the Randstad in the Netherlands, but even in regions where the population is decreasing, urban areas are still growing, notably in Spain, Portugal and Italy and in eastern Germany. Leipzig-Halle is an example of a region which suffers from both the problems of a shrinking city and urban sprawl (Fig. 1.1). The same trend—that urban areas expand approximately twice as quickly the population—can be seen in the United States and China (Fig. 1.2).

As a consequence of these processes, large urban regions have developed which are continuously spreading outwards along major transport routes, connecting major urban centres with smaller settlements and commercial sites as well as recreational areas. Urban development does not happen in neatly delineated

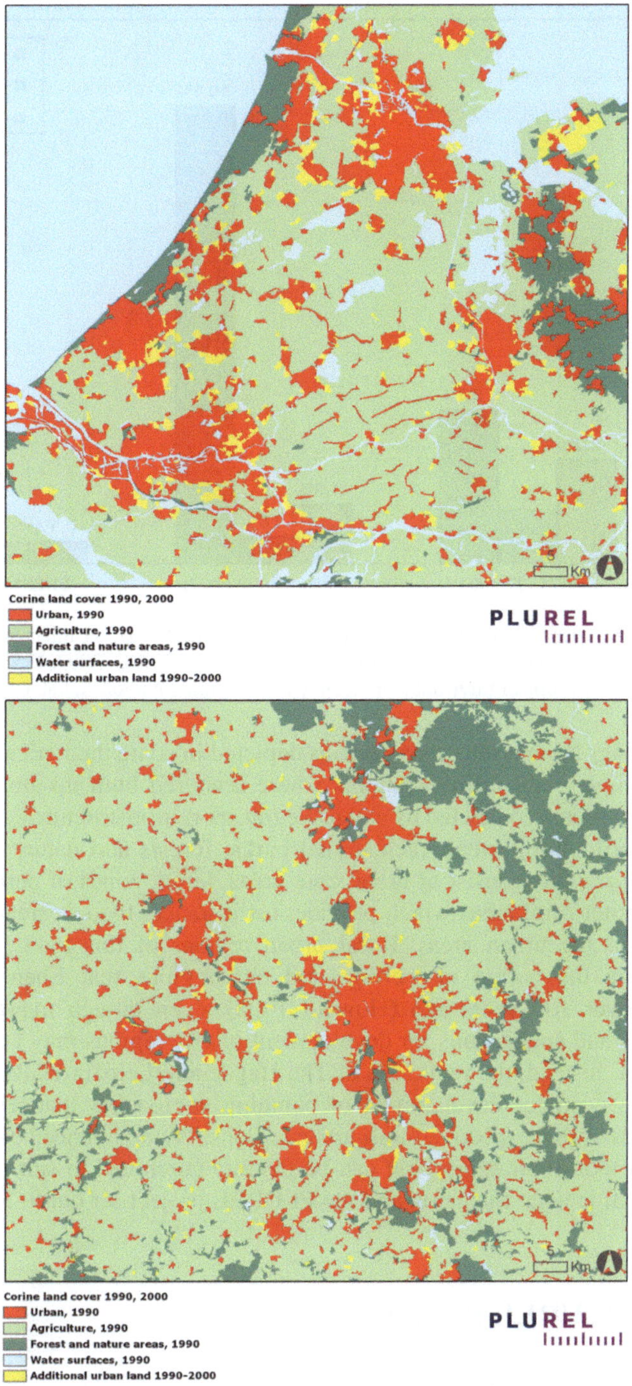


**Fig. 1.1** Urban expansion on the edges of existing agglomeration in the Randstad, Netherlands (above), and the Leipxig-Halle region, Germany (below) (Source: Corine Land Cover Databases 1990 and 2000)

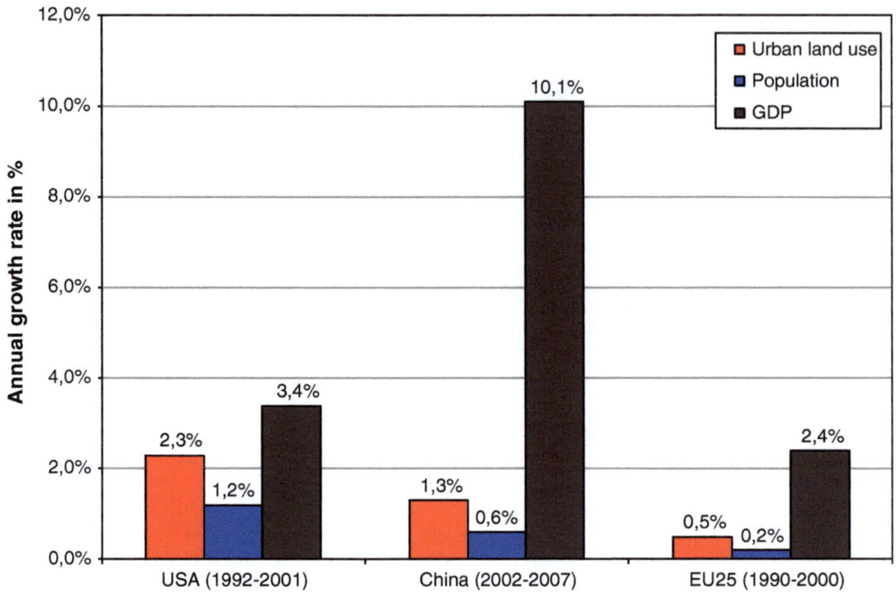


**Fig. 1.2** Growth rates for urban land use, population and GDP in EU, United States and China (Sources: Zhu et al. 2004; The Ministry of Land and Resources P.R.C. 2007; U.S. Department of Agriculture 2003; European Environment Agency 2005; United Nations databases 2009)

concentric rings around the urban core, as depicted in some theoretical models, but in rather complex and diffuse patterns where scattered built up areas of various forms and functions, connected by extensive transport infrastructures, intermix with farmland, horticultural areas, recreational parks, forests and nature areas. These *Rural–urban regions* consist of urban core areas, larger zones of peri-urban areas but also remoter rural areas which provide, for instance, food, recreational spaces and water for the urban areas. Rural–urban regions of this kind have little in common with the notion of well-ordered city regions that Ebenezer Howard introduced over 100 years ago (Howard 1902), even though many of his far-sighted ideas and suggestions for development of city regions may be still worthy of discussion. Instead, the two cases of The Hague and Leipzig-Halle noted above highlight the need for development of planning approaches that respond to the specific needs of the respective rural–urban regions. We do not need one model that fits all but a tool of strategies that can be flexibly combined to suit the particular situation. That is what this book and the PLUREL project set out to achieve.

## 1.2 The PLUREL Project

The focus of the PLUREL project was the sustainable development of land use systems. It tried to answer, at least in part, the question: can resilient land use systems be developed for rural–urban regions that successfully combine compact and resource efficient settlement structures with a multifunctional green infrastructure that provides ecosystem services critical to the quality of life of their inhabitants and which enable them to adapt to various technological, economic, social and environmental changes?

Challenges for the planning of sustainable land use systems in rural–urban regions are particularly pronounced in peri-urban areas where urban and rural land uses closely blend and form a new type of landscape which is neither rural nor urban. Land use dynamics are also highest in these peri-urban areas. While land use change, mainly from rural to urban types, tend to increase the pressure on the environment, e.g. by destruction and fragmentation of natural habitats as well as the loss of valuable agricultural soils, it can also provide the opportunity for establishing new, mutually beneficial relationships between neighbouring land uses, e.g. by providing new sources of income for farmers through production of food, water, energy and recreational spaces for city dwellers. Planning has an important role in enabling such relationships to develop. Therefore, these peri-urban areas have been at the centre of the PLUREL project.

PLUREL was an Integrated Project funded within the Sixth Research Framework Programme of the European Union. Thirty six partners from 14 European countries and China participated in the project which was coordinated by the University of Copenhagen. The project started in 2007 and finished in 2011. It aimed to achieve a deeper understanding of the changing relationships between urban and rural land use with an emphasis on the most dynamic portion, that of peri-urban areas. The project:

- Developed methods and tools to assess the environmental, social and economic impacts of land use changes in rural–urban regions.
- Identified and assessed potential strategies and good practice examples in order to promote the sustainable development of land use systems in rural–urban regions, especially the peri-urban.

The project was interdisciplinary and used multiple methods. Quantitative modelling was based on qualitative land use scenarios and used to develop sustainability impact assessment tools in support of qualitative analysis and assessment of land use strategies. A two-level approach was adopted, that of the pan-EU/national level where broader issues and patterns were evaluated and that of the regional level where real situations of land use change, governance, ecosystem services and quality of life were tested with the participation of local actors and stakeholders.

The project synthesis report *Peri-urbanization in Europe: Towards European Policies to Sustain Urban–Rural Futures* (Piorr et al. 2011) emphasized the pan-European dimension. It concludes with an outline of EU policy options for the integrated development of rural–urban regions. While the European Union has an increasing influence on territorial development, the most important decisions are still made at the local level. In this book, therefore, we offer a complementary perspective by exploring in more detail the land use dynamics within rural–urban regions and the strategies that can be applied to promote their sustainable development. Six European case studies and one reference study in China were selected to explore in depth the land use relationships between rural and urban areas. The case study regions were Warsaw (PL), Leipzig-Halle (DE), The Hague Region (NL), Greater Manchester (UK), Montpellier (FR), Koper (SL), and Hangzhou (CN).

While not statistically representative of all possible models of rural-urban region, the case studies nevertheless reflect the variability of geographic, economic and social conditions prevailing in Europe and are also characterized by different cultures of government and governance. Population trends differ widely between the case study regions and range from growing areas (poly-centric The Hague Region and Montpellier, mono-centric Warsaw) to a region with significant and ongoing shrinkage in population and in city fabric, leading to land use perforation (Leipzig-Halle). The general trends are translated in different ways into rural–urban development patterns within each of the study regions. The Hangzhou reference study explores a very different case: it is a very rapidly urbanizing area of 6.6 million inhabitants in Asia within a top-down political and planning system.

## 1.3 The Structure of the Book

The book is comprised of three parts. The first part introduces the dynamics of peri-urbanization and the theoretical foundations of the project, in particular the concept of *rural–urban regions*. Chapter 2 introduces the definitions of rural urban regions and especially the peri-urban, it identifies the main dynamics found there and introduces the scenarios upon which the modelling of lands use change at a European level was based. It provides the basic context for the later examination of the case studies. Chapter 3 focuses on the breakdown of Europe into different classes of rural–urban regions, presenting a statistical method and the results of the analysis. This further sets the context for the case studies which can be seen as belonging to different classifications, thus enabling the broad sweep of the PLUREL research to be appreciated. Chapter 4 describes the main methods used in the research presented within the individual case study chapters so that readers can focus on the results of each and if they want to find out how the work was done they know where to look.

The second part of the book presents the seven case studies. They follow a common approach to analysis but nevertheless they represent seven individual stories of local trends of peri-urbanisation and response strategies. This part also contains a short introduction so no more will be said about Part 2 at this point, except that it is where the main results of the PLUREL project at the city region level can be found.

The final part of the book then provides a synthesis of the results of the case study work arising out of the PLUREL project. In chapter 13 a comparative analysis of the planning strategies selected and analysed in each case study region is carried out and, while it is not possible to offer solutions which can be applied generally across the whole of Europe, nevertheless a number of lessons can be learnt from this and it is hoped that the results of the PLUREL work will be put to good use by planners and policy makers who read about them in this volume. Chapter 14 examines the government/governance issues in more detail, also making interesting and revealing comparisons of the systems found in each case study. Chapter 15 concludes with an overview and assessment of the achievements of the PLUREL project and the challenges still facing rural–urban regions and the peri-urban landscape.

Coming back to the key question of how to manage urban growth a set of policies for the development of sustainable land use systems in Europe's rural–urban regions is proposed by the PLUREL project as a whole. These aim at better coordination between transport, land use and open space planning, policy integration, containment of urban development by promoting dense cities which, however, also require a green infrastructure that provides essential ecosystem services and improves quality of life in the city. Overall, the need to develop integrated, territorial policies is emphasized which acknowledge the reality of *rural—urban regions* instead of devising segregated urban and rural policies. This also applies to European policies such as the Common Agricultural Policy which does not recognize the specific challenges of peri-urban agriculture. Important directions for future research to support policy making for sustainable rural–urban regions are outlined.

With the publication of this book, the PLUREL project has finally come to an end. It has been a very challenging project, not only because of the theme of this research, but also through the task of integrating 36 partner organisations and many more researchers as well as practitioners very different backgrounds, interests and personalities. It has been an intellectually intriguing and rewarding project for us. We gained new insights into the complexities of rural–urban development. We identified strategies that are applied in rural–urban regions across Europe to steer this development but we also learned about the current barriers and failures to develop more sustainable land use systems in peri-urban areas. Importantly, the concept of the *rural–urban region* has been coined and it has proven in the PLUREL project to be a very useful unit for analysis. Moreover, the methodology applied in this project may be considered to be innovative in linking qualitative work—on scenarios and policy analysis—with quantitative modelling of various types and working across scales and levels of decision-making. By this, we hope that the book can provide insights, tools and inspiration for tackling the daunting challenges of sustainable development in Europe's rural–urban regions.

## 1.4 Other PLUREL Outputs

In addition to this book and a large number of publications in various academic and practice orientated journals, further results from the project can be accessed via the project's website: www.plurel.net. These include in particular:

- Peri-urbanisation in Europe. Towards European Policies to Sustain Urban–Rural Futures. Synthesis Report (Piorr A., Ravetz J., Tosics I, eds.), and a series of newsletters.
- PLUREL XPLORER: a web-based information platform that condenses and presents the knowledge and various products of PLUREL into a form that supports planning and policy discussions on rural–urban land use interactions at a European and regional level. It provides information for planners, practitioners and professionals on processes, problems and places of peri-urbanisation in Europe and its regions.
- IIAT—Integrated Impact Analysis Tool is an internet-accessible tool enables an easy and holistic perception of changes in sustainability indicators, as positive or negative trends according to different scenarios in form of "spidergrams".
- QoLSim, a simulator tool that enables the effects of land use change on residents' perceived quality of life and residential choice decisions to be tested.
- MOLAND-Light—a simplified version of the MOLAND land use modelling software. It can be used to easily and quickly set up a simulation for a region in Europe.

## 1.5 Who Is this Book For?

At the outset of the PLUREL project we kept asking ourselves the question: who is it for? Is it for a handful of European policy wonks? Is it aimed at policy-makers and planners at national, regional or local government level? Does it offer insights of use to developers, infrastructure providers, NGOs or even interested individual residents concerned about where they live? We concluded that at the pan-European level—the level for which the Synthesis Report is aimed—is an important but relatively small target audience in Brussels and in national governmental ministries for spatial planning or environment, for example. The main readership of this book we feel will be the planners and policy makers at the different regional and local levels who are daily faced with complex problems and for whom the time to take a longer strategic view is often a luxury. Finding out what strategies and policies work and which do not is time consuming and often fruitless so that the material offered in this book arising out of the case studies should provide a lot of inspiration as well as practical tools. The fact that the research was carried out with not only the help but also the active participation of local actors and stakeholders should, we feel, give it an added level of credibility among our target readership.

Of course there is also the educational market to be considered and for the students of urban planning, government and governance, land use planning and policy, urban geography, landscape planning, urban ecology and many other fields, the research presented here as well as the academic papers which have also been published should provide valuable theoretical and empirical material of great value not only in Europe but beyond, since, as we have seen, urban sprawl and the peri-urban are in many ways far more pressing problems in city regions outside the borders of the EU.

## References

Adams JT (1931) The epic of America. Little Brown, Boston
Carlisle J (1999) The campaign against urban sprawl: Declaring war on the American dream. National Policy Analysis 239, The National Center for Public Policy Research, Washington DC
Cullen J (2003) The American dream: a short history of an idea that shaped a nation. Oxford University Press, Oxford
Engels F, Marx K (1848) Manifest der kommunistischen Partei. Gedruckt in Office der "Bildungs-Gesellschaft für Arbeiter" von J.E. Burghard, London
EEA (European Environment Agency) (2005) Land cover accounts, LEAC. Based on Corine land cover changes database, 1990–2000. Accessed at http://www.eea.europa.eu/
EEA (European Environment Agency) (2006) Urban sprawl in Europe. The ignored challenge. EEA report 10/2006, Copenhagen
Girardet H (2004) CitiesPeoplePlanet. Wiley Academic, Chichester
Howard E (1902) The garden cities of tomorrow. Swan Sonnenschein & Co., Ltd, London
Ministry of Land and Resources P.R.C. (2007) Communique on land and resources of China 2007
Newman P, Kenworthy JR (1989) Sustainability and cities: overcoming automobile dependence. Island Press, Washington, DC
Piorr A, Ravetz J, Tosics I (eds) (2011) Peri-urbanisation in Europe. Towards European policies to sustain urban–rural futures. Synthesis report. Academic Books, Life Sciences, Thorvaldsensvej 40, DK-1871 Frederiksberg C. A pdf version can be downloaded from: www.plurel.net
Reckien D, Karecha J (2007) Sprawl in European cities—the comparative background. In: Couch C, Leontidou L, Petschel-Held G (eds) Urban sprawl in Europe: landscapes, land-use change & policy. Blackwell, Oxford, pp 39–68
Schmidt C (1998) The specter of sprawl. Environmental Health Perspectives, Vol. 106, No. 6
United Nations Databases (2009) Accessed at http://data.un.org/
U.S. Department of Agriculture, Natural Resources and Conservation Service (2003) Annual National Resources Inventory
Zhu H, Lu C, Li X (2004) Land use studies in China. J Geogr Sci 14:69–73

# Part I
# Concepts

# Chapter 2
# The Dynamics of Peri-Urbanization

**Joe Ravetz, Christian Fertner, and Thomas Sick Nielsen**

## 2.1 Introduction

The peri-urban (sometimes also called the urban fringe) may be the dominant urban form and spatial planning challenge of the twenty-first century. In older industrial or post-industrial countries the peri-urban is a zone of social and economic change and spatial restructuring, while in newer industrializing countries, and most of the developing world, the peri-urban is often a zone of chaotic urbanization leading to sprawl. In both cases the peri-urban can be seen as not just a fringe in-between city and countryside, a zone of transition, rather it is a new kind of multi-functional territory. While it resists simple definitions, there are common features wherever such areas are found, such as a relatively low population density by urban standards, scattered settlements, high dependence on transport for commuting, fragmented communities and lack of spatial governance. Many global challenges arise from the ways that cities grow and change, especially the emerging mega-cities in developing countries where massive social and environmental problems can be found in their peri-urban hinterlands.

This chapter is a broad review of the peri-urbanization phenomenon, in its global and European context. We explore physical, socio-economic and political dynamics of change and we set out a framework for understanding different

J. Ravetz
Centre for Urban & Regional Ecology, University of Manchester, Oxford Road, Manchester M6 9PL, United Kingdom,

C. Fertner (✉)
Department of Geosciences, Natural Resources and Planning, University of Copenhagen, Rolighedsvej 23, 1958 Frederiksberg C, Denmark,
e-mail: chfe@life.ku.dk

T.S. Nielsen
Department of Transport, Technical University of Denmark, Bygningstorvet 116B, 2800 Kgs. Lyngby, Denmark

K. Nilsson et al. (eds.), *Peri-urban futures: Scenarios and models for land use change in Europe*, DOI 10.1007/978-3-642-30529-0_2,

levels of change in the peri-urban system. We outline the methods and results of the European scenario and modelling work from PLUREL, and draw out the implications for the rest of this book.

### 2.1.1 Global Context

The peri-urban zone may become the most common type of living and working situation in the world in the twenty-first century. In some parts of the world it is characterised by affluence and conspicuous consumption. In others it is where poverty and social displacement are more common, a front line between the problems of the city and the countryside. Underlying this is the changing nature of the city itself – as well as the physical expansion of urban or suburban form, there are wider economic, social and cultural dynamics of change. Therefore, we need to look beyond the conventional divide between 'urban' and 'rural', to a new kind of territory where the 'peri-urban' is the central feature – one which is not fixed, rather being in continuous flux and transition. The peri-urban is where the urban structure transitions into the rural landscape so that it can be a significant territory in area terms that must be looked at in the context of the wider transitioning between dense urban cores and rural hinterland – to examine it in the context of the wider city or rural–urban region. The peri-urban acts as a litmus test of change and transition, not just locally at the interface of urban and rural, but in the shape of the whole city-region, or as described below, the 'rural–urban-region'.

Over-shadowing the European peri-urban agenda is the challenge of the global urban system. In one 2005 World Bank study, cities in developing countries were found to have three times the population density of cities in industrialized countries (Angel et al. 2005). However, the current trend is for density to reduce by 1.7% per year; so if this trend continues to 2030, the built-up area of these cities will triple to more than 600,000 km$^2$, while their population doubles. A global remote monitoring study found that there are four main types of urban growth: low-growth cities with modest rates of infilling; high-growth cities with rapid, fragmented development; expansive-growth cities with extensive dispersion at low population densities (generally North American); and 'frantic-growth' cities with very high land conversion rates and population densities (generally found in developing countries) (Schneider and Woodcock 2008). Each of these types had different spatial patterns, whether dispersed or constrained and scattered or contiguous development. To this could be added a fifth type, that of negative growth, more commonly referred to as shrinking cities (Bauer et al. Chap. 7 in this volume). Generally, these definitions of 'urban' include what we term here the 'inner peri-urban' or urban fringe, directly adjacent or within the shadow of the denser urban area. The 'outer peri-urban' – i.e. where the rural areas are in a transition and responding to urban fringe pressures – has not yet been studied at this scale.

Generally the process of urban expansion should not only be seen as a negative change, but also one with positive benefits, particularly for the majority of the world's population, who occupy on average a space of 3.5 m$^2$ per person (Hardoy et al. 2001). However, the implication is that cities in both developed and developing countries should be making realistic plans for large scale physical expansion, building capacity for governance, investing in basic infrastructure and managing sensitive or hazardous areas. Each of these applies in particular to the fast changing peri-urban areas which are the frontiers of expansion.

### 2.1.2 European Trends

Europe is a highly urbanised continent. Over 75% of the population lives in urban areas today, with a projection for this to reach 80% by 2020 (EEA 2006). The dense urban network contains almost 1,000 cities with more than 50,000 inhabitants, but only a few very large cities. In the EU only 7% of the population live in cities bigger than 5 million inhabitants, compared to 25% in the USA (CEC 2008). In recent decades, the most prominent result of the ongoing urbanisation in Europe has been the development of 'functional urban regions' (Nordregio 2005). This process includes the integration of even relatively peripheral areas into the urban system, the connection of neighbouring cities to form polycentric networks and the formation of large-scale metropolitan regions.

Urbanisation in Europe is, however, extremely unevenly distributed. The metaphor of the 'Blue Banana' (Brunet 1989) illustrated the concentration of economic and population development in some core regions in western Europe, while for regions outside the area, in the former CEE countries and in the European periphery, it becomes more difficult for them to compete economically. Since the publication of the European Spatial Development Perspective (ESDP 1999), 'territorial cohesion' became the key concept for counter-balancing this trend, promoting a harmonized development across the continent. The challenge of territorial cohesion is, however, not limited to this macro scale but is also an issue within countries, regions and even cities.

Areas close to cities have historically been subject to high development pressures strongly linked to an increasing per capita consumption of urban land. From the 1950s to around 1990, urban areas expanded their surface area by 78% while the population increased only by 33% over the same period (EEA 2006). This trend continued to 2000 where the population in the EU25 increased by 2% while the urban area increased by more than 5%, mainly as a result of increased numbers of households and decreasing size of households (Jansson et al. 2009). The low level of population growth in Europe in recent decades suggests that the growth of urban areas is slower than in other regions of the world. However the ratio between urban and population growth is comparable to other regions in the world such as the USA and China, creating a trend towards continued de-concentration and urban sprawl. Urban sprawl in the form of low density, discontinuous and dispersed urban development is now a common phenomenon throughout Europe (EEA 2006).

## 2.2 The Nature of the Peri-Urban

### 2.2.1 Theoretical Concepts

'Peri-urban' and peri-urbanisation are generally loose definitions. Often they are used to describe newly urbanised zones at the fringes of cities, especially in developing countries, which are then called the 'peri-urban interface' (Adell 1999; McGregor et al. 2006). From a European perspective, peri-urban areas are often understood to be mixed areas under an urban influence but with a rural morphology (Caruso 2001). The Council of Europe (CEMAT 2007) defines the peri-urban as a transition area moving from strictly rural to completely urban, related to a high pressure towards urban development (Bertrand 2007). Conversely, peri-urban areas can be far from ephemeral, but instead can form a new kind of permanent landscape. Furthermore, the development is not necessarily limited to purely physical development with urban characteristics, but is often marked by the emergence of urban activities in rural areas like hobby farms and second homes (Briquel and Collicard 2005; Caruso 2001). The fact that the residents can be considered urbanised even if they do not live in a strictly urban spatial type, because of their lifestyles and social focus on the urban, for example, emphasises the uniqueness of the zone. These urban transformations which take place outside the urban cores can be summarized by the term peri-urbanisation.

The peri-urban is something between, neither urban nor rural. The historical dichotomy of urban and rural space started to blur in Europe with the formation of nation states, industrialisation and the liberalisation of the economy in the nineteenth century (Bengs and Schmidt-Thomé 2006). However, firstly with the introduction of mass commuter transport systems such as suburban railways, and finally with the increased affordability of the car, the countryside close to towns became a potential place for living, recreation and sometimes also working for former urbanites. This development led to an expansion of cities not only in physical terms with low density housing but also in terms of functional relationships, creating an area of urban influence around cities, also called the urban field (Friedmann and Miller 1965). In this urban field a variety of places developed, characterised by a mixture of urban and rural features.

The blurring of the urban–rural boundary inspired research into the idea of an urban–rural continuum. Bryant et al. (1982) illustrated this by a model where the urban–rural region ranges from core city through inner and outer fringe, a zone of an urban shadow and out to the rural hinterland. However, in reality, while this model works in general, the complex pattern of actual cities and their surroundings, with all their different spatial structures that emerged through geographical and historical as well as political precursors, is often difficult to fit completely if at all. This is the case regardless of the fact that the idea of the continuum includes several dimensions (or several continua) of urbanisation in the urban–rural space, which can result in complex spatial patterns (Robinson 1990). Most recently the term *urban–rural interface* appeared in research, emphasizing the mixed character of these areas without fixing them on a single, simple gradient.

The factors leading to this multifaceted character are many. Several different concepts have been used to try to account for it. One popular concept is ex-urbanisation, originally coined as 'ex-urban' by Spectorsky (1955), who described the development of a ring of wealthy rural communities around New York City, characterized by urban professionals living there but commuting to the urban core for work. Today many of these areas could also be called suburban, and Nelson and Sanchez (1999, p. 689) argued that ex-urbanisation does not differ from suburbanisation, but that exurbia 'is simply the latest incarnation of the continued suburbanisation of American cities.' Ex-urbs are nowadays found in a different manifestation in places like southern Spain, where they form specially built estates for retired people from northern Europe (Zasada et al. 2010).

Another widely used concept describing a form of urban–rural dynamics is 'counter-urbanisation'. This implies an opposite trend to urbanisation, i.e. an increase in migration from the city to the countryside, and was observed in the 1960s and 1970s in the United States and Western Europe (Robinson 1990). Besides the relocation of services and industry into rural areas, the development of part-time farming, second homes and retirement migration play an important role in this process. Champion et al. (1989) emphasized that it is not a unidirectional movement but a tendency towards de-concentration, resulting from a complex pattern of flows.

However, peri-urbanisation also includes other transformations, ones not necessarily dependent on the migration of people. These include movements for commuting or recreation as well as other behavioural changes by old and new rural residents caused by the further integration of the rural area in the system of an urban region. Also, linkages related to human-ecosystem interactions are decisive for peri-urban land use relationships, and will be further discussed below. The impact and significance of these processes, as well as how to turn them towards sustainable development, was the major issue in the PLUREL project and the focus of the case studies presented in Chaps. 4, 5, 6, 7, 8, 9 and 10.

It is clear from this brief overview that one of the main challenges of the PLUREL project was how to develop a research approach to understand the interactions among such multiple forces and complex patterns.

### 2.2.2 Geographic Definitions of the Peri-Urban as a Basis for Research

The PLUREL project used the term 'rural–urban-region' (RUR) as the main unit of analysis, with a range of area types, shown below as nesting circles (Fig. 2.1). This was based on a wide literature review (e.g. Bryant et al. 1982; Champion 1999; Loibl and Toetzer 2003; Gallent et al. 2006: Leontidou and Couch 2007). It was

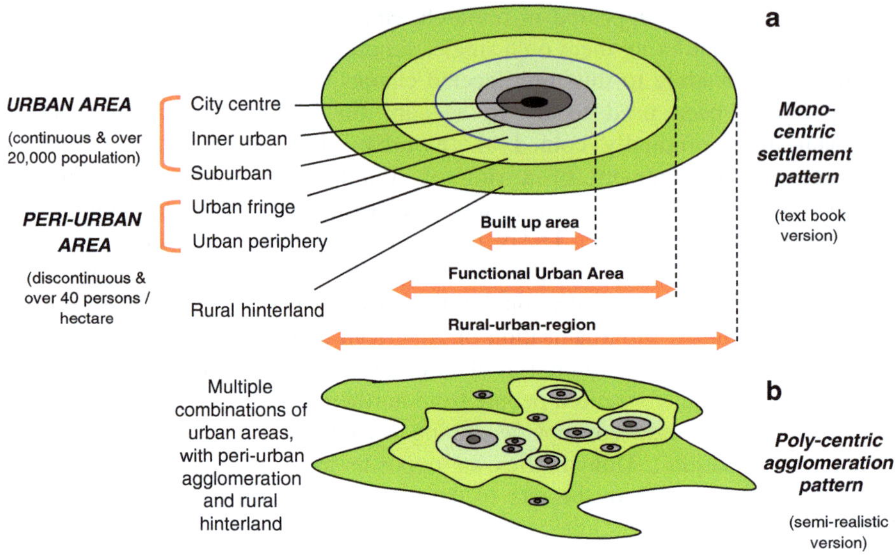


**Fig. 2.1** PLUREL concept of peri-urban areas and rural–urban-region

also found that the meanings of each of these terms could vary between different countries and languages. The basic spatial types which define the RUR include:

- Urban core: including the Central Business District and the site of many other civic and cultural functions and some public spaces associated with these;
- Inner urban area: generally higher density built development (built-up areas) including residential, commercial and industrial types of uses and some public open and green space;
- Suburban area: generally lower density contiguous built-up areas, which are attached to inner urban areas, and where houses are typically not more than 200 m apart, with local shops and services, parks and gardens;
- Urban fringe: a zone along the edges of the built-up area, which comprises a scattered pattern of lower density settlement areas, urban concentrations around transport hubs, together with large green open spaces, such as urban woodlands, farmland, golf courses and nature reserves;
- Urban periphery: a zone surrounding the main built up areas, with a lower population density, but belonging to the Functional Urban Area, as below: this can include smaller settlements, industrial areas and other urban land-uses within a matrix of functional agriculture;
- Rural hinterland: rural areas surrounding the peri-urban area, but within the rural–urban-region and accessible within a practical commuting time and so their rural character is affected by residents with urban incomes and lifestyles.

The peri-urban area therefore includes both the urban fringe and urban periphery segments of the above description: defined for the PLUREL project as 'discontinuous built development, containing settlements of each less than 20,000 people, with an average density of at least 40 persons per km$^2$ (averaged over 1 km cells)'.

Each of these area types is part of a larger 'urban system'. There are different concepts in use to describe urban systems with different boundary definitions – some of them flexible, some specific. The PLUREL project focussed on two main levels:

- Functional urban area: (FUA): 'an urban core and the area around it that is economically integrated with the centre, e.g. the local labour market. Belonging to a commuter catchment area, FUAs represent common local labour and housing markets' (Nordregio 2005). This overlaps with the statistical unit of the 'Larger Urban Zone', as used in the European Urban Audit.
- Rural–urban-region (RUR): spatial clusters of three interrelated regional sub-systems – the urban core, the peri-urban surroundings and the rural hinterland. Areas of recreational use, food supply and nature conservation located in predominantly rural areas are also part of the rural–urban-region. (see Chapter 1). Rural–urban-regions were the overall territorial unit of analysis for the PLUREL project. They include both the 'Functional Urban Area' (zone of daily commuting), and the surrounding rural hinterland. For modelling and analysis, PLUREL defined rural–urban-regions in statistical terms (see Chap. 2) but in policy (and indeed in practical) terms, the boundaries are necessarily more flexible, in order to respond to changing problems and opportunities (see case study chapters).

Figure 2.1 shows two interpretations of this scheme. The upper picture shows a simple text-book version of a mono-centric settlement pattern, surrounded by nested circles. The lower picture is a little more realistic: this shows a poly-centric agglomeration of settlements with different sizes and patterns, surrounded by a rural hinterland with a complex boundary. In the poly-centric version, the peri-urban areas are not only surrounding the urban, they become a geographical type and territory of their own, and the reality on the ground is often complex and fast changing. FUAs overlap and merge to form urban agglomerations, existing settlements change their shape and function and in larger FUAs there are many areas with a combination of infrastructure, housing, industry, open space, and land in transition – a challenge for any kind of definition.

In each of these examples, similar questions arise, in particular, that the conventional measure of radial commuting patterns is only one of a range of factors in urban–rural relationships and fields of influence. This is why the 'rural–urban-region' is important and topical, but also challenging for analysis.

Perhaps the most important definition is that of 'urban sprawl' – generally seen as a land use pattern with lower density, inefficient or wasted land-use, car dependency, and so on. This raises many questions: – which scale or spatial unit is to be counted? Is an airport or industrial complex to be defined as part of urban sprawl, or as economic development or both? A more technical definition looks for low values in one or more of eight measures: density, continuity, concentration, clustering, centrality, nuclearity, mix of uses, and proximity (Galster et al. 2001). In simple terms we use two definitions for sprawl – 'unplanned incremental urban development, characterised by a low density mix of land uses on the urban fringe' (EEA 2006) and also: 'Low density, scattered urban development, without systematic large scale or regional public land-use planning' (Bruegmann 2008, p. 18; Reckien and Karecha 2007). Such definitions can be explored further by looking at the key factors in sprawl, as discussed below.

### *2.2.3 The Peri-Urban Metropolis and Megalopolis*

The context for the 'rural–urban-region' is the broader picture of growth and change in human settlements. The process starts with the expansion of free-standing cities into their rural hinterland, absorbing villages into the urban fabric. At some point, if these separate cities are reasonably close together, a regional agglomeration process takes over. If the agglomeration process scales up, the result can be an 'extended metropolitan region' or 'megalopolis' of over 100 million population, containing within it many types of peri-urban areas, rural areas, or whole rural–urban-regions.

Mumford (1938) defined a megalopolis (also called 'megacity', 'megapolis' or 'mega-city-region') as an agglomeration of adjacent metropolitan areas. Gottmann (1961) used this term to describe the continuous urbanization of the north-eastern seaboard of the USA. Such thinking was then used to inform 'Ekistics', the science of human settlements, and a hierarchy of scales was proposed, including a 'metropolis' of 4 million, a 'small megalopolis' of 25 million, and a 'megalopolis' of 150 million population (Doxiades 1968). Interestingly, the largest examples from around the world (Pearl River Delta, Yangtze Delta, Gangetic Plain) are each in this range (Lacqiuan 2005).

The role of the peri-urban and the rural hinterland in these cases may be different from the typical EU situation. In the Asian megalopolis type there is a focus on the rapid transition from peasant agriculture towards a globalized economic development pattern (Jones and Douglass 2008; Ginsburg and Koppel 2004). By contrast, in the North American type, there is a focus on the 'edge city' as a new kind of CBD, and the rural as a zone of enterprise and opportunity (Garreau 1991; Daniels 1998). The peri-urban and rural hinterland is not so much a fixed thing 'out there', but highly inter-dependent and inter-woven with urban areas.

## 2.3 Dynamics of the Peri-Urban

To explore the dynamics of peri-urban change and the peri-urbanization process, we have to look at more than one aspect, not only in physical scales, but also in understanding the complexity of the system. This is not only an academic question, but a practical question for policy makers, who need to understand the peri-urban in order to work with it. The 5-dimensional framework here has developed out of the PLUREL research, together with an extensive review of literature. The aim is to provide practical routes into a complex situation and to identify the main processes which are visible at different levels. The effects of these in practice can be seen in the regional case study examples in Part 1.

These five dimensions also represent a kind of generic 'story' of how peri-urbanization takes place. The first aspect is where urban expansion occurs as a direct result of growth in population, economics and space demands. Next, as cities expand further, they form regional agglomerations, with step-changes in economies of scale taking place, and a new type of peri-urban territory developing. Thirdly, underlying these developments are the effects of various deeper political and cultural forces which shape the peri-urban territory. The fourth factor is where the whole urban system can go through rapid transitions, with radical change and restructuring. The final aspect concerns policy responses to these changes and transitions which often feed back into the mix, and become 'dynamics' themselves. In practice, the situation is rarely clear or simple, and each aspect will overlap and inter-connect with the others.

Below we set out the main features of each of these five aspects.

### 2.3.1 Direct Factors of Urban Expansion

In simple terms, peri-urban change is a direct result of urban expansion, the peri-urban area spreading outwards into rural areas. This is firstly a result of population and economic growth, which result in demand for housing and commercial areas. The location of housing is then determined in part by transport accessibility to employment and services and in part by the attractiveness of the environment as well as land values. There are usually physical and policy constraints to expansion. Each of these interacts with the others, and each raises further questions, e.g. how do the physical or policy constraints work? What drives the growth in demand for land and buildings, and enables the supply of development? (Fig. 2.2). The following sections outline some of these drivers:

*Demographic and social dynamics* are driven by population change due to fertility and mortality rates and migration. While fertility and mortality are relatively slow to change, over several decades some very different demographic profiles can emerge. International and inter-regional migration flows are more volatile and dependent on political factors and global economic swings among others (Bell et al. 2010). The continuing decrease in average household size as the population ages will also affect housing demand. Urban–rural migration – inwards or outwards – is dependent on spatial policy, the relative attractions of cities or rural areas for jobs and quality of life, as well as transport and communications (Loibl and Bell 2011).

*Economic and employment growth* drives the rate of urbanization. The rate of savings and capital investment feeds into expansion of the building stock and land-use conversion, as both a supply-side push factor of production, and a demand-side pull by consumers. Economic structures and employment patterns also affect

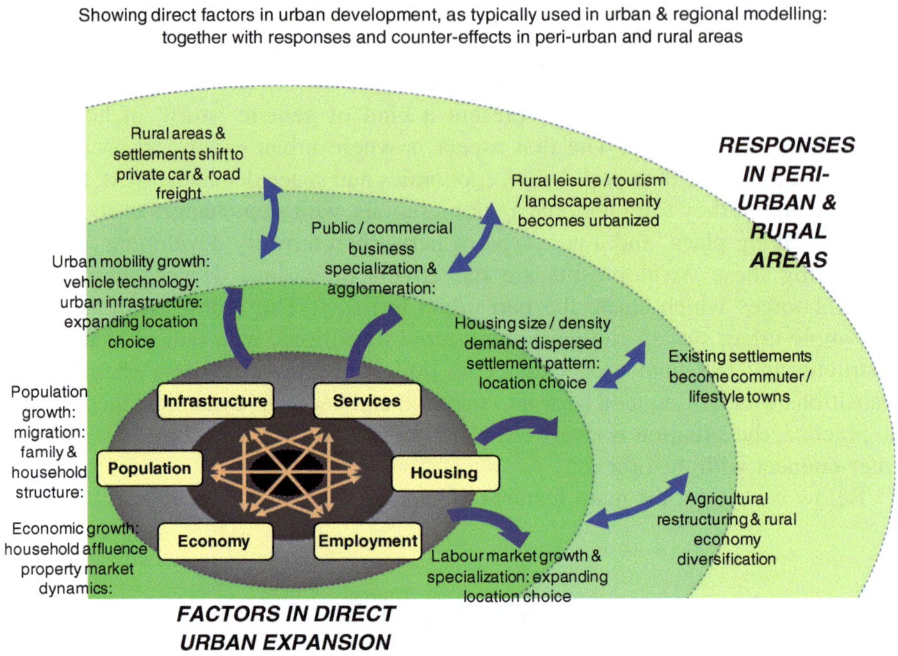


**Fig. 2.2** Dynamics of the peri-urban: (**a**) urban expansion

the trends of peri-urbanization, e.g. if home-working becomes more popular for the service sector. Business technology affects not only employment but supply chain logistics, the distribution of production, services and consumption (Korcelli et al. 2011).

*Environmental dynamics* and constraints include fixed geographic features such as rivers, coasts, wetlands or mountains, which shape the pattern of urban development. Beyond that is a wide range of dynamic factors which are more complex to analyse. Local urban climate factors such as the Urban Heat Island effect may make urban environments less comfortable in summer and broader climate change may possibly affect sea-levels or flooding, for example. As a result, some urban environments may become more unpleasant and hazardous, which may encourage out-migration. Meanwhile increases in sealed surfaces in larger urban territories will affect local hydrological systems, necessitating better water resource and flood management which in turn are likely to put pressure on peri-urban development such as to maintain open flood storage areas along the edges of rivers. Also, there may be pressure on peri-urban land use for the production of bio-mass and other forms of renewable energy (Zasada and Berges 2011).

*Urban built structures and infrastructure* are the components of the physical urban system itself. The floor space per person for living and working, and the land intensity of such floor space, are the primary determinants, especially when household size is taken into account. Then, the pattern of housing investment, housing form, community services, settlement density and morphology, are each relevant to the growth and pattern of peri-urban development. Transport and communications are the other key factors, as infrastructure can encourage or inhibit urban/rural migration, counter-urbanization, or re-urbanization. Transport is not only a matter of direct expansion: there has been a systemic change from a public transport-based radial pattern, to car and highway based network pattern (Ristimäki 2011).

### *2.3.2 Regional Agglomeration and Urban–Rural Linkages*

The urban expansion scheme described above is not a simple one-way process; it also generates responses and changes in the surrounding peri-urban and rural areas. Nor are these only local-scale responses, but they involve a more inter-urban and regional scale in the reshaping of spatial relationships. These peri-urban and rural responses are important feedback loops, which over time can lead towards major peri-urban changes.

Such feedback causes the dynamic of change to evolve towards the inter-urban or 'regional agglomeration' effect. The concept of the free-standing city in rural surroundings is replaced by a wider regional urban system of inter-connected and polycentric settlement forms (Hall and Pain 2006). As the urban markets and peri-urban 'shadows' expand, there comes a point where the agglomeration effect takes over in economic-financial terms, in labour markets, shopping markets, and housing location choices. The rural processes of economic restructuring, land market changes and agricultural modernization can also lead to a more rapid shift towards agglomeration. In order to gain access to larger consumer and labour markets there are clear incentives for new business parks, shopping malls and airport zones to find new locations, not within but between major cities, in order to serve larger populations more efficiently.

The result is that previously separate peri-urban areas can become the linking spaces, forming continuous, functional, low density zones, which provide for most living/working/shopping needs for the majority of suburban or ex-urban car-based residents (see also Soja 2000).

In reality this inter-urban or regional scale agglomeration is not a homogenous space, but more like a diverse territory shaped by many types of 'land use relationships':

- Urban to peri-urban links: urban demands and pressures on suburbs and peri-urban surroundings; space for housing, business, infrastructure. In the other direction, the peri-urban requires urban markets, services and innovations.
- Peri-urban to rural: the relationship of people to landscape, in functional-economic terms such as food, water, minerals, or tourism. There are also services which are more socio-cultural, such as aesthetics, amenity, recreation and cultural identity. In the other direction are relationships of employment, investment, and access to services.
- There are also relationships within peri-urban communities: the potential for economic and social development within settlements and across the peri-urban zone.

This extends the 'ecosystems services' approach to a wider view on 'services, linkages, functions, values': for integrated development policy in the rural-urban-region (based on Ravetz, 2011).

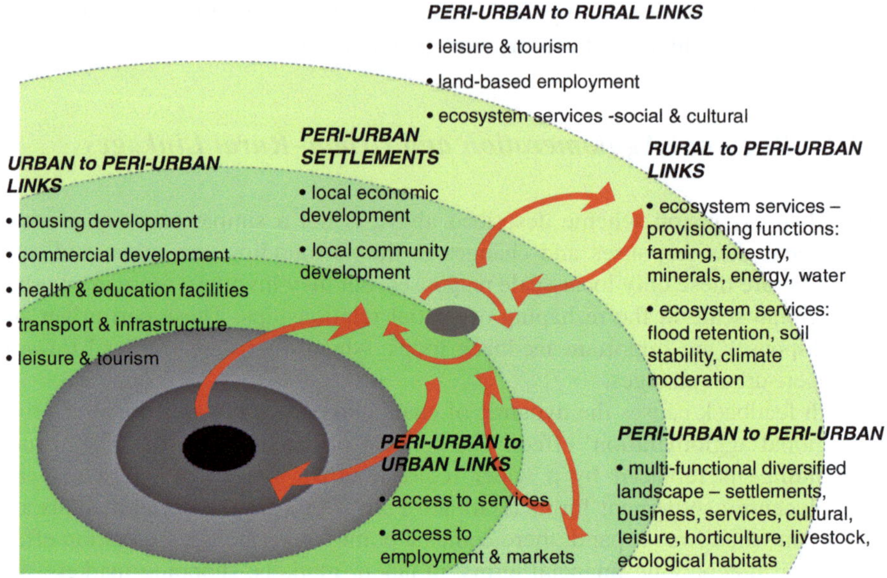


**Fig. 2.3** Dynamics of the peri-urban: (**b**) agglomeration and linkages

Managing such relationships is at the core of the policy agenda for 'integrated development for territorial cohesion in the rural–urban-region' (Ravetz 2011a). This is a broad sustainable development agenda aiming to identify the economic, social or environmental functions and services between different areas, to identify their values, monetary or otherwise and then to construct spatially-based policies for obtaining the best balance of services and values. The implication is that chaotic urban sprawl and unplanned agglomerations can be improved by establishing such relationships, in order to enhance the internal 'territorial cohesion' within and between rural–urban-regions (CEC 2008) (Fig. 2.3).

### *2.3.3 Global–Local and Structural Dynamics*

Behind the physical processes of expansion/agglomeration are powerful social, economic and political forces at work. The first of these is *globalization*, exerting economic effects on the structure of business and finance, political effects on the urban hierarchy and cultural effects through the media and information and communications technology (ICT). There is also a counter-trend of *localization*, where the cultural identities of people and places are being re-invented in new ways. The *liberalization* process involves privatization, franchising and cost recovery, with far-reaching effects on urban governance and public services. Meanwhile the *consumption* culture affects the identity and perceptions of peri-urban places through leisure, tourism and other location decisions. The tension between *risk* and *security* is also a powerful shaper of places and spaces. In search of security, the affluent may seek gated enclaves away from city centres, in a peri-urban 'archipelago' (Beck 1995; Borsdorf and Salet 2007).

All these start with very topical questions – what is the peri-urban landscape for, and who should decide? Who gets the benefits or bears the costs of investment or restructuring? Whose land is it, and why? (Shoard 1983). Large parts of the peri-urban are open to a wide range of possible uses – tourism, high-value housing, business sites, agriculture, nature conservation, flood mitigation or energy production and distribution. There are economic, ecological, historical, and residential aspects in competition with each other and many of the critics of sprawl are the more affluent peri-urban residents, seeking to maintain their own quality of life by limiting further development which might affect their environment negatively.

A structural approach looks beyond physical land use, at the underlying dynamics of power, wealth and ideology. Urban expansion can be seen as one of the paths of the 'dual circuit' in the capitalist system: investment in land and buildings as an alternative to investment in production and consumption (Harvey 1985, 1987). In an alternative version, the peri-urban space is firstly a financial commodity, where the imperative for global capital accumulation drives a system of ideology, as well as the physical system of infrastructure, construction and property speculation (Maciocco 2008). We can also see the peri-urban as a place of dependency and colonization: alongside power stations and waste landfills are the service zones of low income housing for workers and dependents or public housing in 'peripheral estates' (Davis 2005). There is also a discourse on the peri-urban as a kind of frontier capitalism – promoted by business/science park investors with images of green fields and fast road connections to the airport. Favoured parts of the peri-urban are framed as a cultural-cognitive-capitalist zone of creative enterprise, which attracts global investors and entrepreneurs, alongside local utopians and free-thinkers (Scott 2000). This also points the other way, towards many kinds of informal or illegal activity such as farmers trading in scrap vehicles, unlicensed waste dumping or alternative free festivals in the forest (Farley and Roberts 2011).

The 'urban archipelago' concept sees this in terms of a series of increasingly disconnected islands of wealthier and poorer populations, another phase in the so-called 'splintering' of cities (Borsdorf and Salet 2007; Graham and Marvin 2001). By contrast a 'spatial ecology' perspective sees a peri-urban territory with many types of relationships and connections (although not all are positive). A diversity of typical peri-urban land uses and locations is shown in Fig. 2.4, as a 'spatial ecology' (Ravetz 2011b). A wide range of land use relationships as above, is shaped by structural forces, and overlaid on a diverse landscape of 'real places' (Clay 1994).

One axis in this scheme concerns globalization versus localization. Cities become hubs in a geo-political hierarchy, a 'global urban system' (Knox and Taylor 1995). There are also counter forces of localization when commuters seek 'fields at the end of the garden', or citizens participate in new forms of local enterprise. The result can be a diversification of land uses and enterprises; it can also be the NIMBY

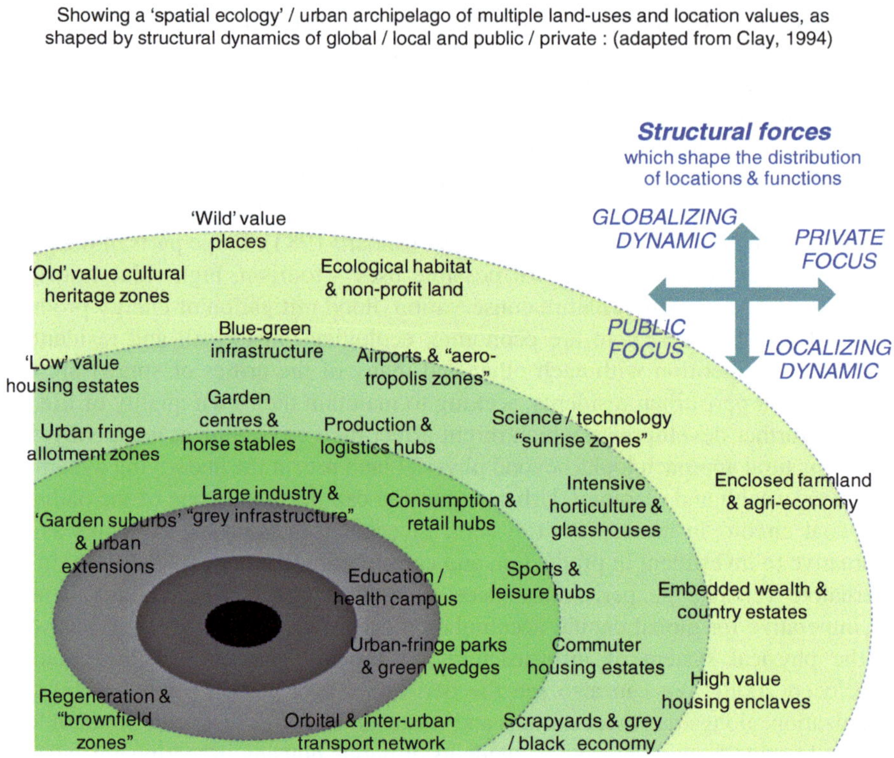


**Fig. 2.4** Dynamics of the peri-urban: (**c**) global–local and structural

('not in my back yard') response of those living there already. The other axis in the scheme focuses on the political-economic aspects of public versus private sector, and the tension between public policy and market processes.

Such questions are crucial for the future of the peri-urban space, where governance is often fragmented and where coordination across boundaries can be more difficult. The diverse land uses each compete for space, access and investment and the line between public and private objectives is closely debated.

## *2.3.4 Complexity, Transition, Resilience*

Each of the above factors – urban expansion, regional agglomeration, and structural effects – can work in combination, with many feedback loops between them, amplifying the processes of change. The result is not always smooth and predictable, but may be a rapid transition, a 'discontinuity', or in some cases a catastrophic failure. Transitions are system-wide changes, which can involve economic systems, social structures, political systems, spatial patterns, technology and infrastructure systems (Geels 2005). There are often parallel transitions in socio-cultural roles, identities, perceptions, and the city's 'reason for being' – a common question in many cities which no longer retain their original function industrially, economically or politically. This is not only a physical and functional transition, but one of overall role and identity as areas evolve from a supporting role as commuter suburb, to a fully functioning low density urban system – a 'post-metropolis' (Soja 2000), or 'metro-scape' (Kraffczyk 2004; Giannini 1994). Some typical transitions are:

- *Rural transition*: away from primarily agricultural production towards a more diverse multi-functional landscape and settlement pattern;
- *Peri-urban transition*: re-structuring for globalized systems of production and consumption;
- *Urban transition*: restructuring towards a networked economy with different patterns of green/grey infrastructure.

Each of these transitions can be micro-, meso- or macro-level in their effects, ranging from individual places, to whole urban systems (De Roo and Silva 2010). Each then generates some kind of response from the system affected. In some cases there is negative reaction and resistance, while in others there are positive responses, to build resilience to shocks, or creative innovation for new roles in a new environment. Such responses may then become incorporated as an objective of policy. For example, a typical response to an economic transition in the Manchester case can be seen where rural development policy aims for new kinds of skills and business models (see Chap. 6). A range of responses focused on economic, governance, community or spatial planning are shown in the diagram as circles; the overlaps reflect the policy links between multiple objectives.

Complexity and transition theory draws on current ecological thinking on 'complex adaptive systems'. These show evolutionary, 'non-linear' and self-organizing behaviour in systems of multiple relationships at multiple scales (Waltner-Toews et al. 2009). The implication for policy is crucial – i.e. managing a complex adaptive system is a very different task to managing a linear system, where outcomes are a direct result of inputs. Similar views on complexity and evolution come from the 'fractal city', composed of endless levels of self-organizing complexity (Allen et al. 1986; Batty and Longley 1994). At the root of such complex cities are the self-organizing patterns of individuals, households, firms and other units – not only economic but social, cultural and political (Portugali 2000). Overall, a 'human' complex adaptive system (in contrast to biological) will aim to build capacity for collective knowledge and social learning, and focuses on the 'shared intelligence' which enables strategic thinking and creative innovation (Ravetz 2011b).

### 2.3.5 Spatial Governance and Policy Responses

The final piece in this framework concerns the responses of policy, spatial planning and the governance system itself. These come with the aim of solving problems, but they can also become part of the problem which they aim to solve. Beneath the surface the role of policy is often to try to turn the 'problems' of transition and restructuring, into 'opportunities'. The systemic responses to transitions – building resilience, adaptive capacity, shared intelligence etc. – then become the objectives of policy, which again becomes part of the system to be addressed.

An example is the Green Belt policy in the UK, seen as successful in its objective of solving one problem – preventing sprawl – by urban containment: but also shaping or distorting the land and property market, and so generating other problems. In the Manchester case-study in Chap. 6, these second-order problems of vacant or neglected land then become the objective of further layers of policy, and so on.

There are questions concerning the overall scope and effects of 'spatial governance' – the system of territorial government, spatial planning, and the policy system. Main factors in direct urban expansion, with alternative modes of spatial governance are:

- *High or low land intensity*, i.e. the amount of land area required per unit of housing or business. This factor could be further analysed into economic intensity, in terms of value or production or social intensity in terms of welfare measures. Environmental land intensity is also relevant, where the urban system demands land for ecosystem services, material supplies, waste management, leisure and amenity. The land intensities are normally assumed to be on a growth trajectory which is similar to that of economic GDP. In the scenario modelling (next section), land intensity is assumed to be correlated with economic growth and capital investment.
- *Strong or weak spatial governance*. This ranges from active and coordinated spatial planning and governance in the public interest to ad hoc and fragmented governance for short term enterprise and private profit.

The analysis is set out in Table 2.1 below. This shows the different effects of population growth and land use growth, mapped on to strong or weak governance. It also shows a third category of 'partial governance', to describe governance which is patchy, or stronger/weaker for different social groups. It also includes 'urban shrinkage', which is a powerful trend in many older cities (see Chap. 7 on Leipzig/Halle).

The foregoing analysis raises questions about the overarching goals of policy, generally as assumed to be *sustainable development*. In principle, sustainable development combines economic, social and environmental 'pillars' or goals, to be achieved both locally and globally and in both the short and longer term. In this case, the principles must begin with the complex, messy reality of peri-urban areas. One approach is to look at the European policy on 'territorial cohesion', and the implications for urban/rural policies and local/regional/national planning (Duhr et al. 2009). Another approach is to explore the 'sustainability tension', between competing urban/rural, and development/conservation agendas (Ravetz 2000; CURE 2003):

- Urban development: a growth and modernization perspective;
- Urban conservation: a containment and regeneration perspective;

**Table 2.1** Combination of growth, expansion and governance effects

| | Strong spatial governance | Partial spatial governance | Weak spatial governance |
|---|---|---|---|
| Urban growth + expansion: (population + space per person) | Hi-growth poly-centric 'social city-region' model (lower density) | Planned agglomeration with enclaves | Hi-growth urban sprawl (lower density) |
| Urban growth: (population) | Poly-centric 'social city-region' model (higher density) | Planned higher density enclaves | Urban sprawl (higher density) |
| Urban expansion: (space per person) | Poly-centric 'social city-region' model (lower density) | Planned lower density enclaves | Urban sprawl (lower density) |
| Urban stability | Compact city and urban containment | Restructuring city within spatial envelope | Ad hoc low-growth or stagnation |
| Urban shrinkage | Planned transition, focused on green infrastructure | Growth in selected zones/enclaves, surrounded by decline | Chaotic decline with derelict/vacant land and buildings |

- Rural development: more local, rural-focused enterprise;
- Rural conservation: an environmental protection approach.

The peri-urban can be seen as being pushed and pulled by these competing aspects. In a dynamic rural–urban-region each of these will be evolving and can be shaped by spatial governance processes. The result can be complex and inter-dependent.

There are basic questions about the degree to which an urban system can be sustainable given its heavy reliance on resources imported from beyond its boundaries – a city region is difficult to consider as a 'unit of sustainability' in its own right and this fact should be recognised.

## 2.4 Scenarios and Modelling Results

The discussions above show that the dynamics of the peri-urban and of land use change are complex and multi-level; they are also the subject of conflict and competition between different social and political groups. To describe this accurately is beyond the scope of any technical analysis or modelling system. One solution to this problem as a route into the exploration of the complexities of processes and dynamics is to work with scenarios. The first thing to make clear is that scenarios are not 'forecasts' or 'predictions'. Instead, scenarios are useful tools in asking a wider set of 'what-if?' questions, and comparing the results of policy options. Technical analysis and modelling, using starting parameters and decision rules based on these scenarios, can be combined with broader exploration of possibilities in terms of social, economic, cultural and political change. Technical modelling forecasts are useful but never good enough to describe complex environmental, economic or human problems fully. Scenarios are not new tools – they are used all the time to test how policies or plans could deal with sudden unforeseen emergencies of all types. War games, earthquake responses, economic modelling and foreign policy analysis all use them in one form or another – a 'what if?' story is prepared and different models used to test what might happen in the event of the storyline taking place. Scenarios are most useful for helping to prepare governments for what experts might consider to be the least likely but most devastating contingencies and so for that reason in the PLUREL project a number of 'shock scenarios' were used (see below).

### 2.4.1 Scenario Method and Framework

PLUREL developed a scenario framework for the peri-urban research agenda. As a source for these, given that the project was funded under the 'Climate change and ecosystems' programme, was the IPCC Special Report on Emissions Scenarios (SRES) (IPCC 2000) (Fig. 2.5).

- Adapting the IPCC global context scenarios to the EU space, up to the years 2025 and 2050.
- Focusing these on the peri-urban issues.
- Building in a set of 'shocks', i.e. rapid changes, which were relevant to the scenario and the theme.
- Translating the scenarios into economic, demographic and land use modelling parameters.
- Completion of EU-wide scenario model results, including spatially explicit maps.

The scenarios were used mainly as the basis for 'top-down' modelling work on economic, demographic, and environmental and land use changes. In the study of spatial typologies and governance systems, they were applied to different urban types and governance types. In each of the case study regions, the focus of this book, the top-down scenarios became the starting point for modified, localised 'bottom up' scenarios within each of the regional case studies. These took inputs from regional policy-makers and stakeholders, in the form of issues, questions, and policy options which were then modelled in the MOLAND system and the implications of these results were fed back to policy-makers and stakeholders for their reaction and response. More detail on the technical methods can be found in Chap. 3.

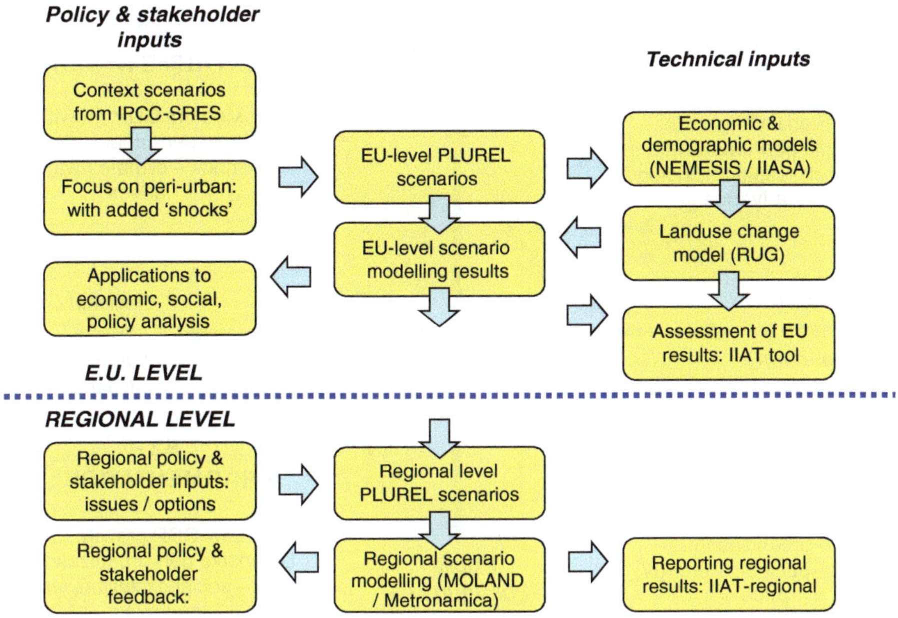


**Fig. 2.5** Outline of scenario and modelling process

The result of the initial scenario development was four main scenario types aligned along two conceptual axes as illustrated in Fig. 2.6. 'Storylines' to flesh out the bare bones of the types were then produced as follows.

A1 – 'Hyper-tech' scenario (globalizing and privatizing dynamics)

This visualises a future world of rapid economic growth, a global population that peaks by mid-century and the rapid spread of more efficient technologies. Investment in research and development is high and nations share knowledge and pool resources in a global research market place. Energy prices decline because supply is driven by new developments in renewable energy production and nuclear fission. The shock concerns the rapid acceleration of information and communication technology (ICT), which transforms home and work as never before.

For peri-urban areas in Europe, this scenario would be expected to result in a range of smaller 'polycentric' towns and cities becoming even more popular. New transport technologies would lead to more rapid journeys and the expansion of the commuting distances around towns and cities. New forms of ICT would enable people who prefer country life to work from home or a neighbourhood centre, and this would lead to peri-urbanisation and 'metropolization' of rural areas on a widespread scale.

A2 – 'Extreme water' scenario (localizing and privatizing dynamics)

This imagines a more diverse world of self reliance by people and the preservation of local identities. While the population increases, economic development would be primarily regionally-orientated, and per capita economic growth and technological change would be more fragmented and slower than in the other storylines. The shock

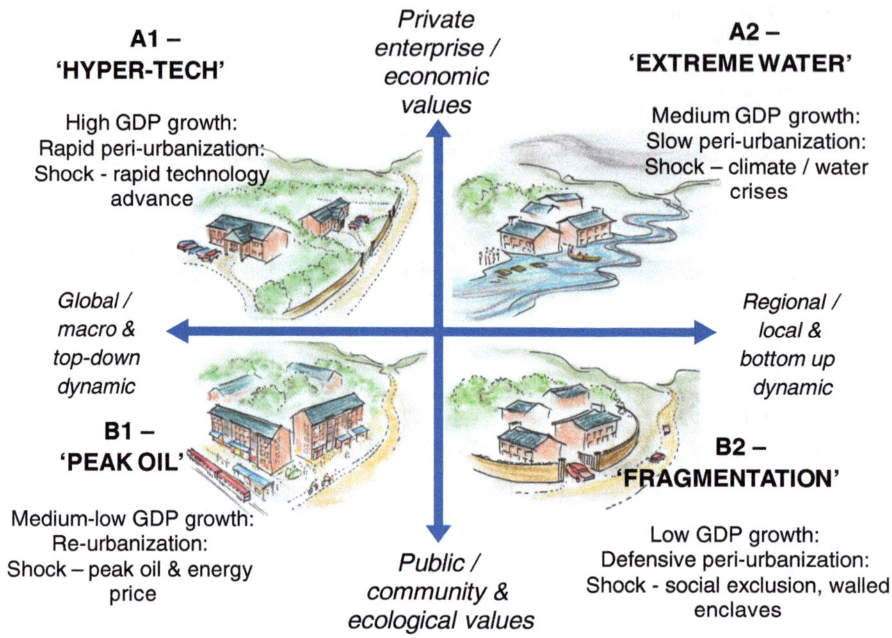


**Fig. 2.6** PLUREL scenario framework and images

here is subtitled 'extreme water', and is based on rapid increase in sea level rise accompanied by flooding in some areas and drought in others.

Peri-urban areas would be expected to be strongly affected; affluent yet vulnerable city-regions such as London or the Dutch Randstad would spend large sums of money on flood defence and adaptation strategies. Population growth due to climate-change induced immigration would put more pressure on urban infrastructure and services.

B1 – 'Peak oil' scenario (globalized community dynamics)

This scenario envisages a future of environmental and social consciousness – with a global approach to sustainable development involving governments, businesses, media and households. Economic development would be more balanced with major investment in resource efficiency, social equity and environmental protection. The 'shock' in this scenario is driven by the arrival of the point of 'peak oil', that is, a decline in global oil availability following maximum production levels, leading to rapid increases in energy prices and many resulting social and economic effects.

For peri-urban areas, rising energy prices would have an enormous impact on residential location choices because transport costs limit commuting distances. Although tele-working would be encouraged, most people would attempt to return to larger cities and towns, and many of the more remote rural areas would be expected to decline.

B2 – 'Social fragmentation' scenario (localized community dynamics)

Here, the vision is of a fragmentation of society, in terms of age, ethnicity and international distrust. The voter-strong older population would become increasingly dependent on the younger generation, but the working-age population would be disinclined to transfer their resources, leading to growing intergenerational conflicts. Cities would become more socially fragmented as younger migrants would tend to dominate city centres and older 'natives' to favour the outskirts and enclaves outside the cities – so that peri-urban areas would also become 'peri-social' areas. New development would slow down but much existing urban form would change its function.

The scenarios shown here are only four possibilities out of an almost infinite number. What is interesting is that each scenario suggests not only a very different spatial development path but a different direction for the research and modelling focus. For instance, one of the first technical parameters in spatial development is the gravity function, i.e. the field of attraction of urban areas. Each scenario suggests not only variations in the gravity function, but also different kinds of functions which take priority. The A1 scenario suggests a network effect; the A2 suggests disaster management; the B1 suggests an energy/climate policy focus; and the B2, a localized repulsion effect which replaces attraction.

Again, these differences pose not only academic questions but ones which concern practitioners, as seen in the case study chapters to follow.

## 2.4.2 Scenario Applications and Assumptions

The scenarios were applied to the issue of land use change through modelling and statistical analysis based on assumptions for economic growth, development and employment as well as demographic growth and change. Environmental issues such as resource consumption and climate change were not modelled at the European level. The overall modelled assumptions are summarized in Table 2.2.

Using the macro-economic model NEMESIS, various assumptions on economic growth and employment restructuring were modelled at national level for each EU27 country. A limited 'downscaling' provided a breakdown at regional level (NUTS-x[1]), together with analysis of housing and land use demand factors. Demographic projections were modelled using the IIASA model (Samir et al. 2008), also at national and regional levels. The base assumptions for each scenario included estimates for fertility and life expectancy, as well as internal and cross-border migration.

[1] NUTS = Nomenclature of Units for Territorial Statistics, a geocode standard for subdivisions of countries for statistical purposes. For modelling and analysis of rural–urban-regions across Europe, we used NUTS-x units, a combination of NUTS2 (regional) and NUTS3 (sub-regional) units. In our analysis 510 NUTS-x regions are included, each with an average population of 900,000 inhabitants.

**Table 2.2** Summary of scenario assumptions

| | A1 | A2 | B1 | B2 |
|---|---|---|---|---|
| | 'Hyper-tech' | 'Extreme water' | 'Peak oil' | 'Fragmentation' |
| Population growth | Medium-high | Medium | Low | Medium |
| Fertility | Medium | Medium | Low | Medium |
| Mortality | Low | Medium | High | Medium |
| International migration | Medium | Medium | Low | Medium |
| GDP growth | High | Medium-high | Medium-low | Low |
| Urban population growth | Low | High | Medium | Medium |
| Peri-urban/rural population growth | High | Low | Very low | Medium |
| 'Shock' storyline | Rapid technology advance | Extreme water events | Peak oil | Fragmentation, social exclusion |

Urban transport/land use interactions were modelled using a combination of the Regional Urban Growth model (Rickebusch and Rounsevell 2009) and, in case study regions, application of the MOLAND model (Barredo et al. 2003). These models are described in detail in Chap. 3. As above, the technical inputs and results need to be seen in combination with a wider set of driving forces and assumptions including technological change, infrastructure investment, socio-political and cultural issues.

### 2.4.3 Scenario Modelling Results

The key result of the scenario modelling is the calculation and allocation of built development (i.e. 'artificial surface') growth for the period 2000–2025, for each of the four scenarios. The definition of artificial surface includes urban residential, industrial and commercial areas, as well as transport infrastructure, leisure and non-green public spaces in urban areas. These types and the baseline state for 2000 were derived from the CORINE land cover database, managed by the European Environmental Agency (EEA 2000). The projected growth was based on the scenario assumptions and the respective projections for economic growth and population change, as downscaled to NUTS-x regions.

## 2.5 European-Wide Development Trends

As described at the beginning of this chapter, there is a gap between population growth rates and the growth of urban areas in Europe. During the post-war period European cities grew between 40% and 300% in size but considerably less so

**Table 2.3** Annual growth by scenario 2000–2025, EU27[a]

| | Projected annual increase by scenario | | | |
|---|---|---|---|---|
| | A1 (%) | A2 (%) | B1 (%) | B2 (%) |
| Population | 0.16 | 0.14 | 0.13 | 0.15 |
| GDP/capita[b] | 2.22 | 1.92 | 1.53 | 1.43 |
| Artificial surface | 1.86 | 1.55 | 1.10 | 1.09 |

[a]Without Bulgaria.
[b]Data for 2005–2025. Prices 2000.

**Table 2.4** Growth of artificial surface in urban, peri-urban and rural areas, 2000–2025, EU27[a]

| Sub-region | Artificial surfaces area in 2000 (with share of total area)[b] | Annual increase by scenario | | | |
|---|---|---|---|---|---|
| | | A1 (%) | A2 (%) | B1 (%) | B2 (%) |
| Urban | 48,765 $km^2$ (79.1 %) | 0.65 | 0.61 | 0.50 | 0.48 |
| Peri-urban | 47,532 $km^2$ (8.3 %) | 2.46 | 2.06 | 1.44 | 1.44 |
| Rural | 72,182 $km^2$ (2.5 %) | 2.13 | 1.75 | 1.24 | 1.24 |
| Total | 168,478 $km^2$ (4.7 %) | 1.86 | 1.55 | 1.10 | 1.09 |

[a]Without Bulgaria.
[b]The remaining areas which are not classified are unpopulated areas as water, rocks and glacier surfaces.

in population (EEA 2006). According to all four scenarios, this gap would increase further up to 2025, meaning increasing per capita consumption of urban land.

There are, however, some differences between the scenarios at a pan-European scale, reflecting the various assumptions taken as the basis for each. Economic development would be 50% higher in scenario A1 than in B2, which would also have a strong influence on the increase of artificial surface. By contrast, population development would be rather similar in all scenarios, changing by only a very low annual increase. Table 2.3 summarizes the key drivers for each scenario.

In the PLUREL project a regional typology was developed, grouping the EU-27 territory into urban, peri-urban and rural areas at a resolution of 100 $\times$ 100 m cells. The typology incorporates population as well as land use data. Urban areas are defined as continuous areas with artificial surface (according to CORINE) and a minimum of 20,000 inhabitants. Peri-urban areas are those with more than 40 inhabitants per $km^2$ and adjacent to or very close to urban areas. Rural areas include all remaining populated areas. Details are explained in more detail in Chap. 2. Table 2.4 illustrates the results of projected artificial surface change broken down into the three sub-regional types.

All the four scenarios suggest a continued growth of artificial surface in Europe. Peri-urban areas would experience the highest growth rates, while areas which are already predominantly urban would experience relatively small changes. Many rural areas are also projected to have strong growth.

## 2.6 Land Use Dynamics

Table 2.4 already gives a first impression of the expected dynamics in peri-urban areas, which are shown as absorbing a large amount of future urban growth. However, these dynamics are, unsurprisingly, allocated very differently across Europe. In the following section some maps showing different aspects of these dynamics will be presented and discussed in terms of the regional variations across Europe. Before that, we will summarize the development of the economy, expressed in GDP, and the projected population development, which both are decisive for urban growth.

The scenario results for GDP growth present a well known picture. The new EU member states would experience a far higher annual growth rate than the rest of the EU, resulting in a relative convergence at country level in all scenarios. The lowest growth rates would be expected in Eastern Germany and parts of Italy. There is though, some difference between the scenarios. In the A1/A2 scenarios a generally higher rate of growth is projected than in B1/B2. Also, in the B1/B2 scenarios parts of Western Europe, especially France, Belgium, German, Austria, Italy and Portugal, would be expected to have a lower growth than other parts of the EU. Furthermore, the increase in GDP would not be linear, but a slowdown for the period of 2010–2020 is suggested in the B1/B2 scenarios and also in A2. Shock events introduced by the scenarios, such as 'peak oil' and a subsequent but delayed technological change, are reflected in this development.

Regarding population development, again, a well known picture emerges, showing a west–east divide, with population loss mainly in Eastern Europe and some peripheral regions due to a negative migration balance. Western Europe and especially the UK and parts of the Netherlands, Belgium, France and Spain, would experience the highest population growth rates in all scenarios. There are only minor differences between them as mentioned above, with slightly more increase of population suggested by scenario A1 but without differences in regional trends.

Against the background of these development trends we will now discuss the projected changes of land use, which is illustrated by the development of artificial surface, and not least the role peri-urban areas would be expected to play. We will only present some key results; more details will be found in Chap. 2.

### 2.6.1 *Loss of Natural Surface*

The maps in Fig. 2.7 show the projected transformation of natural into artificial surface up to 2025. In the economic growth scenarios 'Hyper-tech' and 'Extreme Water' the loss of natural surface would be highest. Geographically most natural surface transformation would be in the economic core area of Europe between London, Hamburg, Munich, Milan and Paris (the so-called 'Pentagon area').

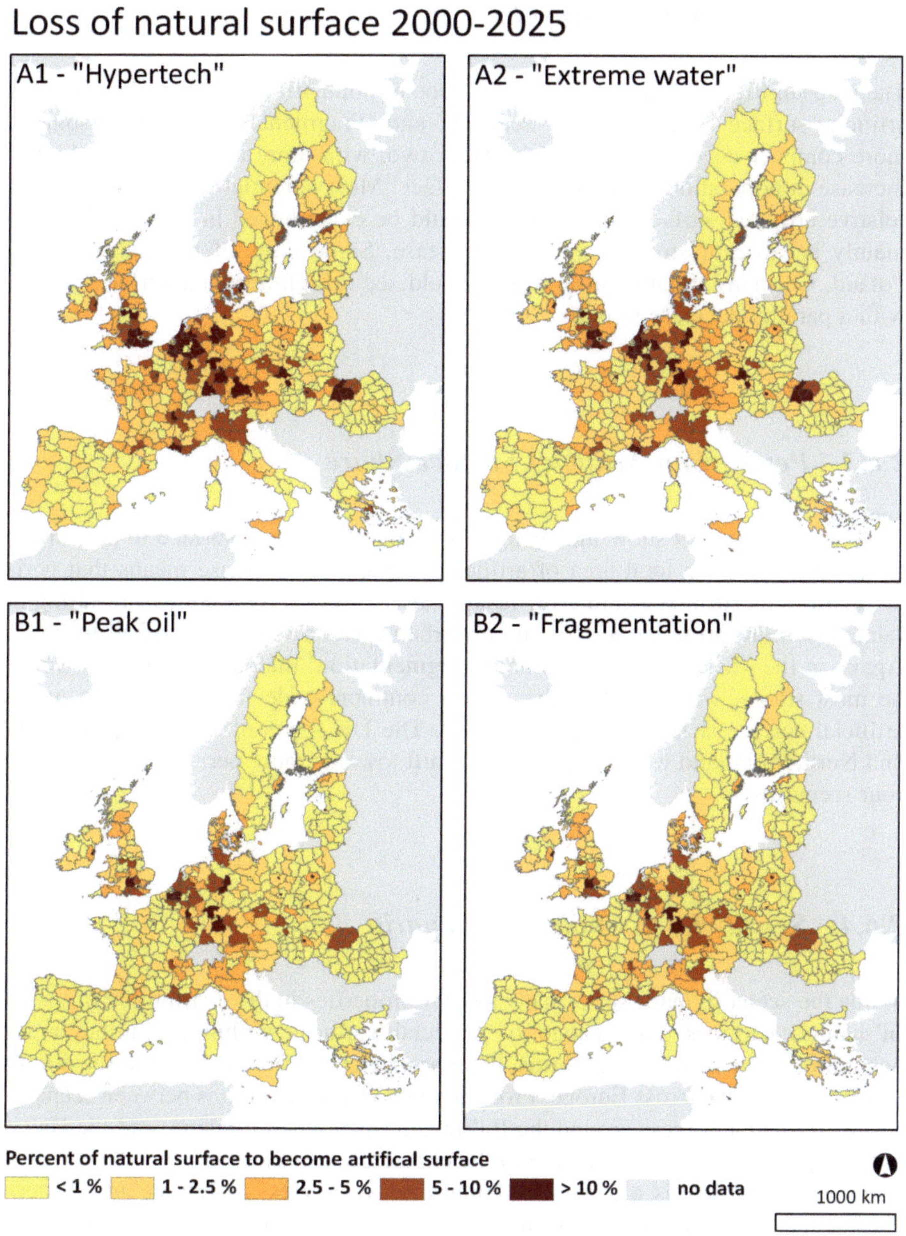


**Fig. 2.7** Loss of natural surface

### 2.6.2 Type of Urban Growth

The maps in Fig. 2.8 illustrate the comparison of population change and changes of artificial surface. The scenarios 'Peak oil' and 'Fragmentation' seem to suggest more compact development than the other two, with more areas having a higher increase in population than in artificial surface. Most potential for this decrease in relative artificial surface consumption would be experienced in Western Europe, mainly because of strong population increase. Some parts of Northern Europe, Poland, Germany, Spain and Portugal would see a high artificial surface growth with a parallel decrease in population.

### 2.6.3 Peri-Urban Artificial Surface Share

The maps in Fig. 2.9 show the changes in share of artificial surface in peri-urban areas relative to the total area of artificial surface. An increase means that peri-urban artificial surface would grow faster than in urban and rural areas, resulting in relatively more artificial surface in peri-urban areas by 2025 compared to 2005. Again, in the scenarios 'Peak oil' and 'Fragmentation' the peri-urban share would be most stable, while in the other two a continent-wide increase of peri-urban artificial surface share would be expected. The UK, parts of the 'Pentagon' area and Northern Poland would experience a shift towards more peri-urban areas in all four scenarios.

### 2.6.4 Sensitivity to Choice of Scenario

While the scenario maps show many similar properties in the distribution of losses of natural surfaces and change of artificial surfaces within peri-urban areas according to the different scenarios, there are general differences in the sensitivity to scenario choice across Europe. From analysis of the deviations between scenario results Ireland and areas around the Baltic Sea (in Sweden, Finland, and the Baltic states) appear to be most sensitive to the choice of scenario.

In addition, there is a general difference between Western Europe and Eastern Europe and the Iberian Peninsula where the sensitivity to the choice of scenario tend to be less. The sensitivity towards the choice of scenario is related to sensitivity towards population growth scenarios – but not reducible to underlying population growth.

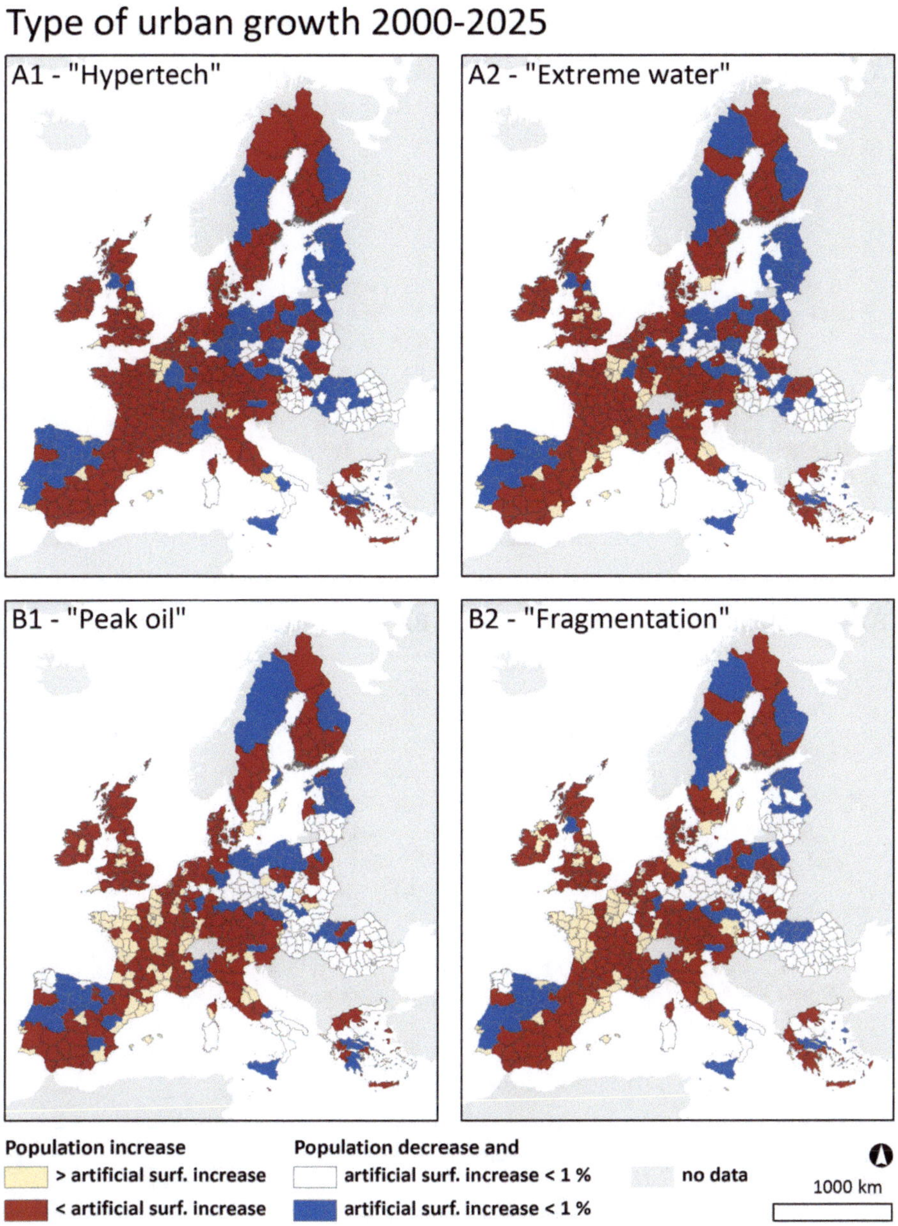


**Fig. 2.8** Type of urban growth

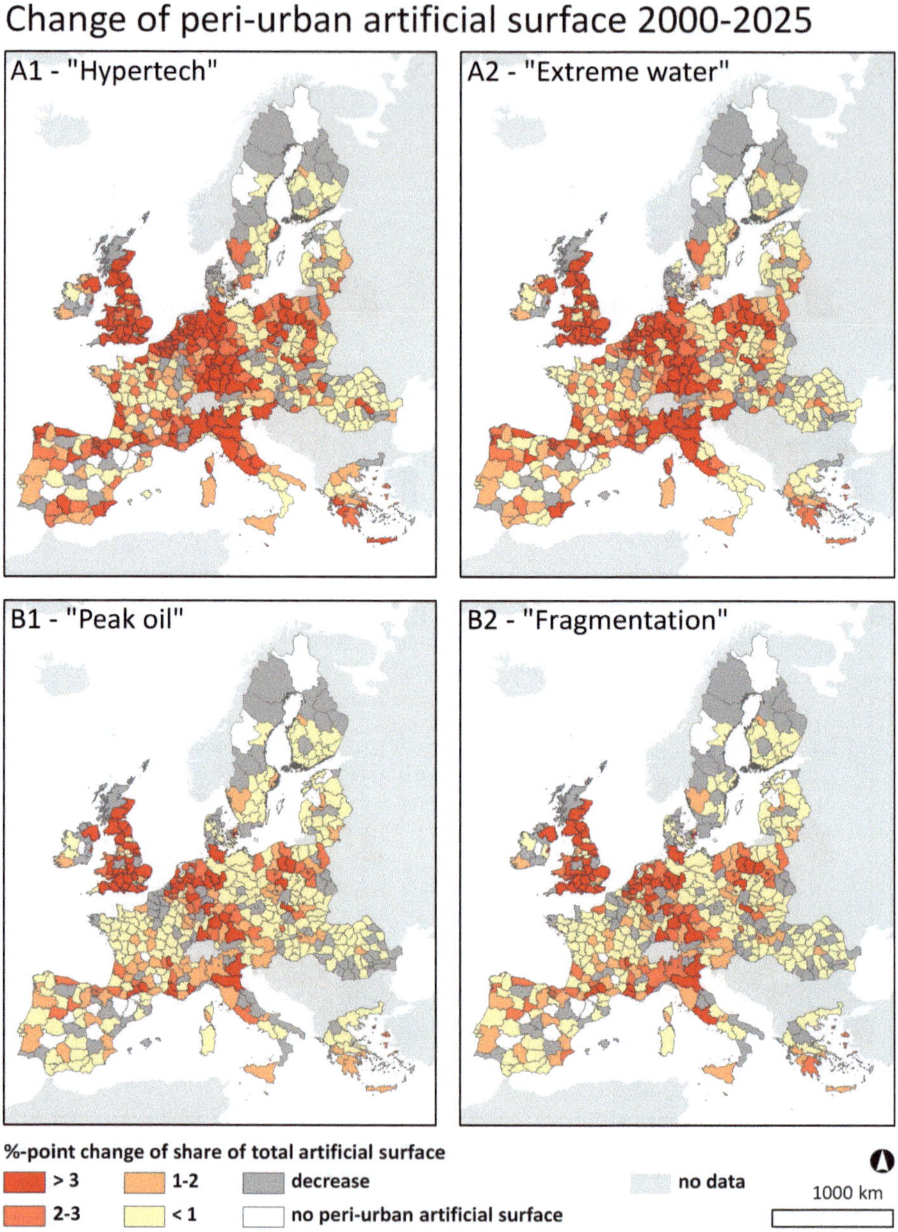


**Fig. 2.9** Change of peri-urban artificial surface share (The maps show the projected change of the share in artificial surface in peri-urban areas relative to the total artificial surface in percent points. In the majority of regions the share of artificial surface in peri-urban areas would increase.)

## 2.7 Implications and Conclusions

This chapter has shown that the peri-urban is, by definition, something in-between, not clearly delineated, a hybrid result of different forces at different scales. It is often defined as a transition zone, on a spectrum from rural to urban, the direct result of urban development and expansion. However, there are other levels of understanding dynamic change such as the regional agglomeration process, the structural effects of globalization and capital accumulation, the transitions and complexity effects and the feedback and collective intelligence from policy and governance. Overall, these urban-driven transitions, which take place in a new kind of territory outside and between the urban cores, can be summed up by the word 'peri-urbanisation'.

The PLUREL scenarios have systematically explored the driving forces and their uncertainties, and translated these into a pan-European modelling scheme. The results highlight some of these wider issues:

- Natural areas would become further marginalised in the 'Pentagon' area, in all scenarios.
- Artificial surfaces would increase the fastest in peri-urban areas, in all scenarios.
- In the B1 and B2 scenarios, with generally stronger governance and planning, the share of artificial surface in peri-urban areas would remain stable in many regions.
- The fastest increase of peri-urban artificial surface would take place in the already highly urbanised regions in Western Europe (Pentagon), UK, the Alpine region, but also northern Poland and northern Spain.

It should be clear from the above, that the peri-urban is a challenging and topical subject in many ways. It is certainly a challenge for scientific research. There are many physical and human processes and interactions to analyse and model and the reality is driven by inter-dependency, uncertainty and complexity. The modelling and analysis tools to describe this fully do not yet exist, so we need to look for other ways of generating useful evidence and insights.

This chapter has fulfilled the purpose of setting the scene for the rest of the book, and providing the larger-scale overview of the European picture, as a context for the case study regions. The scenarios will be used as the basis for the analysis of the issues of urban and rural development in the case study chapters, as a means of developing and testing the strategic thinking of policy makers in each region.

The PLUREL project focussed on the peri-urban, but did so in the context of the rural–urban region. How Europe can be divided into such regions is the next necessary step, and the focus of Chap. 2.

## References

Adell G (1999) Theories and models of the peri-urban interface. A changing conceptual landscape, strategic environmental planning and management for the peri-urban interface – research project. Development Planning Unit, University College, London

Allen P, Engelen G, Sanglier M (1986) Towards a general dynamic model of the spatial evolution of urban systems. In: Hutchinson B, Batty M (eds) Advances in urban systems modelling. North-Holland/Elsevier, Amsterdam

Angel S, Sheppard SC, Civco DL (2005) The dynamics of global urban expansion. World Bank, Transport, Urban Development Department, Washington, DC

Batty M, Longley PA (1994) Fractal Cities: A Geometry of Form and Function. Academic Press, San Diego, Calif

Barredo JI, Kasanko M, Lavalle C, McCormick N, Demicheli L (2003) Sustainable urban and regional planning. The MOLAND activities on urban scenario modelling and forecast, European Commission-Joint Research Centre. http://publications.jrc.ec.europa.eu. Accessed 14 Mar 2012

Beck U (1995) Ecological politics in an age of risk. Polity Press, Bristol

Bell S, Alves S, Silveirinha de OE, Zuin A (2010) Migration and land use change in Europe. A review. Living Rev Landscape Res 4(2). Online article. http://www.livingreviews.org/lrlr-2010-2. Accessed 14 Mar 2012

Bengs C, Schmidt-Thomé K (2006) ESPON 1.1.2. Urban–rural relations in Europe. Final report. ESPON, Luxembourg. Online access. www.espon.eu. Accessed 14 Mar 2012

Bertrand N (2007) Introduction: ESDP ideals and the inheritance of rural planning failures. In: Bertrand N, Kreibich V (eds) Europe's city-regions competitiveness: growth regulation and peri-urban land management. Van Gorcum, Assen, pp 1–35

Borsdorf A, Salet W (2007) Spatial reconfiguration and problems of governance in urban regions of Europe. An introduction to the Belgeo issue on advanced service sectors in European urban regions. Belgeo 1:1–12

Briquel V, Collicard J-J (2005) Diversity in the rural hinterlands of European cities. In: Hoggart K (ed) The city's hinterland – dynamism and divergence in Europe's peri-urban territories. Ashgate, Aldershot, pp 19–40

Bruegmann R (2008) Sprawl: a compact history. University of Chicago Press, Chicago

Brunet R (1989) Les villes europeénnes: Rapport pour la DATAR, Délégation à l'Aménagement du Territoire et à l'Action Régionale, Paris

Bryant CR, Russwurm LH, McLellan AG (1982) The city's countryside: land and its management in the rural–urban fringe. Longman, New York

Caruso G (2001) Peri-urbanisation. the situation in Europe. A bibliographical note and survey of studies in the Netherlands, Belgium, Great Britain, Germany, Italy and the Nordic countries. Report prepared for DATAR, France

CEC (2008) Green paper on territorial cohesion. Turning territorial diversity into strength. SEC (2008) 2550, Commission of the European Communities, Brussels

CEMAT (2007) Spatial development glossary. Council of Europe, Strasbourg

Champion A (1999) Urbanization and counter-urbanization. In: Pacione M (ed) Applied geography. Principle and practice. Routledge, London, pp 347–357

Champion AG, Fielding AJ, Keeble D (1989) Counterurbanization in Europe. Geogr J 155 (1):52–80

Clay G (1994) Real places: an unconventional guide to America's generic landscape. University of Chicago Press, Illinois

CURE (Centre for Urban and Regional Ecology) (2003) Sustainable development in the countryside around towns. Report TX 171. Countryside Agency, Cheltenham

Daniels T (1998) When city and country collide: managing growth in the metropolitan fringe. Island Press, New York

Davis M (2005) Planet of slums. Verso, London

De Roo G, Silva EA (2010) A planners encounter with complexity. Ashgate, Aldershot
Doxiades C (1968) Ekistics. An introduction to the science of human settlements. Oxford University Press, New York
Duhr S, Colomb C, Nadin V (2009) European spatial planning and territorial cooperation. Routledge, Oxford
EEA (2000) CORINE land cover technical guide – Addendum 2000. Technical report no. 40, European Environment Agency, Copenhagen
EEA (2006) Urban sprawl in Europe: the ignored challenge. European Environment Agency, Copenhagen, 10/2006
ESDP (1999) European spatial development perspective: towards balanced and sustainable development of the territory of the European Union, Informal Council of Ministers responsible for spatial planning. European Commission, Luxembourg
Farley P, Roberts MS (2011) Edgelands. Journeys into England's true wilderness. Jonathan Cape, London
Friedmann J, Miller J (1965) The urban field. J Am Inst Planners 31(4):312–320
Gallent N, Andersson J, Bianconi M (2006) Planning on the edge: the context for planning at the rural–urban fringe. Routledge, Abingdon
Galster G, Hanson R, Ratcliffe M, Wolman H, Coleman S, Freihage J (2001) Wrestling sprawl to the ground. Defining and measuring an elusive concept. Hous Policy Debate 12(4):681–717
Garreau J (1991) Edge city: life on the new frontier. Doubleday, New York
Geels FW (2005) Technological transitions and system innovation: a coevolutionary and socio-technical analysis. Edward Elgar, Cheltenham, UK
Giannini E (1994) Metroscape. Dissertation, Royal Melbourne Institute of Technology, Melbourne
Ginsburg NS, Koppel B (2004) The extended metropolis: settlement transition in Asia. University of Hawaii Press, Honolulu
Gottmann J (1961) Megalopolis: the urbanized northeastern seaboard of the United States. Sage Publications, New York
Graham S, Marvin S (2001) Splintering urbanism: networked infrastructures, technological mobilities and the urban condition. Routledge, London
Hall P, Pain K (2006) The polycentric metropolis: learning from mega-city regions in Europe. Earthscan, London
Hardoy J, Mitlin D, Satterthwaite D (2001) Environmental problems in an urbanizing world. Earthscan, London
Harvey D (1985) The urbanization of capital. Blackwell, Oxford
Harvey D (1987) The urban process under capitalism. Int J Urban Reg Res 2(1):101–131
IPCC Working Group III (2000) IPCC special report on emissions scenarios, SRES. UN Environment Programme, New York
Jansson T, Bakker M, Hasler B, Helming J, Kaae B, Lemouël P, Neye S, Ortiz R, Nielsen TS, Verhoog D, Verkerk H (2009) The SIAT model chain. SENSOR report series 2009/2. Leibniz-Centre for Agricultural Landscape Research (ZALF), Müncheberg. http://tran.zalf.de/home_ip-sensor. Accessed 14 Mar 2012
Jones GW, Douglass M (2008) Mega-urban regions in Pacific Asia: urban dynamics in a global era. NUS Press, Singapore
Knox PL, Taylor PJ (1995) World cities in a world-system. Cambridge University Press, New York
Korcelli P, Kozubek E, Piorr A (2011) Economy and employment. In: Piorr A, Ravetz J, Tosics I (eds) Rural–urban regions and peri-urbanisation in Europe. Academic Books Life Sciences, Copenhagen
Kraffczyk D (2004) The MetroScape: a geography of the contemporary city. Harvard University, Graduate School of Design, Cambridge, MA
Lacqiuan A (2005) Beyond metropolis: the planning and governance of Asia's mega-urban regions. Woodrow Wilson Center Press, Washington, DC

Leontidou L, Couch C (2007) Urban sprawl and hybrid city-scapes in Europe: comparisons, theory construction and conclusions. In: Couch C, Leontidou L, Petschel-Held G (eds) Urban sprawl in Europe: landscapes, land-use change and policy. Blackwell, Oxford

Loibl W, Bell S (2011) Population and migration. In: Piorr A, Ravetz J, Tosics I (eds) Peri-urbanisation in Europe. Academic Books Life Sciences, Copenhagen, pp 50–55

Loibl W, Toetzer T (2003) Modelling growth and densification processes in sub-urban regions – simulation of landscape transition with spatial agents. Environ Modell Softw 18(2003):553–563

Maciocco G (2008) The territorial future of the city [electronic resource]. Springer, Berlin

McGregor D, Simon D, Thompson D (2006) The peri-urban interface: approaches to sustainable natural and human resource use. Earthscan, London

Mumford L (1938) The culture of cities. Secker and Warburg, London

Nelson AC, Sanchez TW (1999) Debunking the exurban myth. A comparison of suburban households. Hous Policy Debate 10(3):689–709

Nordregio (2005) ESPON 1.1.1. Potentials for polycentric development in Europe. Final report. ESPON, Luxembourg. www.espon.eu. Accessed 14 Mar 2012

Portugali J (2000) Self-Organization and the City. Springer-Verlag, Berlin

Ravetz J (2000) City-region 2020: integrated planning for a sustainable environment. Earthscan, London

Ravetz J (2011a) Integrated development for peri-urban territorial cohesion. In: Piorr A, Ravetz J, Tosics I (eds) Peri-urbanisation in Europe. Academic Books Life Sciences, Copenhagen, pp 114–123

Ravetz J (2011b) Peri-urban ecology: green infrastructure in the 21st century metro-scape. In: Douglas I (ed) Handbook of urban ecology. Routledge, Oxford

Reckien D, Karecha J (2007) Sprawl in European cities. The comparative background. In: Couch C, Leontidou L, Petschel-Held G (eds) Urban sprawl in Europe: landscapes, landuse change and policy. Blackwell, Oxford

Rickebusch S, Rounsevell M (2009) Where will Europeans live in the future? PLUREL newsletter no. 6, pp 1–3. www.plurel.net. Accessed 14 Mar 2012

Ristimäki M (2011) Mobility and transport. In: Piorr A, Ravetz J, Tosics I (eds) Peri-urbanisation in Europe. Academic Books Life Sciences, Copenhagen, pp 61–64

Robinson GM (1990) Conflict and change in the countryside: rural society, economy and planning in the developed world. Belhaven Press, London

Samir KC, Skirbekk V, Mishra T (2008) Demographic projections for NUTS2 regions in EU countries based on national probabilistic population projection. PLUREL deliverable report 1.2.3. www.plurel.net. Accessed 14 Mar 2012

Schneider A, Woodcock C (2008) Compact, dispersed, fragmented, extensive? A comparison of urban expansion in twenty-five global cities using remotely sensed, data pattern metrics and census information. Urban Stud 45:659–92

Scott AJ (2000) The cultural economy of cities: essays on the geography of image-producing industries. Sage, New York

Shoard M (1983) This land is our land. Paladin, London

Soja E (2000) Postmetropolis: critical studies of cities and regions. Blackwell, Oxford

Spectorsky AC (1955) The exurbanites. Lippincott, Philadelphia

Waltner-Toews D, Kay JJ, Lister N-M (eds) (2009) The ecosystem approach: complexity, uncertainty, and managing for sustainability. Columbia University Press, New York

Zasada I, Berges R (2011) Environment and landscape. In: Piorr A, Ravetz J, Tosics I (eds) Peri-urbanisation in Europe. Academic Books Life Sciences, Copenhagen, pp 72–75

Zasada I, Alves S, Müller FC, Piorr A, Berges R, Bell S (2010) International retirement migration in the Alicante region, Spain: process, spatial pattern and environmental impacts. J Environ Plann Manag 53(1):125–141

# Chapter 3
# Rural–Urban Regions: A Spatial Approach to Define Urban–Rural Relationships in Europe

**Ingo Zasada, Wolfgang Loibl, Regine Berges, Klaus Steinnocher, Mario Köstl, Annette Piorr, and Armin Werner**

## 3.1 Introduction

The previous chapter introduced the range of issues associated with the peri-urban, the subject of this book. The peri-urban as a specific morphological type was defined and the different dimensions of its dynamics were explored. This peri-urban zone is intimately associated with the transition from a dense urban structure to that of a rural character and since it also involves movements into, out of and across it from both these extremes, it is difficult to consider it properly without understanding the broader regional context and dynamics across the urban–rural gradient. Therefore, this chapter will focus on the broader context of urban–rural relationships. Based on recent scientific debates concerning the concept of functional regions and urban–rural relationships, both current and previous definitions and their political implementations are introduced before presenting a new typology to represent Rural–urban Regions (RUR) spatially. Covering the territory of European Union (EU), this typology classifies regions into different types, considering city size, degree of regional mono- and poly-centricity, as well as their urban, peri-urban or rural predominance. The development of the typology includes a further delineation of regions into urban, peri-urban and rural sub-regions, all based on land use patterns and population distribution and density. The typology was subsequently used throughout the PLUREL project and each of the case studies presented in Part Two refers to one of these types, although not all are represented there, since the case studies were unavoidably selected before the typology was developed.

I. Zasada (✉) • R. Berges • A. Piorr • A. Werner
Leibniz Centre for Agricultural Landscape Research (ZALF), Eberswalder Str. 84, 15374 Müncheberg, Germany
e-mail: ingo.zasada@zalf.de

W. Loibl • K. Steinnocher • M. Köstl
AIT Austrian Institute of Technology GmbH, Donau-City-Straße 1, 1220 Vienna, Austria

K. Nilsson et al. (eds.), *Peri-urban futures: Scenarios and models for land use change in Europe*, DOI 10.1007/978-3-642-30529-0_3,

## 3.2 The Region as a Physical, Socio-Economic and Political Entity

With increasing European integration in many spheres, the idea of the region as a spatial entity has been gaining in political, economic and socio-cultural relevance. When facing global problems and issues, such as the recent financial crisis, climate change, an ageing population as well as scarcity and rising prices for food, energy and other resources, it is the *region* which becomes increasingly important as an arena for political strategy- and decision-making. Constant innovations in transport and communications technologies facilitate improvements in accessibility to remote information exchange, permitting, on the one hand, easier home-working and, on the other hand, daily long-distance commuting. With ever-growing flows of goods and people, the predicted "Death of Distance" (Hall 1998) or the "Flat Earth" (Friedman 2006) seems to be taking place. Nevertheless, a majority of economic and social activities remain place-based and most movements occur over relatively short distances. Therefore it is still the local and regional levels which offer solutions to global challenges. The inter-relationships and partnerships between the urban and the rural spheres have attracted political attention particularly through the European Spatial Development Perspective (ESDP) which called for an enhanced spatial balance between the city and the surrounding countryside (COM 1999). However, a lack of spatial coherence between administrative and political decision-making power (e.g. of individual municipal jurisdictions) and the functional dimensions of cities and regions has also been acknowledged (Jensen-Butler 1997; Marvin et al. 2006).

Considering governance inefficiencies resulting from this discrepancy, there have been a number of attempts to delineate and classify regions at both the national and European level in terms of, for example, centrality and urban hierarchy, urban–rural relationships (OECD 2002; Bengs and Schmidt-Thomé 2006) and the degree of urbanity and remoteness (Dijkstra and Poelman 2008; EUROSTAT 2010). However, these approaches often aligned the boundaries of their typologies to administrative and statistical entities, mostly at the NUTS3[1] level, tolerating both the size variances between countries (e.g. the small German *Landkreis* compared to the large Swedish *Län* or Spanish *Provincia*) and the continued neglect of spatial adjacency relationships. Other attempts have aimed at using narrower spatial definitions, such as urban, peri-urban and remote rural areas (Le Jeannic 1997; Marvin et al. 2006). These approaches have provided some insights into the functional relationships and spatial realities and have contributed to the development of appropriate local and regional governance structures, decision-making and policy support mechanisms. In this chapter an alternative method for the territorial delineation of European regions is presented, which aims to address two main objectives missing from the earlier approaches: firstly, the definition of a regional typology which will spatially represent urban–rural relationships at a pan-European

[1] NUTS is an acronym for "Nomenclature d'Unités Territoriales Statistiques" (Nomenclature of Territorial Units for Statistics). EU countries are subdivided into smaller administrative units on three scales. E.g. Belgium: NUTS1 = Regions, NUTS2 = Provinces, NUTS3 = Arrondissements.

level and secondly, the allocation of urban, peri-urban and rural sub-entities explicitly within these regions. Particular challenges to the method derive from the large variety of regions with many degrees of urbanisation, different administrative structures and varying complexities of urban–rural relationships throughout the EU.

## 3.3 Functional Relationships at the Regional Level: Theory and Models

### *3.3.1 Urban–Rural Relationships*

Urban–rural relationships as a subject field are gaining importance in European spatial development research and policy development (Piorr et al. 2011). In their work on urban–rural relationships in Europe, Bengs and Schmidt-Thomé (2006) distinguished the types of relationships as being structural and functional. They defined structural properties as including both the relatively stable physical structure of land use and settlement and the distribution of population. In relation to the physical environment, the functional aspects of urban–rural relationships encompass the modes of production and consumption causing interactions, like commuting and communication, which change over time. Traditionally, urban–rural relationships refer to the capacity of cities' rural hinterlands to supply them with food, fibre, water and energy resources. These roles became marginalised by the growing systems of inter-regional and global resource provision. Nevertheless, the population of the rural surroundings still make use of the functions which only the central city can efficiently provide—mainly a broad spectrum of goods and services, such as specialised medical care, higher education, shopping and cultural activities. Furthermore, relationships between various enterprises also connect rural and urban areas, such as production, trading, consulting or research and development. Modern industrialised agriculture is highly dependent on industrial goods (machinery, fertilisers, etc.) frequently produced in urban areas and the transfer of knowledge delivered by city-based companies (Kalantaridis 2010). Food and resources used by the city are still produced, often in more specialised or niche ways and these include functions like recreation associated with quality of life.

Relationships between land use and spatial configuration were first described by Alonso (1964), who defined the 'centrality' of a place by locational factors such as accessibility or penetration by infrastructure. His land use allocation model was based on a balance between land bid rents, transport and commuting costs. Thus, a hierarchical urban system as described in Christaller's (1933) central-places theory emerged. Regions are defined in this theory as an area of a certain market size distributed around a central place, representing a common catchment area for goods and services available in the central place. The commuting movement between rural living and urban working places represents one of the most obvious relationships linking both areas. Suburbanisation at the edge of the city and socio-economic changes in the rural hinterland driven by influxes of new residents and increasing numbers of the rural population working in the urban labour market also reinforce the pattern of home-work relationships between the urban and rural spheres (Antrop 2004). These activities lead to increasing commuter and goods flows between the rural, the peri-urban and urban.

However, it has been argued more recently that the patterns of flows of people and materials have become more complex and dynamic as the countryside around the cities increasingly provides rural public goods and services (Buciega et al. 2009; Bengs and Schmidt-Thomé 2006). Land use types such as forests, farmland, water bodies and nature areas provide ecosystem services, such as drinking water, soil protection, flood control, outdoor recreation or the moderation of the urban climate. Local self-subsistence, the provision of biotic and abiotic resources and the toleration of changing framework conditions has also received more attention as part of the so-called "metabolism and resilience" debate (Alberti and Marzluff 2004). A post-productive, consumption-orientated adaptation and transition of peri-urban agriculture has led to an increased provision of multiple goods and services to the nearby urban area. Hobby farming, horse-keeping, farm-gate purchase and social farming are recent land use phenomena mainly found at the urban fringe, where farmland is recognised as part of the landscape setting and as recreational space for urban dwellers (Zasada 2011; Zasada et al. in press).

### *3.3.2 Functional Urban Regions*

Since urban–rural relationships are based on socio-economic interactions, they require either spatial adjacency or accessibility provided through the transport infrastructure. Therefore most models try to capture the spatial extent of these relationships in geographical terms by using commuting and travel time distances (e.g. 1 h commuting time) to define the outer limit of the functional urban region. The OECD (2002) defined a *Functional Urban Region* (FUR) as

> A territorial unit resulting from the organisation of social and economic relations in that its boundaries do not reflect geographical particularities or historical events. It is thus a functional sub-division of territories. The most typical concept used in defining a functional region is that of labour markets. (OECD 2002, p. 11)

A FUR thus represents a common regional market for labour (where maximum marginal commuting times represent the outer limits of the labour catchment), land, housing and intra-regional trade. First attempts to delineate the catchment areas of urban agglomerations date back to the 1950s, when in the United States the concept of the Standard Metropolitan Statistical Area was adopted. Later, Hall and Hay (1980) developed the concept of the Daily Urban System in Europe, which was based on critical masses of population amount and density as well as daily commuting levels to a central city. The FUR became a particularly relevant model as part of the new economic geography and its focus on the role of spatial proximity for innovation processes (Porter 1990). At the European and global level, urban agglomerations gain their competitive advantage from their roles as centres and gateways of social and economic development, innovation and creativity, transport and communication (Anderson 2000). In promoting competiveness and economic development, cities require the integration of their surrounding area in order to expand the size of their functional reach, e.g. by improving the efficiency of their transport systems.

Infrastructure improvements which lead to expansion of the FUR, provide important preconditions for peri-urbanisation (see Chap. 1 in this book). Diminishing commuting time-distances, decreasing transport costs and the overall increase of car ownership and accessibility within regions all result in land use transformations. Urban sprawl, low density settlements (Couch et al. 2007), re-concentrations at infrastructural nodes (Bontje and Burdack 2005) and marginal open space segments of different land uses characterise current peri-urban landscapes (Sieverts 2003; Meeus and Gulinck 2008). Depending on the definition, these areas represent a considerable amount of the territory of European countries, e.g. 22% in France (Cavailhes et al. 2004) or 20% in the United Kingdom (UK) (Gallent et al. 2006). The FUR concept embraces a wide range of different regional types, but lacks common definitions and spatially continuous delineations based on criteria such as population density, commuting distance or settlement topology and morphology.

Owing to their relevance for spatial coherence and economic competitiveness, urban–rural relationships and functional urban regions appeared on the political agenda concerning polycentric spatial development a decade or so ago and again recently (COM 1999; ESPON 2005a; METREX 2010). In 1999, the EU initiated the European Spatial Development Perspective, officially formulating the political objective of polycentric development with strong urban–rural relationships and partnerships. Aiming at a balanced pattern of regional growth, making efficient use of endogenous potentials and moderating regional differences, the polycentric development concept represents a cornerstone of current spatial development policy at the European level (COM 1999).

Looking ahead from theory to implementation in development policy and planning action "on the ground", polycentric development and urban–rural relationships require substantial knowledge and strengthened evidence based on the actual situation of spatial structures and processes within regions. In the final report of the Study Programme on European Spatial Planning, the authors state clearly that:

> The development of policies aimed at encouraging co-operation between rural and urban areas would be greatly facilitated if a typology of regions, reflecting different forms of urban–rural relationships, can be discerned. (Nordregio 2000, p. 23)

The next section of the chapter examines some of the current and previous approaches for regional typologies.

## 3.4 Current and Previous Regional Model and Typology Approaches

A number of national and European approaches exist which attempt to differentiate geographical areas and regions along an urban–rural continuum. Some of the models represented in Table 3.1 define entire regions based on their relative urbanity. Others aim at a spatially explicit identification of suburban and peri-urban areas within regions strongly influenced by an urban core, comprising a functional urban area and its spatial expansion as a result of urban–rural dynamics.

In their study on the diversity of the rural hinterlands of European cities, Briquel and Collicard (2005) demonstrated that in different countries diverging concepts of rurality and urbanity exist according to varying criteria of land use, settlement patterns, population size and commuting flows. Reviewing definitions and classifications of FURs, the OECD (2002) also found large differences. The French statistical institute, INSEE, has undertaken several attempts to define areas under urbanisation, including urban core areas (*Unité urbaine*) and the surrounding peri-urban areas (*Couronne périurbaine*) (Le Jeannic 1997). In Germany, settlement structure types are classified at various spatial levels. At the regional level (NUTS2), three types can be identified: agglomeration areas, urbanised areas and rural areas. Depending on the number of inhabitants, counties (NUTS3) are defined as being core cities, highly densified counties, densified counties or rural counties (BBR 2005). A framework for city-regions has also been developed in the UK to identify nodes and catchment areas for functional urban areas based on commuting patterns for different purposes such as work and leisure (Marvin et al. 2006).

At the European level, various approaches have been used to trace the spatial extent of regional functional urban areas, degrees of urbanity and rurality and urban structures. Together with the European statistical agency, EUROSTAT, the European Commission set up the Urban Audit programme to evaluate quality of life in urban areas (COM 2004). For the data analysis a spatial concept of Larger Urban Zones (LUZ) was developed. These encompass urban cores and their surrounding peri-urban region and they mainly cover an area equivalent to the FUR. The European Spatial Planning and Observation Network (ESPON) put forward yet another approach to delineate Functional Urban Areas (FUA). This made use of national FUA definitions and was complemented by population figures and an area radius around the FUA centre defined as accessible in 45 min by car. According to sectoral criteria, including transport, tourism or population, the authors classified FUAs ranging from Metropolitan European Growth Areas, Transnational/national FUAs and regional/local FUAs (ESPON 2005b).

Urbanity and rurality represent important regional properties at the heart of many political and research questions. The OECD (2007) developed a well-known and widely applied typology of European NUTS3 regions based on a

**Table 3.1** Regional classifications of urban regions in the EU

| Name | Types | Criteria | Spatial scale | Source |
|---|---|---|---|---|
| *Single units* | | | | |
| Functional region classification | Depends on countries: e.g. labour market districts (AT), commuting catchment area (DK), NUTS4 (PL) | Commuting conditions; travel time; depends on country | Depends on country | OECD (2002) |
| Urban Audit (UA) | Large urban zones | Approximation of existing functional urban regions to NUTS3/LAU1, country specific definitions | NUTS3 or LAU1 for selected towns/cities | EC/Eurostat (2004) |
| *Different types* | | | | |
| Regional types of urban-rural spatial patterns | Regions dominated by large metropolis; polycentric regions with high urban and rural densities; polycentric regions with high urban densities; rural areas under metropolitan influence; rural areas with networks of medium-sized and small towns; remote rural areas | Share urban population; density rural population; distribution of settlement size; population weighted distance to urban settlements; share of population in biggest city; number of inhabitants | NUTS2/NUTS3 | Moriconi-Ebrard, Geopolis (1994) |
| Typology of rural–urban regional settings | Mono- or polycentric metropolitan areas; Mono- or polycentric regions with medium sized cities; Mono- or polycentric regions dominated by smaller cities or without cities >= 50,000 inhabitants | Size of the city, commuting distance, share of population in biggest city, number of inhabitants, overlapping catchments (1 h travel time) | Partly aggregated NUTS3 | BBR (1999) |
| Typology of functional urban area (FUA) | Metropolitan European growth areas (MEGA), transnational/national FUAs and regional/local FUAs | National FUA definitions, number of inhabitants, travel time to centre, population, tourism indicators, etc. | Depends on FUA definition | ESPON (2005b) |
| Urban rural typology | High/low urban influence combined with high/medium/low human intervention | Population density, Functional Urban Area rank, land use shares | NUTS3 | Bengs and Schmidt-Thomé, ESPON (2006) |

(continued)

**Table 3.1** (continued)

| Name | Types | Criteria | Spatial scale | Source |
|---|---|---|---|---|
| Urban-rural typology of NUTS3 regions | Predominantly urban; intermediate; predominantly rural | Population density, Share of population living in rural areas, population number in urban centre | Territorial level TL 3 (partly NUTS3 level) | OECD (2007) |
| Urban-rural typology of NUTS3 regions | Predominantly urban regions; intermediate regions close to a city; intermediate, remote regions; predominantly rural regions close to a city; predominantly rural, remote regions | Population density, Share of population living in rural areas, population share in proximity to bigger cities | NUTS3 | Dijkstra and Poelman, EC (2008) |
| Urban-rural typology | Predominantly urban; intermediate; predominantly rural | Population density, distribution and numbers in urban centres | NUTS3, partly aggregated | EUROSTAT (2010) |

definition of urban and rural communities (using a population density threshold of 150 inhabitants per $km^2$ in the LAU2[2] as the threshold for distinguishing between urban and rural). Depending on the share of the regional population living in rural areas, regions are classified as "predominantly urban", "intermediate" or "predominantly rural". Since the OECD system does not take relationships between adjacent NUTS3 areas into account, Dijkstra and Poelman (2008) developed a supplementary criterion which considers the proportion of people living within the travel radius of 45 min of cities with more than 50,000 inhabitants. In doing so, they derived two further regional types of "regions close to cities" and "remote regions". More recently this typology has been refined further by EUROSTAT (2010) in order to cope with methodological shortcomings due to the varying sizes of regions and communities. Thus, population density was based on a more precise 1 $km^2$ grid and small NUTS3 regions were merged. A different regional classification was also developed by Bengs and Schmidt-Thomé (2006). Combining their FUA classification with population density and the share of artificial surface, the degree of "urban influence" and "human intervention" were also taken into account.

More narrowly addressing poly-centricity questions, Moriconi-Ebrard (1994) developed a typology of "Regional types of urban–rural spatial patterns" for the NUTS2/3 level. The analysis focused on population distribution and densities in urban and rural areas as well as on distances between centres and settlement size distribution. As a result, he distinguished amongst "regions dominated by a large metropolis", "polycentric regions with high urban and rural densities", "polycentric regions with high urban densities", "rural areas under metropolitan influence", "rural areas with networks of medium-sized and small towns" and "remote rural areas".

However, the main disadvantage of these earlier classifications is the lack of accurate, spatially continuous delineation to allow intersecting of, for example, the CORINE land cover (CLC) maps with administrative borders. Another problem is the frequent incompleteness of the European coverage—many classifications consider only the densely urbanized regions of Europe and ignore the rest.

## 3.5 The Rural–Urban Region: A Spatial Typology

Within the research context of PLUREL, the necessity for a broadly applicable regional typology meeting requirements which existing approaches do not fulfil was identified. The typology was necessary to provide a spatially explicit analysis and modelling framework to be applied, using statistical and geospatial data, for the purpose of regression modelling on urbanisation impacts. The available urban region classifications, as noted above, lacked spatially explicit regional delineation

[2] LAU – Local Administrative Unit.

and only considered the densely urbanised regions of Europe, ignoring the rest. Therefore, a new regional typology of *Rural–urban Regions* (RUR) was developed to meet the following objectives:

1. Since it should cover the full range of urban–rural relationships, including those which evolve in remote rural areas, a system was needed to cover the whole of Europe, filling in the blank spaces beyond the boundaries of existing FURs (see Fig. 1.1 in Chap. 1). Therefore the RUR was defined as a spatial entity which covers the urban core areas, passing the peri-urban commuter belt and reaching the rural hinterland.
2. In order to provide the spatial framework for pan-European-level modelling, it was necessary to ensure a consistent pan-European coverage in terms of geographical location and degree of urbanisation.
3. Observation and modelling of processes and impacts of peri-urbanisation were to be applied at different spatial levels in the project. Thus, a categorisation of regions defined by population size and density as well as settlement morphology criteria (e.g. mono-centric/polycentric/rural) was required. This also needed to be complemented by a spatially explicit sub-regional division into urban, peri-urban and rural areas representing different influence spheres—the urban cores, the peri-urban surroundings and the rural hinterland.

At the outset, the purpose of the typology was to define classes of RUR which show high spatial similarity, being characterised by population size, population density and settlement distribution (mono-centric vs. polycentric). These types of different RURs were to be used as the bases for the sustainability impact assessment of land use change, particularly urban growth. Based on this analysis, planning and policy advice aiming at sustainable development for the different RURs was to be derived. In order to meet these objectives, certain preconditions were required:

1. Spatial comprehensiveness: Not only the large urban regions but the entire EU should be covered by RURs in order to explore the variety of regional impacts of urban development projections.
2. Data limitation: The RUR typology and sub-regional delineation should be based on a limited but robust set of statistical and spatial data with European coverage.
3. Application at different geographic scales: a pan-European coverage would require an adequate set of regional types as well as robust threshold indicators for an EU-wide RUR classification and sub-regional delineation, whereas for the case study analysis a further level of detail would be needed at a sub-regional scale.

On a pan-European level, the smallest unit for which data is available is NUTS3. These regions are usually large administrative and statistical units covering urban centres and their peri-urban fringe, but in some countries (such as Germany or Austria) they consist of either urban cores or peri-urban or rural areas. Thus, as a first step in developing the typology, NUTS3 regions which cover only parts of a

**Table 3.2** Characteristics of regional morphology types

| Morphology type | Characteristics |
|---|---|
| 1. Monocentric | Regions with a core city area without notable peri-urban sub-centres, three sub-types by core city size: very large (+metropolitan), large, medium |
| 2. Urban polycentric | Regions with one or more core cities and peri-urban sub-centres |
| 3. Dispersed polycentric | Regions with several (medium-sized) peri-urban centres |
| 4. Rural | Regions without notable centres but dispersed smaller low density settlements |

RUR had to be aggregated into clusters, thus generating a functional urban region of urban core, peri-urban fringe and remote rural hinterland—related to the respective urban centre of the RUR. As a result 1,278 NUTS3 regions were aggregated into 904 RURs. In this way, a compact set of regional entities was defined as a backbone for applying various forms of analysis, such as regression models, in order to assess sustainability impacts of potential land use changes.

Since the overall objective was to distinguish distinct morphology classes as the spatial framework for urban–rural relationships, and keeping in mind the limited spatial data sets available for the entire EU at NUTS3 level, the classification focused on the distinction between intra-regional mono- and poly-centricity. Table 3.2 provides an overview of the morphological types generated by the method.

As the RUR morphology classification is based on the number and size of urban centres within a region, urban centres which serve as RUR nuclei were identified first. Due to the large number of urban centres, NUTS3 entities and the EU-wide scope of the analysis, the RUR delineation and classification was carried out by combining spatial and statistical data in GIS. Limited data availability at the European level, however, was a key barrier to developing a sophisticated data-driven typology. Data available for the entire EU includes CORINE Land Cover 2000 (CLC2000) provided by the European Environmental Agency (EEA), and total population data from 2000 to 2006 for NUTS3 entities. Additionally population data from the GISCO urban centre point data base (STEU) with population numbers above 10,000 (in Germany and Netherlands above 20,000) were made available to match around 5,000 larger urban settlements with their inhabitant numbers.

At the preparatory stage a preliminary classification of single NUTS3 regions was conducted by ranking them based on the cities' population numbers. To identify settlements, urban fabric patches were extracted from the CLC2000 data set and neighbouring or adjacent settlement patches were merged into the settlement area featuring a "Settlement Morphological Zone" (SMZ). The STEU centre points and the settlement areas were geometrically intersected to allow a link between settlement areas and the related population numbers to be made. Settlements containing a STEU point with population numbers greater than

10,000 were defined as "urban centres" and those exceeding 100,000 inhabitants were termed "core cities". NUTS3 entities featuring a core city were then selected as initial RUR centre regions. NUTS3 regions with no or only small urban centres were treated as catchment areas and merged with one of the neighbouring initial RUR centre regions. For these "centre-less" NUTS3 entities the urban centre to which they mainly related to had to be identified. Therefore, information on interactions (e.g. commuter flows) was required in order to quantify the physical extent of the urban centre spheres of influence. As these data were not available for the EU at NUTS3 sub-region level, population numbers of urban centres served as proxies for the extent of interaction instead. These numbers were weighted through a logarithmic function for defining the extent of the commuting catchment of the respective RUR centres. The NUTS3 regions without a distinct larger urban centre were then merged with the initial urban centre region which showed the largest coverage of influence circles in order to achieve "complete" RURs. Figure 3.1 shows the urban centres (in blue) with their influence sphere around the RUR centre (a geometric buffer around the SMZ) and the initial morphology class.

Figure 3.2 depicts the final six main morphology classes derived by applying this method: the urban mono-centric class is divided into three sub-classes, while the polycentric classes show either urban or rural characteristics, depending on the size of the centres. The resulting typology indicates a majority of urban mono-centric RURs, lacking distinct peri-urban sub-centres. Urban polycentric RURs (orange) are concentrated in the UK (Central and Southern England), large parts of Belgium and the Netherlands, as well as regions in Poland (Silesia), Germany (Lower Rhine) and Northern Italy. The majority of dispersed polycentric RURs (yellow) with some small urban centres embedded in rural areas, are located in the Eastern part of EU (Poland, Slovakia, Hungary, Romania) and scattered over France. The remaining rural RURs (green) with no distinct urban centres are located in mountainous and forested areas in peripheral Southern and Northern Europe, in the Alps and scattered over Germany and Austria.[3]

## 3.6 Sub-Regional Urban–Rural-Delineation

The sub-regional delineation was applied in order to identify urban, peri-urban or rural areas within individual RURs. Therefore, an automated approach, applicable for the whole of Europe, was required. The database used for this consisted of a 100 × 100 m grid cell map for population density provided by EEA (Gallego 2010), municipality boundaries, LUCAS (Land use/cover area frame survey), 4,887 STEU

[3] Here the NUTS3 region-size influence is obvious—even when aggregated to RUR: large regions tend towards polycentric classification, small regions tend towards mono-centric classification. Widening core city buffers to define larger RUR would result in fewer—mainly polycentric—RURs which, however, would mask the smaller cities' influence spheres.

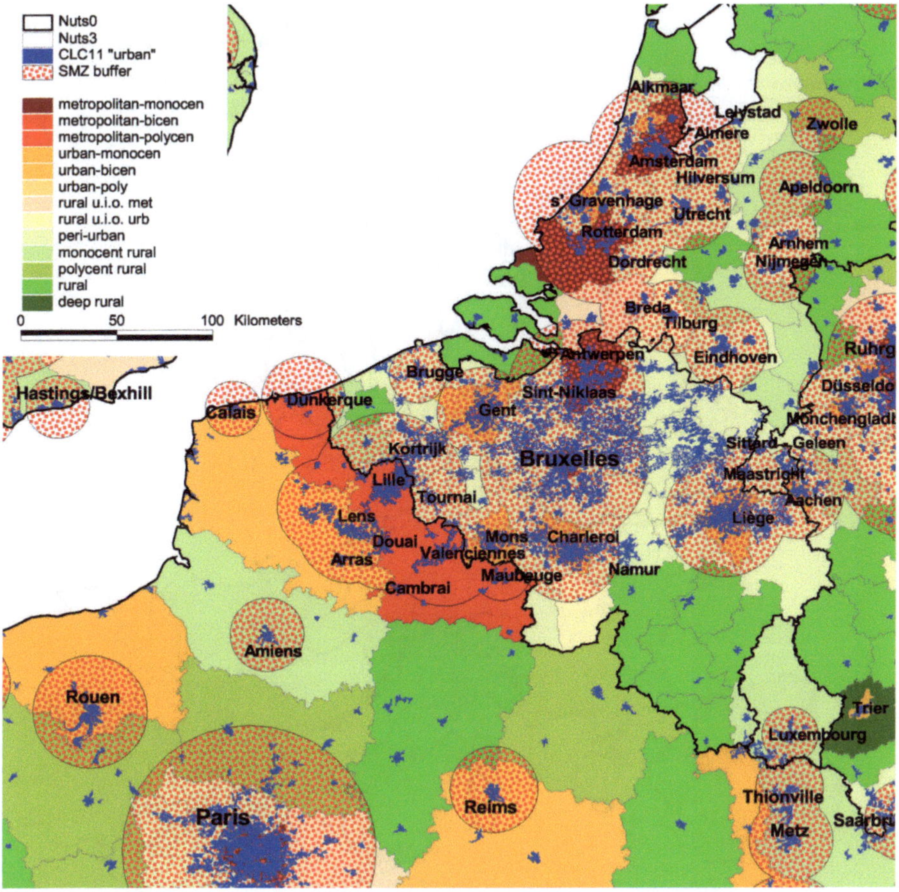


Fig. 3.1 Clustering process of NUTS3 entities to rural urban regions (detail)

points with population numbers greater than 10,000 compiled by EUROSTAT for the year 2000 and CLC2000 data. The approach makes use of population density in certain land cover classes applied via a logistic regression model. The delineation was conducted stepwise, applying map algebra functions to establish GIS-based Boolean decision rules for extracting the RUR sub-regions. Finally, all sub-region layers were extracted from the input data sets and were merged into one RUR sub-region layer. The delineation resulted in six (populated) sub-region entities:

– *Urban high density areas (1)* were defined by CLC class 111 (continuous settlement area), if inside the general urban sub-region. No population density values were applied as threshold for delineation, because high density urban cores are not necessarily inhabited, but also contain business, manufacturing and commercial areas with service and retail workplaces.

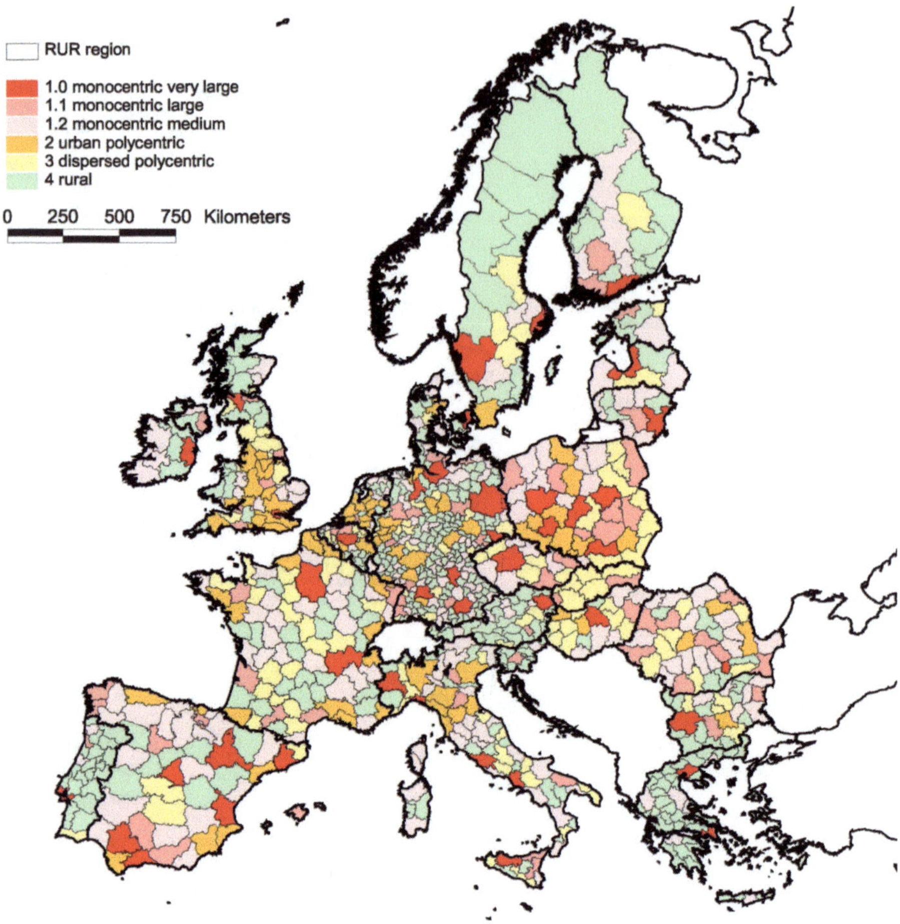


**Fig. 3.2** RUR delineation and morphology classification for EU

- *Urban low density areas (2)* are CLC class 1 (artificial surfaces), excluding CLC class 13 (mining) and with population numbers in the respective centres above 20,000 inhabitants.
- *Peri-urban high density areas (3)* are defined by a population density greater than 75 inhabitants per km$^2$ or by CLC class 11 (settlement area) and STEU-population numbers above 10,000 inhabitants within the low density regions.
- *Peri-urban low density areas (4)* are defined by population density greater than 40 inhabitants per km$^2$ when located within 300 m of urban areas. The density threshold avoids the need for complex rules for including or excluding certain land cover classes. The 300 m distance criteria assumes a spatial connection between urban core regions and avoids the exclusion of areas separated from urban areas by rivers or small open space corridors.

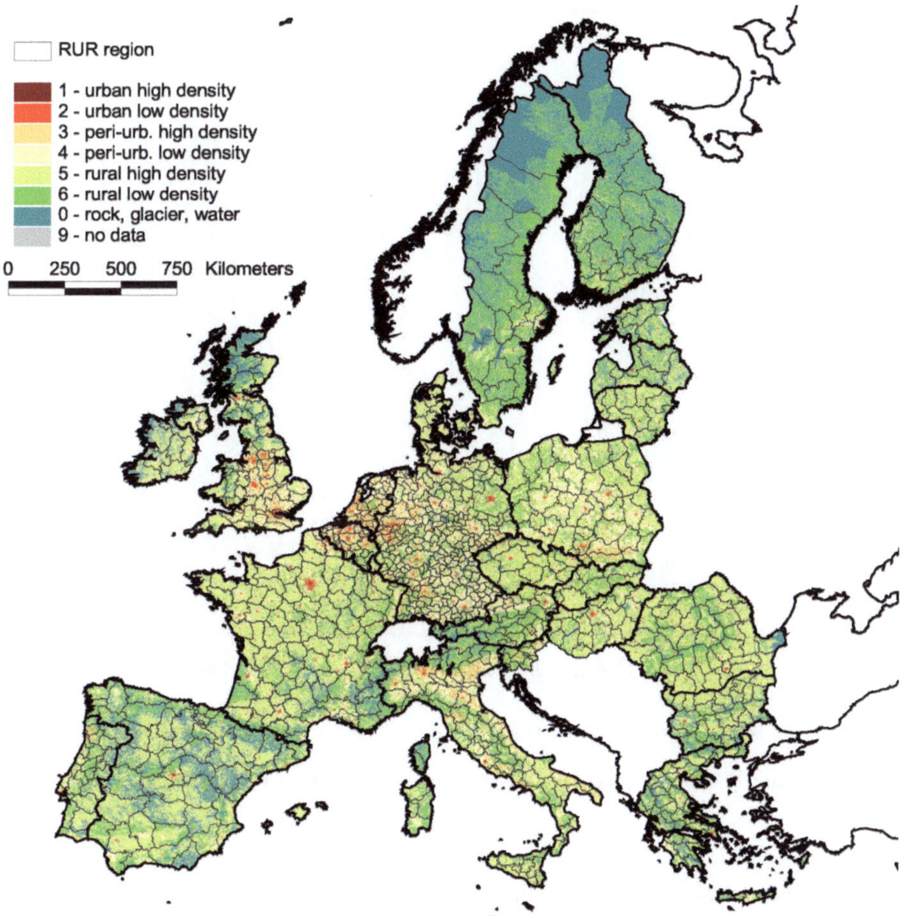


**Fig. 3.3** Sub-region delineation into urban, peri-urban and rural areas of high and low population density

- *Rural high density areas (5)* are defined by a population density greater than 10 inhabitants per km$^2$.
- *Rural low density areas (6)* are defined by a population density greater than 0 inhabitants per km$^2$. This classification includes all remaining areas which host any population at all, even if very little.
- The remaining unpopulated areas are natural areas, military territories or similar. They were kept separate to avoid "immigration" during the later modelling tasks to simulate population growth and land use change at a regional to local level.

The RUR sub-regional delineation in the EU shown in Fig. 3.3 is quite comprehensible: the concentration of the urban high density area classes can be identified

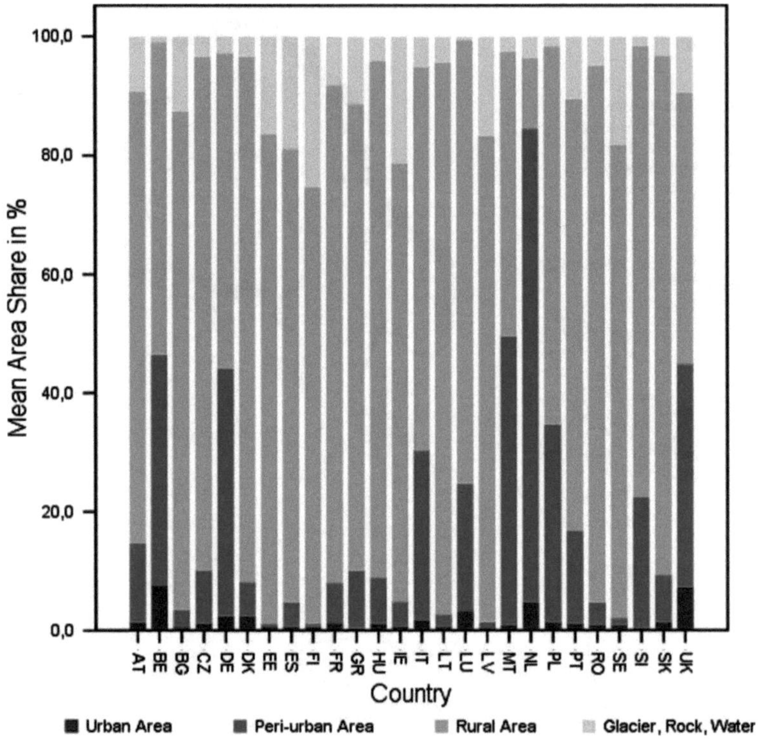


**Fig. 3.4** Share of urban, peri-urban and rural areas per country EU27 (no data available for Cyprus)

in the European capital regions. Peri-urban areas mainly occur in RURs in the UK, the Netherlands, Belgium, northern Italy, western and southern Germany along the Rhine valley and in southern Poland.

## 3.7 Rural–Urban-Regions and Their Sub-Regional Composition

The distribution of urban, peri-urban and rural areas across the EU is characterised by significant differences between countries (see Fig. 3.4). Belgium (7.7%) and the UK (7.3%) possess extensive amounts of urbanised areas. The Netherlands accounts for the highest share of peri-urban area (79.7%) particularly with high densities (43.5%), followed by Malta (48.7%), Germany (41.7%), Belgium (38.8%) and the UK (37.5%). By contrast, rural areas dominate in the Nordic, Baltic and southern European countries (73.4–92.8%), with even partly uninhabited and depopulating areas. Generally, the differences in the composition of urban, peri-urban and rural areas in the countries of the EU arise due to different urban development and planning traditions which have led and continue to lead to different urbanisation patterns. In the southern European countries, many urban areas with no real growth dynamics are much more compact, often lacking urban green areas and are similar to many cities in neighbouring dry climate regions in the Mediterranean. Conversely, in the rest of Europe, the differences are strongly related to the population density in the various countries, with extremely low densities in the Nordic countries and extremely high ones in the Netherlands and Belgium.

Even though Belgium and the Netherlands share similar spatial planning traditions and population densities, when looking at the detailed patterns for peri-urban high and low densities, strong differences become apparent. Whereas the Dutch regions are characterised by a significantly higher amount of densely populated peri-urban areas (43.5%), the peri-urban areas in Belgium tend to be low density (36.3%). Leinfelder (2009) referred to a study revealing that Belgian regions have the highest ratios of settlement area per inhabitant compared to Dutch, French and German regions. It provides evidence that sub-urbanisation is a stronger force in Belgium than in the Netherlands, where specific policies are in place to protect the remaining open space (e.g. the Green Heart in the Randstad region). Apart from urbanisation traditions and population densities, it is also the administrative structure and planning systems over recent decades which influence the peri-urbanisation processes of a particular country.

Figure 3.5 provides an overview of the average shares of urban, peri-urban and rural areas in relation to the six different RUR types. Predictably, the mono-centric RURs with very large centres encompass comparably high shares of urban area, while urban polycentric RURs, while comprising a similar amount of urban area show a much higher share of peri-urban area. Urban and peri-urban areas further decrease while rural as well as uninhabited areas increase in the following sequence: Large and medium mono-centric, dispersed polycentric and finally rural. The mono-centric medium-sized RUR is quite similar to the dispersed polycentric RUR. The former has a higher share of peri-urban low density areas, whereas the latter is mainly characterised by rural high density areas. This aspect also differentiates the mono-centric medium type from the rural equivalent, which has the highest share of area defined as low density and uninhabited rural areas. Thus, there is a generic relationship between the morphological rural–urban structure of a region and the distribution of urban/peri-urban/rural sub-entities.

## 3.8 Analysing Regional Patterns of Peri-Urban Agriculture: An Application of the RUR Typology

As one example of urban–rural relationships, local food production has recently come back into fashion, because many sophisticated middle-class urban consumers increasingly prefer locally grown or regionally specific products to those from the

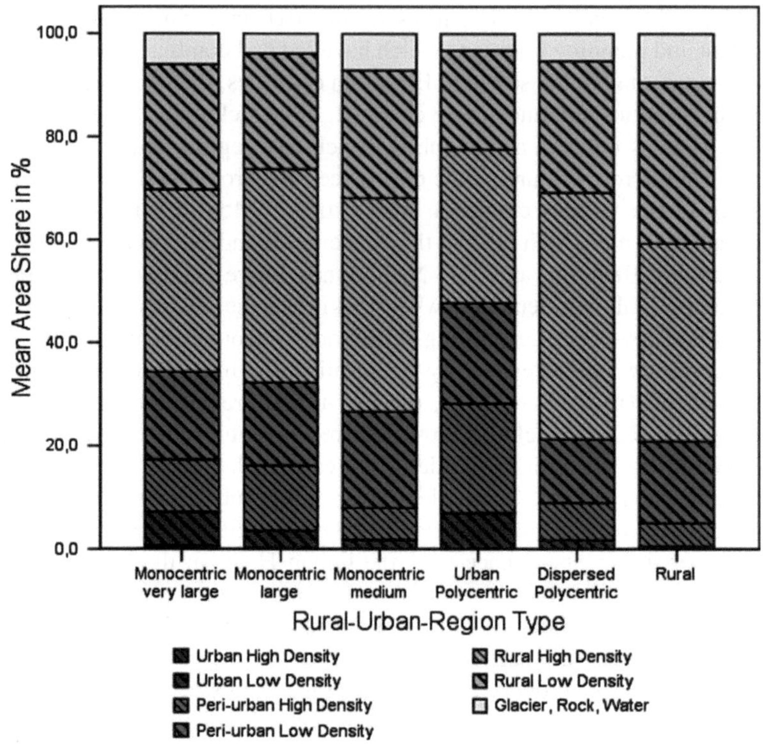


**Fig. 3.5** Share of urban, peri-urban and rural areas per RUR type

global market. Subsequently, farming in urbanised areas has tended to become specialised in high-value fresh products, such as vegetables, berries or flowers. The need for agricultural production in the rural hinterland to be close to the central city, already described in the classical model by Heinrich von Thünen in 1826 (Hall 1966), is found once more as a result of this renaissance with a focus on specialised and high-value products. Von Thünen was the first to discuss the functional relationships occurring between the city and its surroundings on the basis of food supply. He developed the "isolated state" model illustrating the balance between land and transport costs triggering certain land uses and land use intensities forming a pattern of concentric circles around a city. According to the model, the farmer balances the cost of transport, land, production and revenue, resulting in the most cost-effective product for the market. Thus the economic relationships between a centre and the complementary surrounding region were understood for the first time.

Today, owing to their proximity to urban markets and potentially close consumer-producer relationships, peri-urban farmers possess comparative advantages concerning their ability to specialise in urban-orientated production and to adopt direct marketing activities, allowing them to generate higher revenues.

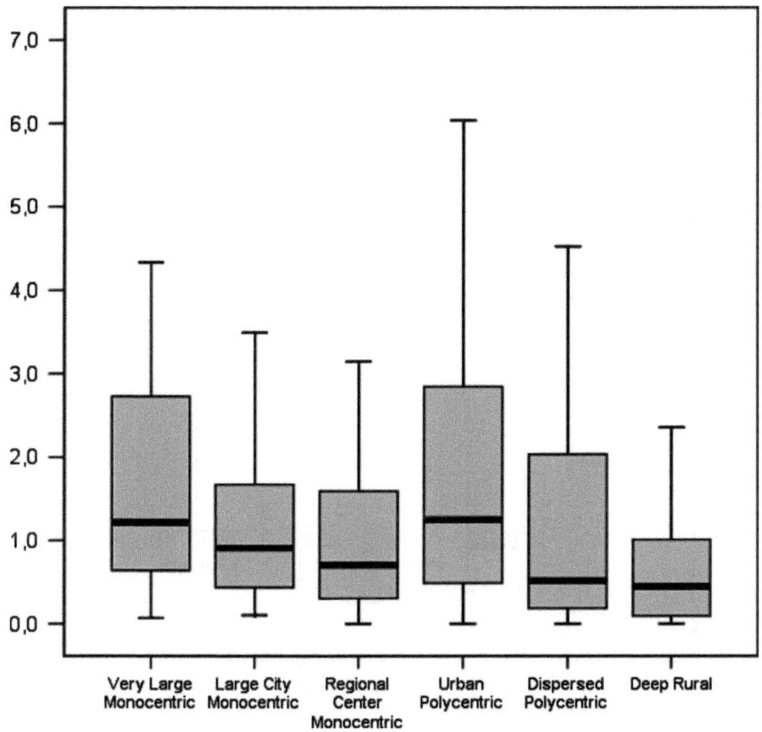


**Fig. 3.6** Share of UAA for horticultural products (per RUR type) in % (EU27, year 2000) (Database Eurostat)

The utility of the RUR typology, described above, can be demonstrated by analysing whether or not the farming sector is more specialised in horticultural products in urbanised regions and peri-urban areas. A cross-regional comparison of all European regions revealed the spatial distribution of (1) the share of the agricultural area (UAA) dedicated to horticulture as well as (2) the share of holdings which are specialised in horticultural production.

Based on the RUR typology, an analysis of variance (ANOVA) was applied, to compare the farming systems in the different RUR types and to examine the influence of urban, peri-urban and rural areas on the prevalence of horticultural production. The results show that the RUR typology significantly contributes to the explanation of the regional value distribution of horticultural area and farm holdings. Figure 3.6 demonstrates that the share of holdings specialising in horticulture is nearly 3.6 times (urban polycentric) and 2.7 times (very large monocentric) greater than in comparable rural regions. In urbanised regions noticeably more farmland is also used for horticultural production (Fig. 3.7). Furthermore, there is evidence that polycentric regions are characterised by a stronger horticultural specialisation and concentration than comparable mono-centric urbanised

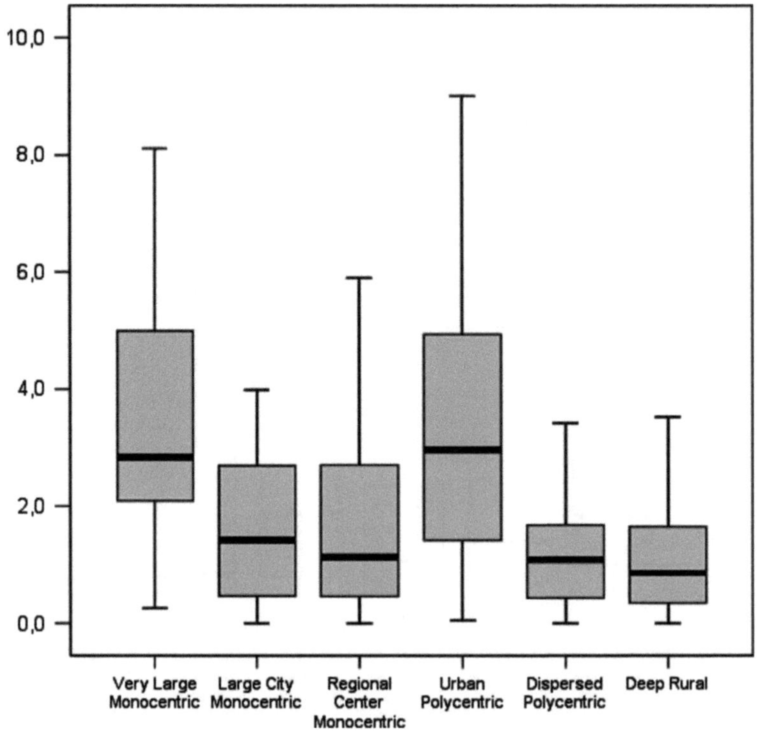


**Fig. 3.7** Share of holdings, specialised on horticulture (per RUR type) in % (EU27, year 2000) (Database Eurostat)

ones. The application of the sub-regional typology revealed that horticultural areas are most prevalent in peri-urban areas, whereas the adjacency to urban, particularly high density areas, brings about high shares of specialised farms (Figs. 3.8 and 3.9).

The RUR typology therefore enables us to differentiate among European regions according to the presence of urban centres of a specific population size. It helps to understand the intensity and potential of urban–rural relationships in a region. The sub-regional, spatially explicit delineation of urban, peri-urban and rural high and low density areas focuses in addition on the specific locational context where agriculture takes place—whether directly within or adjacent to urban agglomerations or more distant from them. Furthermore, it also takes the population density in the particular area into account. Both types are characterised by significant differences between classes, although they suggest the influence of additional factors, such as the geographical location within Europe. Major variations in farming traditions, bio-physical and socio-economic differences need to be taken into consideration for understanding the degree of specialisation and productivity of farming. For instance the production of ornamental plants (e.g. in the Netherlands) and vegetables and fruit (e.g. south-eastern Spain) are characterised by regional concentration patterns.

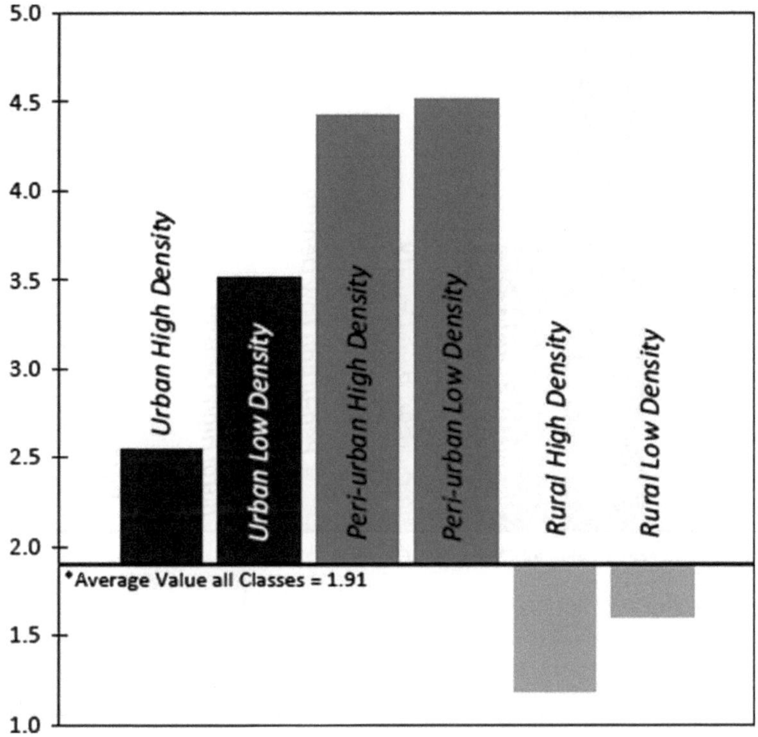


**Fig. 3.8** Average share of UAA for Horticultural Products (per sub-regional type) in % (EU27, year 2000) (Database Eurostat)

## 3.9 Conclusion: Relevance for Research and Policy

This chapter has introduced a novel alternative approach for delineating RUR-regions which derives its strength from the simultaneous application of the issue of urban centrality at the regional level for the issue of peri-urbanity at the pan-European as well as regional scales. Furthermore, the RUR typology extends the scope of the functional urban region by also including the rural hinterland, which is not the case with previous approaches. In this context, the spatially more explicit subdivision of the RUR types enables detailed insights into intra-regional differences to be obtained. The RUR typology and sub-division represent an approximation to describe the complex and diverse situation of RUR regions throughout the EU as a whole. Researchers aiming to provide estimations of rural–urban development for the whole EU can thus make use of it. It is also a way to present the pattern more simply while, at the same time, capturing the quite complex issue of urban–rural regions. Thus it is also a tool for policy support, enabling policy makers at the EU level to evaluate regional developments.

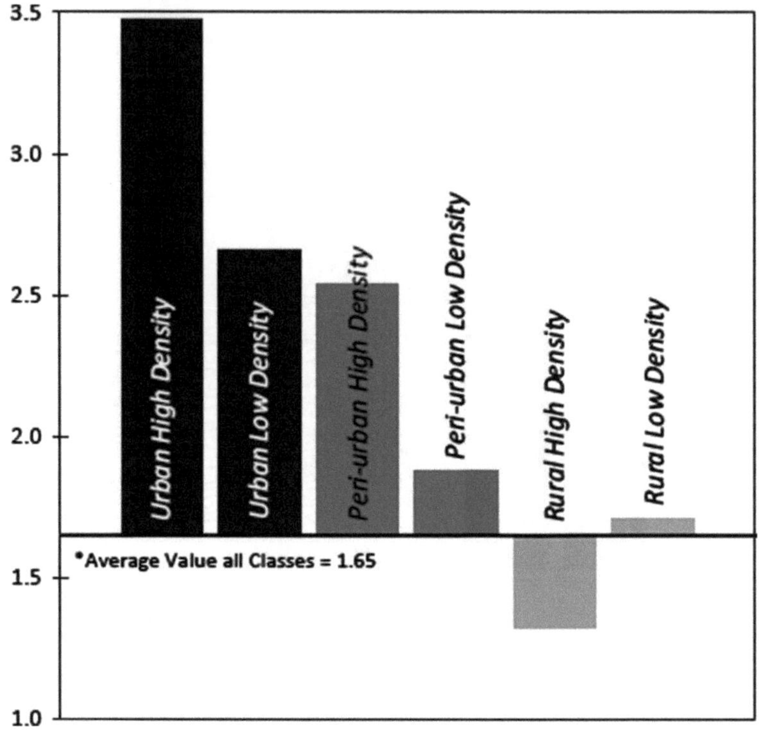


**Fig. 3.9** Average share of holdings, specialised on horticulture (per sub-regional type) in % (EU27, year 2000) (Database Eurostat)

**Acknowledgements** The authors would like to thank Stefanie Lange and Petri Kahila from Nordregio, Stockholm for their kind reviewing work helping to improve this chapter.

# References

Alberti M, Marzluff JM (2004) Ecological resilience in urban ecosystems: linking urban patterns to human and ecological functions. Urb Ecosys 7(3):241–265

Alonso W (1964) Location and land use. Harvard University Press, Cambridge

Anderson AE (2000) Gateway regions of the world—an introduction. In: Andersson AE, Andersson DE (eds) Gateways to the global economy. Edward Elgar, Cheltenham, pp 3–16

Antrop M (2004) Landscape change and the urbanization process in Europe. Landsc Urban Plan 67:9–26

BBR (2000) Urban development and urban policy in germany. Bundesamt für Bauwesen und Raumordnung, Bonn

BBR (2005) Raumordnungsbericht 2005. Bundesamt für Bauwesen und Raumordnung, Bonn

Bengs C, Schmidt-Thomé K (2006) ESPON Report 112: urban–rural relations in Europe. ESPON, Luxembourg. www.espon.eu. Accessed 14 Mar 2012

Bontje M, Burdack J (2005) Edge cities, European-style: examples from Paris and the Randstad. Cities 22(4):317–330

Briquel V, Collicard J-J (2005) Diversity in the rural hinterlands of European cities. In: Hoggart K (ed) The city's hinterland: dynamism and divergence in Europe's peri-urban territories. Ashgate, Aldershot, pp 19–39
Buciega A, Pitarch MD, Esparcia J (2009) The context of rural–urban relationships in Finland, France, Hungary, The Netherlands and Spain. J Environ Pol Plan 11(1):9–27
Cavailhes J, Peeters D, Sekeris E, Thisse JF (2004) The periurban city: why to live between the suburbs and the countryside. Reg Sci Urban Econ 34(6):681–703
Christaller W (1933) Die zentralen Orte in Süddeutschland. Eine ökonomisch-geographische Untersuchung über die Gesetzmäßigkeit der Verbreitung und Entwicklung der Siedlungen mit städtischer Funktion. Gustav Fischer, Jena
Couch C, Leontidou L, Petschel-Held G (2007) Urban sprawl in Europe: landscapes, land-use change & policy. Blackwell, Oxford
Dijkstra L, Poelman H (2008) Remote rural regions: how proximity to a city influences the performance of rural regions. In: COM—Commission of the European Communities (ed) Regional policy 1/2008. European Commission, Brussels, pp 1–7
ESPON (2005a) ESPON 1.1.3: enlargement of the European Union and the wider European perspective as regards its polycentric spatial structure. ESPON, Luxembourg, www.espon.eu. Accessed 14 Mar 2012
ESPON (2005b) ESPON 1.1.1: potentials for polycentric development in Europe. ESPON, Luxembourg. www.espon.eu. Accessed 14 Mar 2012
COM—Commission of the European Communities (1999) European spatial development perspective. Towards balanced and sustainable development of the territory of the European Union. Office for Official Publications of the European Communities, Luxembourg
COM (2004) Urban Audit - Methodological Handbook. Commission of the European Communities, Luxembourg
EUROSTAT (2010) Regional yearbook 2010. Chapter 15: a revised urban–rural typology. EUROSTAT, Luxemburg, pp 240–253
Friedman TL (2006) The world is flat. Penguin, London
Gallego FJ (2010) A population density grid of the European Union. Pop Environ 31:460–473
Gallent N, Anderson J, Bianconi M (2006) Planning on the edge: the context for planning at the rural–urban fringe. Routledge, London
Hall P (1966) Von Thunen's Isolated State: an English edition of Der Isolierte Staat. (transl: by Carla MW). Pergamon, Oxford
Hall P (1998) Cities in civilisation. Phoenix, London
Hall P, Hay D (1980) Growth centres in the European urban system. Heinemann Educational Publishers, London
Ingo Zasada, Regine Berges, Annette Piorr, Armin Werner: Leibniz Centre for Agricultural Landscape Research (ZALF), Eberswalder Str. 84, 15374 Müncheberg, Germany
Jensen-Butler C (1997) Competition between cities, urban performance and the role of urban policy: a theoretical framework. In: Jensen-Butler C, Shachar A, van Weesp J (eds) European cities in competition. Avebury, Aldershot, pp 3–42
Kalantaridis C (2010) In-migration, entrepreneurship and rural–urban-interdependencies: the case of East Cleveland, North East England. J Rural Stud 26(4):418–427
Leinfelder H (2009) Formalisation of 'open space' as 'public space' in zoning: the Belgian experience. In: Van der Valk A, van Dijk T (eds) Regional planning for open space. Routledge, London, pp 225–247
Le Jeannic T (1997) Trente ans de périurbanisation: extension et dilution des villes. Econ Stat 307:21–41
Marvin S, Harding A, Robson B (2006) A framework for city-regions. Office of the Deputy Prime Minister (ODPM), London
Meeus SJ, Gulinck H (2008) Semi-urban areas in landscape research: a review. Living Rev Landsc Res 2:3

METREX (2010) Intra-metropolitan polycentricity in practice: reflections, challenges and conclusions from 12 European metropolitan areas. METREX, Glasgow

Moriconi-Ebrard F (1994) Geopolis, pour comparer les villes du monde. Anthropos, Paris

Nordregio (2000) Study programme on European spatial planning. Final report. Stockholm, Nordregio

OECD (2002) Redefining territories: the functional regions. Organisation for economic co-operation and development (OECD), Paris

OECD (2007) OECD regional typology. Organisation for economic co-operation and development (OECD), Paris

Piorr A, Ravetz J, Tosics I (eds) (2011) Peri-urbanisation in Europe: towards a European policy to sustain urban–rural futures. Synthesis report. Academic Books, Frederiksberg

Porter M (1990) The competitive advantage of nations. Free Press, New York

Sieverts T (2003) Cities without cities: between place and world, space and time, town and country. Routledge, London

Wolfgang Loibl, Klaus Steinnocher, Mario Köstl: AIT Austrian Institute of Technology GmbH, Donau-City-Straße 1, 1220 Vienna, Austria

Zasada I (2011) Multifunctional peri-urban areas: a review of societal demands and agricultural provision of goods and services. Land Use Policy 28(4):639–648

Zasada I, Berges R, Hilgendorf J, Piorr A (in press) Horse-keeping and the peri-urban development in the Berlin Metropolitan Region. J Land Use Sci. DOI: 10.1080/1747423X.2011.628706

# Chapter 4
# Tools for Modelling and Assessing Peri-Urban Land Use Futures

**Dagmar Haase, Annette Piorr, Nina Schwarz, Sophie Rickebusch, Franziska Kroll, Hedwig van Delden, Affonso Zuin, Tim Taylor, Marco Boeri, Ingo Zasada, Carlo Lavalle, Roel Vanhout, Alessandro Sarretta, Felix Müller, Mark Rounsevell, and Simon Bell**

D. Haase (✉)
Humboldt University Berlin and Helmholtz Centre for Environmental Research – UFZ, Leipzig, Germany
e-mail: dagmar.haase@ufz.de
A. Piorr • I. Zasada
Institute of Socio-Economics, Leibniz Centre for Agricultural Landscape Research, Müncheberg, Germany
e-mail: apiorr@zalf.de; ingo.zasada@zalf.de
N. Schwarz
Helmholtz Centre for Environmental Research – UFZ, Leipzig, Germany
e-mail: nina.schwarz@ufz.de
S. Rickebusch • M. Rounsevell
David Kinloch Michie Chair of Rural Economy and Environmental Sustainability, Geography and the Lived Environment Research Institute, Edinburgh, UK
e-mail: sophie.rickebusch@ed.ac.uk; mark.rounsevell@ed.ac.uk
F. Kroll • F. Müller
Institute for Natural Resource Conservation, Department of Ecosystem Management, University of Kiel, Germany
e-mail: franzikroll@googlemail.com; fmueller@ecology.uni-kiel.de
H. van Delden • R. Vanhout
Research Institute for Knowledge Systems (RIKS), Maastricht, Netherlands
e-mail: hvdelden@riks.nl; rvanhout@riks.nl
A. Zuin • S. Bell
Edinburgh College of Art, Openspace Research Centre, Edinburgh, UK
e-mail: a.zuin@eca.ac.uk; s.bell@eca.ac.uk
T. Taylor
University of Bath, Department of Economics, Bath, UK
e-mail: t.taylor@bath.ac.uk
M. Boeri
Fondazione Eni Enrico Mattei, Venezia, Italy
e-mail: mboeri01@qub.ac.uk
C. Lavalle • A. Sarretta
Institute for Environment and Sustainability of the Joint Research Centre (JRC), Ispra, Italy
e-mail: carlo.lavalle@jrc.it; alessandro.saretta@jrc.it

K. Nilsson et al. (eds.), *Peri-urban futures: Scenarios and models for land use change in Europe*, DOI 10.1007/978-3-642-30529-0_4,

## 4.1 Introduction

Chapters 1 and 2 set the scene for the scenarios of possible futures, presented modelling results and identified a typology of rural urban regions. The next set of chapters in Part 2 focuses on the case studies and what can be learned from them. Before examining each case study it will be useful to provide an overview of the various models and tools used to simulate potential land use futures of peri-urban regions and to assess their impacts across Europe. The focus of this chapter is at two scales: the pan-European (EU27) and the regional scale, as produced by the typology in Chap. 2 and exemplified by the case studies presented in Chaps. 4, 5, 6, 7, 8 and 9.

## 4.2 The Need for Tools for Impact Assessment

The transformation of natural, open or agricultural land into urban land dominated by built structures and sealed surfaces causes many major environmental impacts in most urbanized countries and regions (OECD 1997). Moreover, along the urban–rural gradient the results of this land consumption are often characterized by dispersed fragmented developments, mono-functional and low-density land uses and a reliance on private car ownership as the main mode of transport (Nuissl et al. 2008; EEA 2006a).

Land consumption is a problem not only because it contradicts the normative ideals of spatial planning and compact development. Many studies have shown that peri-urban land consumption is usually detrimental to the environment in a number of ways (Johnson 2001). Its impact reduces the ability of nature to satisfy many human requirements and thus reduces the functioning of various ecosystem services (De Groot et al. 2002; Curran and Sherbinin 2004). Land consumption in and especially around cities frequently creates impervious surfaces, infills fragile wetlands, fragments ecosystems and has disproportional impacts on the carbon cycle through changes in the net primary productivity of affected ecosystems (Nuissl et al. 2008). Projections for the future suggest that the number of people living in urban areas is very likely to continue to increase (UNESA 2007) and that urban expansion and peri-urbanisation is occurring faster than rates which might be expected by population growth alone (Liu et al. 2007). Hence, there is a need to search for sustainability pathways for global and regional urban development, which can sustain both quality of life and environmental resources (Lambin and Meyfroidt 2010).

Following this argument, it becomes clear that a quantitative evaluation of the major impacts of potential future land use changes in rural–urban regions and particularly their peri-urban zones is crucial for an overall assessment of sustainable development in Europe (Ravetz 2000; Lambin and Geist 2006). Decision-making about future core areas of residential and commercial development as well as

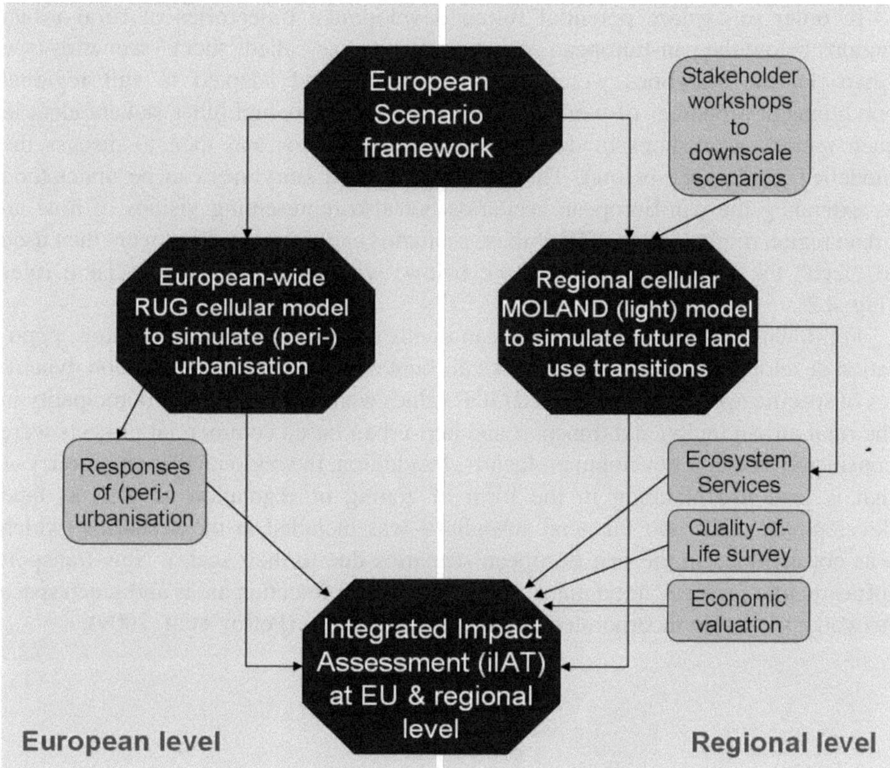


**Fig. 4.1** Logic of the tools developed and applied to assess peri-urban land use changes

planning for nature protection and open space requires spatially explicit, comprehensive and comparable data about ongoing and anticipated land use changes around cities in Europe (EEA 2006a). Therefore, the major objectives of the model and tool development were (1) to generate the land use scenarios for urban regions in Europe (as presented already in Chap. 1), (2) to model the relationships between land use change and the provision of resources and functions (e.g. residential, transport, environmental and recreational) at both the pan-European and regional level and (3) to assess the related environmental, social and economic impacts of land use changes induced by both global and regional driving forces, as well as by regional land use strategies, using a multi-criteria Impact Assessment (IA) approach. The respective models and tools were developed as shown in Fig. 4.1.

The tools to be presented here were developed in order to understand the drivers of each modelled scenario. The scenarios are those described in Chap. 1 and applied at a regional level in each of the case studies, as described in the respective chapters in Part 2. The methods used in the regional scenario workshops will be presented before the modelling methods and impact assessment tools are described.

In order to explore potential future development trajectories of rural–urban regions below the pan-European scale based on the so-called 'shock' scenarios (see Chap. 1), the storylines were further developed and adapted to suit regional conditions at a number of workshops held with planners and other stakeholders in each region—workshops to develop the local scenarios and then to discuss the modelled results (see below). The resulting regional storylines can be understood as extending the pan-European scenarios but also representing visions of how an urban region might look in 2025 if these scenarios came to pass. They were then used to "feed" the modelling systems (see below) with initial data for decision rules (Fig. 4.2).

For downscaling of the pan-European trends to single rural–urban regions, population development was adapted to take account of the in- and out-migration dynamics of specific age groups. Localized GDP values were added for each municipality in the rural–urban region and transport and peri-urban-based commercial projects were considered as major development factors. In addition, the regional planning policy—that is land use planning in the form of zoning or regulation as well as land development plans and financial subsidies—was included in the scenarios (which was not included in the pan-European scenarios due to their scale). New transport infrastructure projects, flood maps, nature and other protection areas and green space networks were also incorporated as suitability variables (Petrov et al. 2009).

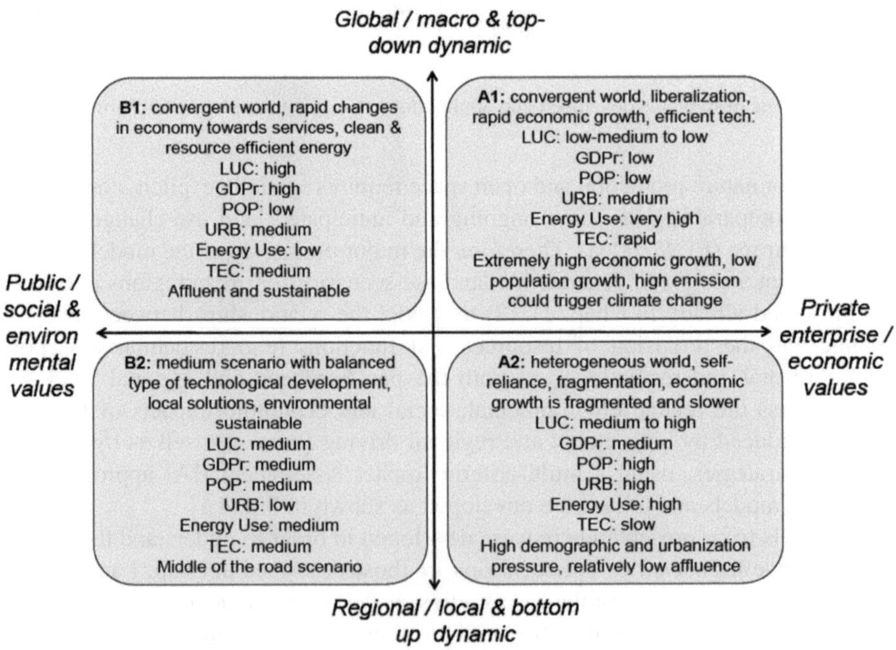


**Fig. 4.2** European scenario framework (Ravetz and Rounsevell 2008). *LUC* land use change, *GDPr* growth domestic product, *POP* population growth, *URB* urbanisation, *TEC* technological development

**Fig. 4.3** Example of the agenda of a regional stakeholder workshop—here shown for the urban region of Leipzig-Halle

The main purpose of the first scenario workshop in each case study region was to let representative stakeholders, such as planners and other practitioners, create their own potential land use maps framed by the possible land use options relevant to the area in question and based on the adaptation of the main storylines. The researchers were interested in the locations where the stakeholders placed future housing and transport development and where and why they assumed land abandonment and brownfield sites would occur. In addition, the stakeholders listed the instruments they would wish to use to steer their land use maps (Haase et al. 2009). Compared to the pan-European scenario framework the regionalized storylines which emerged reflected development and decision-making processes within rural–urban regions. Their limitations have to be seen firstly in their non-transferability to other regions and secondly the time and resources they require in order to be effective.

The second scenario workshops' main objective was to discuss the maps produced by the MOLAND model (see below) for each of the European case studies with local stakeholders, experts and practitioners. Such workshops help to contribute to the dialogue between research and practice on spatial and land use development in urban regions. The land use scenarios and the MOLAND land use modelling were presented by researchers in order to stimulate a debate on the implications of potential future land use development. In addition, policy options for regional planning and governance were discussed (Fig. 4.3).

An innovative element of one of the stakeholder workshops in the urban region of Leipzig-Halle was a planning game for jointly developing future land use maps under different assumptions (Haase et al. 2009). Using the same driving forces, scenarios were developed for the following planning options by the stakeholders: "Unrestrained growth", "Managed growth" and "Managed shrinkage". In practice, according to the scenario, each planning game was supplied with a certain amount of "land use" in the form of coloured paper and a storyline. The stages and the results of the planning game were documented in notes, photographs and films (Haase et al. 2009).

## 4.3 Land Use Change Simulations

In order to determine future land use changes, two cellular models were developed and applied in PLUREL: the RUG (Regional Urban Growth) urbanisation model was applied to all urban regions across Europe and the constrained cellular automata models MOLAND and MOLAND-Light, respectively, for the case study scale. MOLAND-Light in particular is a specific stakeholder-orientated development of the PLUREL project: it is a tool that makes integrated dynamic modelling available to stakeholders since growth rates (land use pressures) are predetermined by expert-based regional estimates using targets for the future. Further, infrastructure and zoning as the spatial expression of planning and governance strategies were incorporated based on the regional stakeholder workshops and a joint "regionalization" of the European scenarios. The model outcomes can be used to assess the effectiveness and applicability of the strategies by planners.

### *4.3.1 Regional Urban Growth Model (RUG)*

The Regional Urban Growth (RUG) model gives projections of the percentage of artificial surfaces (using CORINE land-cover level 1, class 1 categories) for Europe or a selected region, following the assumptions in the four European-level scenarios (as described in Chap. 1). It covers 25 European countries using a $1 \times 1$ km grid and includes two time steps, 2015 and 2025. This is done by first evaluating the potential for settlement in each grid cell, based on the cell's characteristics and the scenario storyline. On this basis, the model calculates the new percentage of artificial surfaces in each grid cell, which depends on the total amount of artificial surfaces expected in the region (NUTS2) for the scenario and year considered.

The main inputs to RUG are the projected proportions of artificial surfaces per region, for each time step and scenario. These are derived from projections of population (KC et al. 2010) and gross domestic product (GDP) per capita (Brécard et al. 2006) using a linear regression model (Reginster and Rounsevell 2006), with country and urban type (large city vs. smaller city/rural region) as additional factors.

Grid cell characteristics include the current proportion of artificial surfaces, Euclidian distance to the coast (sea or large lake), flood risk classification and

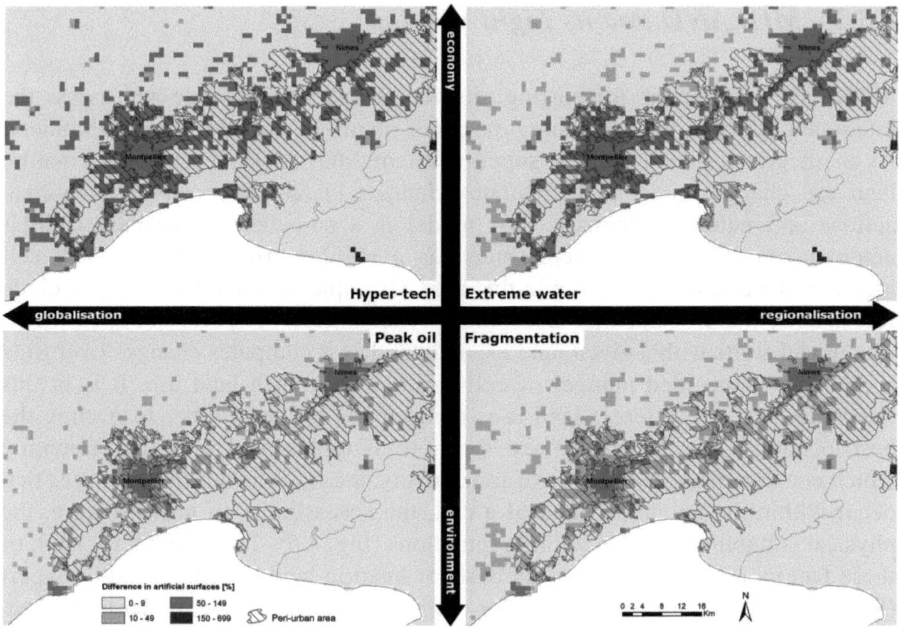


**Fig. 4.4** Results of the RUG model for the Montpellier urban region. The maps show the increase in artificial surfaces in 2025, as a proportion of baseline (2005) values, for the four pan-European scenarios (cf. Chap. 2). The darker the color the higher is the increase in artificial surface

"remoteness" relative to the nearest medium (≥100,000 inhabitants) and large (≥500,000 inhabitants or national capital) city. The "remoteness" index combines the travel time cost (Verburg et al, 2008) and the proportion of commuters (Helminen et al. 2012), relative to the region. The travel time costs for 2025 are modified to reflect the technological changes in the European scenarios. For 2015, current values are used, because these changes do not occur gradually but in discrete steps and transport projects take years to gain approval and be implemented.

The scenario storylines determine the location preferences of residents relative to the various cell characteristics and the type of planning policy. The scenarios therefore define the parameter values which shape the relationship between the location preferences and the cell characteristics. The model can be run for a single region or for all 25 countries, in which case it will run for each region in turn.

The RUG model only simulates one land use class (urban) against several in MOLAND/MOLAND-Light, but it gives a quantity within each cell, rather than presence/absence. RUG's limited thematic resolution makes it lighter in processing requirements, which allows simulations on a larger scale. It is therefore possible to compare a case study region to the whole country or to the whole of Europe, as well as comparing different case study regions. The two models are therefore complementary (Fig. 4.4).

### *4.3.2 MOLAND and its Light Version*

MOLAND is a generic forecasting tool for planners to simulate and assess the integrated effects of their planning measures on urban and regional development. As an integrated spatial decision support system, MOLAND incorporates a mature land use change model that simulates changes in land use caused by human actions and natural processes. This model is a so-called constrained cellular automata model. Space is represented as a regular grid, in which each cell indicates the dominant land use in that area. Examples of land use classes include forest, agriculture, residential areas, industry, commerce, airports and open water. The model starts with a given land use map and as it computes changes over time it allocates a specified number of cells annually for each land use. It is in this allocation that the different land uses compete for space in order to occupy the best locations. The notion of "best location" is represented by a total potential which consists of a combination of four factors: the influence of the land uses that occur within the neighbourhood of a cell, the accessibility to infrastructure, the physical suitability and the zoning regulations (Fig. 4.5). The weights for each of these factors need to be set on the basis of historic land use maps and/or expert (local) knowledge and are specific for each region to which the model is applied, which is also a limiting factor in terms of the model transferability and objectivity (which depends on the experts involved).

MOLAND-Light is a simplified, web-based version of MOLAND (an overview of the main differences can be found in Table 4.1). It includes the same model, but data, land use classes and parameter settings are generic for each RUR region in

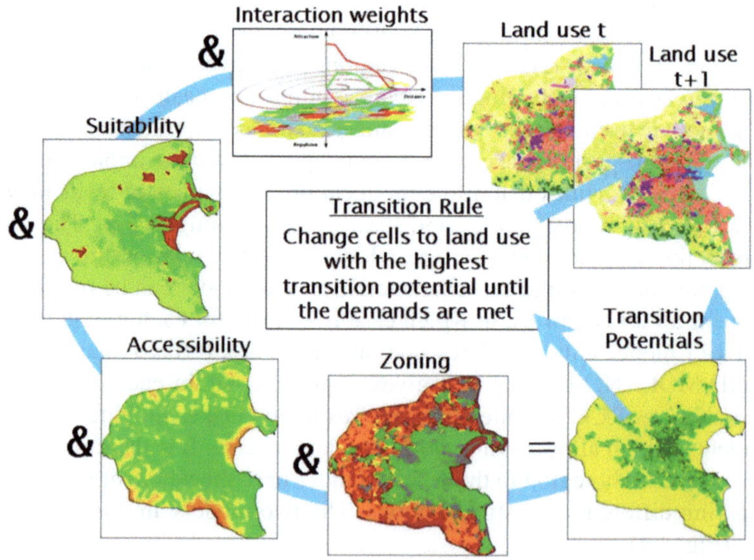


**Fig. 4.5** Simulation step of the land use model showing the important drivers at local level (RIKS 2009)

**Table 4.1** Overview of the main differences between MOLAND and MOLAND-Light

| MOLAND | MOLAND-Light |
|---|---|
| Uses region specific data | Uses generic datasets |
| Full desktop application | Desktop component (Light part) and web component (link to MOLAND and data stored on JRC server) |
| Full functionality, training in application, calibration and use required | Limited functionality, to be used without training |
| Tool can be—and is—used for policy support in regions worldwide | Not directly suitability for policy analysis, main focus is quick scan |
| Region-specific calibration | Generic parameter set for RUR regions in Europe |

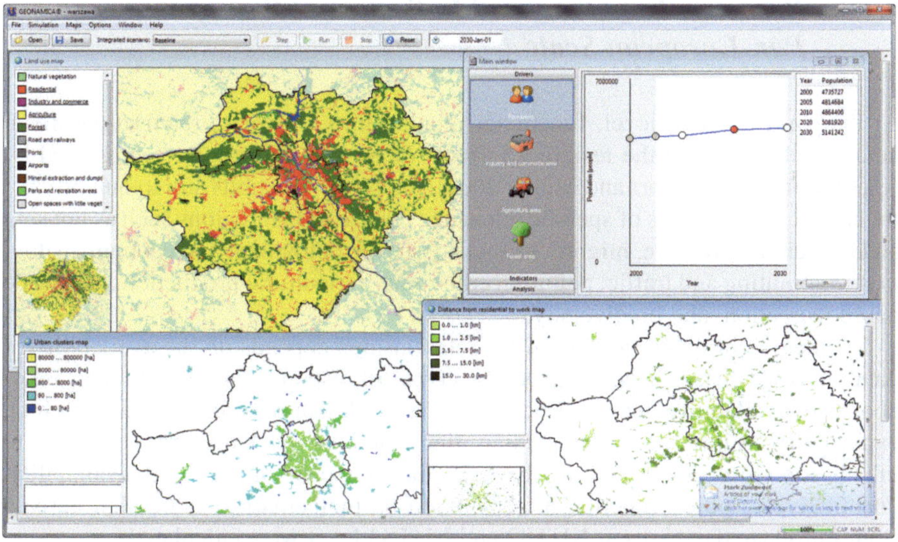

**Fig. 4.6** Screenshot of the user interface of MOLAND-Light showing the drivers, the calculated land use map and the urban clusters indicator

Europe. Users can download a desktop front-end application (Fig. 4.6) and select for which RUR region they would like to run a simulation. The desktop application then connects to a server based at the Joint Research Centre of the European Commission (JRC) in Ispra, Italy where the model simulation is executed. Model results for each year in the simulation are shown on the desktop application. Results are provided as an animation showing yearly land use development and as maps for selected years, which can be opened in common GIS packages or the freely available Map Comparison Kit (www.riks.nl/mck).

Drivers available to the user are limited to the projections for regional population and land use area demands for industry and commerce, agriculture and forest. Besides the land use developments, selected indicators are, among others, urban clusters, soil sealing, abandoned land and distance from residential to work locations.

## 4.4 Impact Assessment Models

Similarly to the two land use change models which were applied at two different scales, namely the EU27 and the regional, PLUREL also used two different approaches to model the impacts of land use change in rural–urban regions: firstly, statistical response functions and indicators at the NUTSx scale based on aggregate Corine Land Cover (CLC) and secondly, impact assessment models with higher spatial grain at the case study scale.

### *4.4.1 Response Functions and Indicators at the Pan-European Scale*

At the pan-European level, functional relationships between urbanisation-driven land use changes and the resulting impacts on sustainability were represented by response functions (Piorr and Müller 2009). Response functions aim to quantify the functional relationships of specific relevance for peri-urban regions such as housing and household structure, migration, social issues, employment and economic situation, commuting and traffic, pollution emissions, agricultural production, landscape based recreation, and biodiversity.

The degree of quantification of the relationships depends on the availability of data. Thus we distinguished between two types of response function. In the case of poor data availability generic functions were used, that describe curves with high explanatory value for understanding the reactions of land use functions to urbanisation processes, and in the case of high data quality and availability, quantitative response function models based on mathematical regression derived from European databases at NUTS2 and NUTS3 level were calculated.

Quantitative response models make use of regression analysis with independent and dependent variables being represented by quantitative indicators. Independent variables describe urbanisation with land cover-related and socio-economic indicators. In PLUREL, three independent variables were used: the change in artificial surface, in population density and in Gross Domestic Product (GDP). The dependent variables were indicators describing the strength of anthropogenic impacts on the sustainability potential. Examples of dependent variables are the indicators used in the PLUREL Integrated Impact Assessment Tool (iIAT) (Table 4.2). The data sources for both kinds of variables were EEA (2006b), EEA (2006c) and others such as Bird Life International. Land use data was derived from Corine Land Cover (EEA 2009) and, as it was also for the dynamic modelling, also from EEA (2009).

The spatial scale of response function modelling is NUTSx (a combination of NUTS2 and 3 as noted in Chaps. 1 and 2), chosen to create more equally-sized regions. Derived from databases that assign a value pair of the dependent and independent variables to each NUTSx region and each indicator, the response

**Table 4.2** Response function (RF) models used for the analysis of urbanisation impacts at EU scale

| Sustainability issue | RF on relationship between change of artificial surface or of GDP or of population density |
|---|---|
| Social | 1 person households |
| | 2 person households |
| | 3 person households |
| | 4 person households |
| | Educational level 1 |
| | Educational level 2 |
| | Educational level 3 |
| | Educational level 4 |
| | Population $<$ 15 years |
| | Population $>$ 60 years |
| Environmental | Number of endangered bird species per 100 $km^2$ |
| | Interspersion and juxtaposition index |
| | Landscape shape index |
| | Effective mesh size |
| | Green background index |
| | Heavy metal and other emissions |
| | $NO_x$ emissions |
| | CO emissions |
| | HC emissions |
| Economic | GDP |
| | GDP per purchasing parity |
| | Total employment |
| | Employment sector I |
| | Employment sector II |
| | Employment sector III |
| | Employment sector IV |
| | Artificial area |
| | Agricultural area |
| | Part-time farm holders |

function, as a result of the modelling process of all European NUTSx regions, provides the opportunity to calculate future values for each indicator. The aim of the regression analysis was to derive regression models with the best possible explanatory power. Out of some 75 response functions this was the case for around 25 (see Table 4.2).

For social issues, where the focus was on migration patterns e.g. of inter-European labour migration, extra-European immigration or international retirement migration, mainly qualitative as opposed to quantitative response functions were developed due to poor data availability and in any case could only be resolved to NUTS2. The response functions for commuting were derived from highly detailed and spatially explicit datasets from good practice cases and by using more specific models that considered data which do not underlie a continuous monitoring at European scale. In Fig. 4.7 the generic response function on commuting probability is shown in relation to increasing distance to the urban centre, as a function of

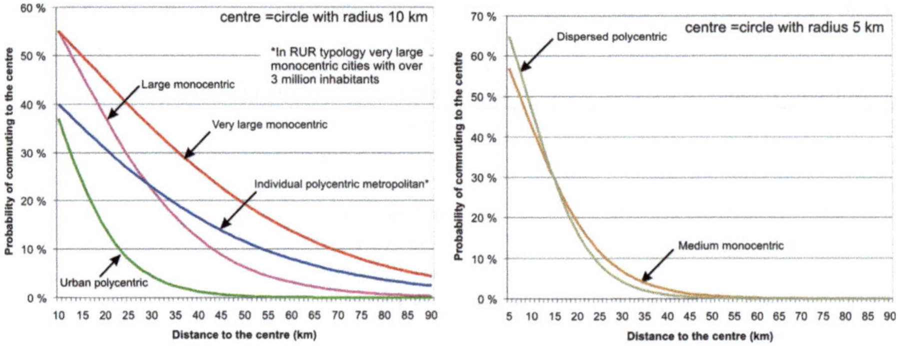


**Fig. 4.7** Example for generic response functions: archetypical curves for commuting RF per settlement morphological type of rural–urban region (RUR) (Helminen et al. 2010)

different settlement morphological types of rural–urban regions. Overall, regression functions represent the statistical way of identifying relationships between regional variables and they are clearly limited in their value to explain causes.

## *4.4.2 Impact Assessment at the Regional Scale*

At the regional scale, no new indicators were developed, so a sub-set taken from already existing sustainability indicator sets was refined for the purpose and especially for assessing the impact of land use change across the gradient of a rural–urban region (RUR). None of the existing local or regional sustainability indicator sets focuses on the integration and relationships of both rural and urban areas, so this was an area for further development. The selected indicator set also reflected the stakeholder-orientated participatory research that was carried out and so was able to function as a tool to integrate the results. The method of developing the indicator set for sustainability impact assessment was first built on a theoretical model of the role and function of indicators, the derivation of a preliminary indicator list was then based on the theoretical model and following discussions in workshops. The "long list" of potential indicators was put into matrix and tested for their integrative power and finally the selection of key indicators was made from this "long list" based on the matrix results and stakeholder feedback (Kroll et al. 2010). Table 4.3 provides an overview over the indicators and the respective impact models.

For the detailed land use change impact assessment, a smaller sub-set of key indicators was chosen to cover the three pillars of sustainability. For the environmental aspects, the ecosystem services concept was chosen as a framework. The concept of ecosystem goods and services (which are jointly called ecosystem services) has recently demonstrated a rapid conceptual development. Based on the idea of "functions of nature" (De Groot 1992), ecosystem services are "the benefits people obtain from ecosystems and the processes that support the production of ecosystem goods" (MA 2005). They result in "the benefits of nature to households,

**Table 4.3** List of all impact models and indicators used

| Concept | Indicator | Impact model |
|---|---|---|
| ESS[a] | Local climate regulation | Land surface emissivity is computed as a proxy, with case-study specific satellite images (Schwarz et al. 2011). |
| ESS | Recreation potential | Recreation of natural and semi-natural areas is computed using the proxy of the per-capita green space. |
| ESS | Carbon mitigation | Estimated by a quantification of carbon storage in trees, based upon extrapolation of field data. |
| ESS | Water quantity regulation | Quantified with evapotranspiration (ETP) using the empirical model of the TUB-BGR approach by Wessolek et al. (2004). |
| ESS | Biodiversity potential | Potential of land use types for serving as habitat for urban, agricultural and forest bird species (regression model; Strohbach & Haase, 2012). |
| ESS | Food production | Empirical additive model of the produced food in comparison to yields in tons per hectare out of statistical data. |
| ESS | Energy provision | Summary of all types of energy, computed out of statistical values per land cover type. |
| ESS | Fresh water provision | Calculated using the proxy of groundwater recharge (TUB-BGR approach by Wessolek et al. 2004). |
| QoL[b] | Quality of life aggregate | Weighted sum value of all quality of life indicators measured for a case study using the Quality-of-Life Simulator (see Sect. 4.6). |
| QoL | Air quality | Single variable to determine the quality of life for a case study using the Quality-of-Life Simulator (see Sect. 4.6). |
| QoL | Access to green space | |
| QoL | Public transport | |
| QoL | Shops in the neighborhood | |
| QoL | Noise pollution | |

[a]Ecosystem services
[b]Quality of life

communities, and economies" (Daily et al. 1997) as "components of nature, directly enjoyed, consumed, or used to yield human well-being" (Boyd and Banzhaf 2007).

Generally, three classes of ecosystem services can be distinguished: provisioning services (goods produced or provided by ecosystems such as food and clean water), regulating services (benefits obtained from regulation processes in ecosystems such as local climate regulation) and cultural services (non-material benefits obtained from ecosystems such as recreation; Kroll et al. 2012). The social aspects were framed by the concept of quality of life and the economic aspects were calculated using GDP and external costs of land uses. Both methods and tools of the quality of life and economic impact assessment are discussed in more detail below.

## 4.5 Integrated Impact Assessment Tool (iIAT)

The integrated Impact Assessment Tool (iIAT) synthesized the modelling results from all aspects of the impact assessment into one tool. It is a tool for presenting the integrated results of a broad impact assessment (IA) and is multipurpose and

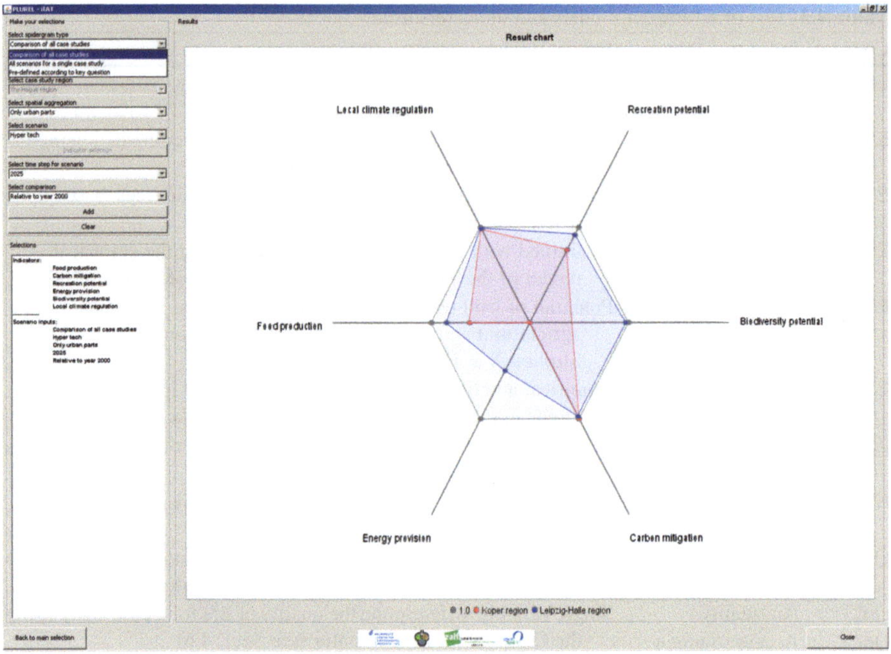


**Fig. 4.8** Graphical user interface (GUI) of the iIAT-EU

interactive in nature. It allows for the integration of many aspects of problems associated with land use and its functions and services related to urbanisation. It considers conflicts of interest of different stakeholders (e.g. residents, planners, developers) within a planning process. It also covers all dimensions of sustainability, namely the economic, the social and the environmental. Based on the land use change maps produced by the RUG and MOLAND (Light) simulations, economic, social and environmental impacts were evaluated using the range of indicators developed as described earlier.

The iIAT is an internet-based tool that displays results in the form of "spidergrams", providing a graphic presentation enabling an easy and holistic perception of multilevel information. The interactive nature lies in the possibility to obtain an in-depth view of different thematic aspects and different scales, chosen according to the requirements of individual users. The iIAT uses the indicator database and generates the desired outputs, displaying them in the graphical user interface (GUI). Via the GUI it is possible to explore the effects of global drivers, national planning policies and local governance on land use change in rural–urban regions and, consequently, their impacts on sustainability (Fig. 4.8).

The iIAT consists of two modules—the iIAT-EU and the iIAT-Region, since it covers two spatial levels: the EU27 and the regional (rural–urban region) level. Being user determined, e.g. by displaying comparative spidergrams, the tool facilitates discussions for different end-user groups ranging from EU policy assistants to local planners. In addition, collaborative working with the iIAT could provoke learning processes about different views of land use development.

The iIAT-EU shows how the impacts of urbanisation under future scenario conditions could differ from the current situation. The initial situation 2000 (baseline) can be compared to the four PLUREL scenarios of future development for the two time slices of 2015 and 2025. The main purpose of the iIAT-EU is to create awareness of how sustainability trends develop at different scales for different types of regions and where policy action might be necessary, thematically and spatially. It also allows users to distinguish between the impacts of trends in predominantly urban, peri-urban and rural regions. The underlying data are derived from modelling results at the spatial NUTS3 (administrative unit across the EU comparable with districts or counties) or NUTSx scale.

The iIAT-approach is limited in terms of expressing value uncertainties for the single indicators displayed. Although the spidergram values can be compared to thresholds given for specific social, economic or environmental issues, the opportunity to assess the value validity/representativeness is limited.

## 4.6 Quality of Life in Rural–Urban Regions

For the assessment of quality of life a novel approach was used in PLUREL. Instead of taking a set of indicators and applying them as a technical exercise the opportunity offered by the project was taken to collect some new data from some of the case study regions. This enabled the effects of land use change on perceived quality of life to be based on real data and related to real places in the sampled rural–urban region. The theory behind this is that of “affordances” (Nefs et al. (in press)) where an environment presents possibilities of different functions to people. This may, however, vary from person to person according to their lifestyle, life stage, socio-economic status etc. How people might react to a perceived deteriorating change in the quality of their living environment could be by moving away to a better or more preferred location (if they can afford to do so), to one with more affordances. The quality of life in rural–urban regions can be subject to improvement or deterioration as a result of many factors of which some are specifically associated with land use change, the focus of the indicator sets used in PLUREL.

In this study a series of indicators were selected, derived from pre-existing sets and which were sensitive to land use change. Each of these could be measured quantitatively but could also be understood perceptually by residents. For example, air quality can be measured by levels of particulate matter or NOx in the atmosphere but is perceived by people as being relatively good or bad as a result of the amount of smog present at any one time. This enables links between quantitative and qualitative perceptions to be made.

The data was collected from samples of residents in a number of European city regions—several case study examples plus others (to enable a pan-European model to be constructed as well as the case study regional level). The residential choice decision (to stay or move) was used as the threshold trigger of an indicator but

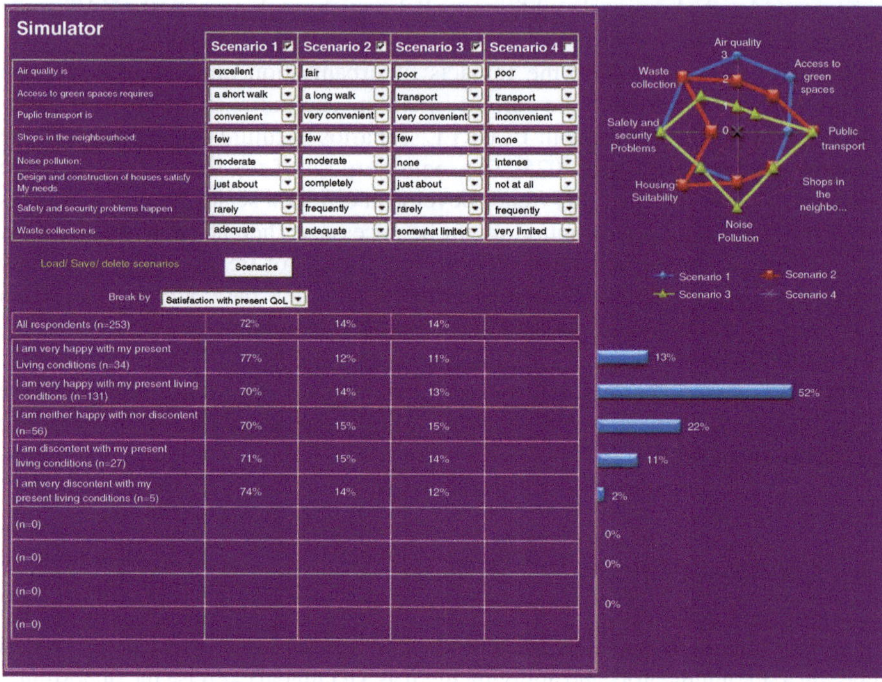


**Fig. 4.9** Graphical user interface (GUI) of the quality-of-life simulator QoLSim displaying resident responses of an arbitrary RUR in Europe

because tradeoffs among many factors have to be made in real life when choosing a preferred environment, the methodology used was Adaptive Conjoint Analysis (ACA), one of a family of choice experiment methods. The method is based on offering respondents choices between different scenarios of residential quality based on randomly selected combinations of levels of indicator. The selection of the preferred choice along a 10 point scale for a number of combinations enables the preference levels and threshold tolerances of individual respondents to be calculated and aggregated according to different classes of demographic variable. The data was analysed using Sawtooth Software (www.sawtoothsoftware.com) and fed into the a modelling system developed for the project, based on market simulator systems, called the Quality of Life Simulator (QoLSim) (Fig. 4.9), which allows the effects of land use change derived from the land use modelling scenarios to be converted into predictions of behaviour by residents, using residential choice as the vehicle for measuring the threshold. The effects of different land use changes on predicted quality of life were also added into the iIAT by inferring what effect on the quality of life indicators would be expected by specific land use changes, such as more roads leading to shorter commuting times but worse air quality and more noise.

By developing a tool for modelling behaviour it is possible to predict what changes to land use may affect people for the better or worse. This can provide valuable information for planners and policy makers who can use the QoLSim tool themselves. However, the linkage of the measured quality of life perception values with the land use change values of the regionalized MOLAND scenarios remains challenging. Here, still more emphasis is needed to prove the rule-set developed so far of how land use change quantitatively affects quality of life in a specific region. The QoLSim Europe software and data can be downloaded from the PLUREL website www.plurel.net.

## 4.7 Economic Assessment of Peri-Urban Land Use Changes

Most public policies affecting land use decisions entail a cost to society, in terms of higher taxes, higher prices, or reduction in the provision of other public goods and services. When considering new land use policies, it is therefore useful to gain an insight of the public's preferences for the effects of those policies, as well as assessing how much citizens are willing to pay to implement those changes. A number of methods exist to estimate the value of land use change in peri-urban areas. These include stated preference and revealed preference methods (for more detail on these methods see Markandya et al. 2002).

Stated preference methods include contingent valuation and choice experiments. People are asked directly about the values they attribute to different types of land use change either in terms of a willingness to pay for a defined change in land use (e.g. the development of green/blue services) in the case of contingent valuation or in terms of their preferences between projects with different attributes in the case of choice experiments (including a cost attribute, which can be used to impute the willingness to pay). Both of these methods require carefully designed questionnaires to be answered by a sample of the population. Bateman et al. (2002) gives an excellent introduction to both contingent valuation and choice experiments.

Revealed preference methods involve the examination of actual behaviour in terms of buying houses (hedonic pricing) or travelling to recreational sites (travel cost method). Using the hedonic pricing approach it is possible to analyse the value placed on different attributes in peri-urban areas. Here, data on house prices (ideally sales data) is used together with factors that affect the price—such as the characteristics of the house (e.g. number of bedrooms, size of garden), neighbourhood characteristics (e.g. crime rate, access to good schools) and environmental characteristics (e.g. green space, pollution). The travel cost method uses data on costs for those travelling to a site (e.g. a forest) as a proxy for their minimum willingness to pay.

These methods facilitate policy making in terms of both the identification of priorities for action and in allowing environmental impacts to be considered in monetary terms in cost-benefit analysis. In the peri-urban context, a stated preference method using choice experiments was applied to identify how much people value changes in public transport, cultural heritage, housing developments and green spaces. An example of a "choice card" is given in Table 4.4. For PLUREL, as part of the assessment of sustainability, questionnaire surveys were carried out in selected case study regions—The Hague, Leipzig and Warsaw. Respondents were presented with different alternatives—all involving a cost in terms of a one off tax or fee.

**Table 4.4** Example of a choice experiment

| Characteristics | Policy A | Policy B | Policy C |
|---|---|---|---|
| Transport | **A lot of improvements** in public transport lead to a large reduction in air pollution and greenhouse gases emissions | **Some improvements** in public transport lead to a large reduction in air pollution and greenhouse gases emissions | **Some improvements** in public transport lead to a large reduction in air pollution and greenhouse gases emissions |
| Cultural heritage preservation | **Most** cultural heritage monuments are preserved | **Most** cultural heritage monuments are preserved | **Some** cultural heritage monuments are preserved |
| Housing and green open spaces | New apartment complexes are **developed in green open areas** | Partly abandoned and idled buildings are **refurbished** to accommodate new flats | **Keep** buildings and green open areas **as** they are **now**, with limited efforts to accommodate the requests for new flats |
| Cost of the project to the respondent as additional obligatory fee to household for 2011 only | **175** | **25** | **0** |
| Which policy would you choose? | □ | □ | □ |

## 4.8 Summary and Conclusions

This chapter has presented the range of modelling and impact assessment tools used in PLUREL. Some use existing modelling systems (e.g. MOLAND) while others have been developed specially for the project (iIAT, QoLSim). Some also use existing data sets and others collected new data. They demonstrate that such tools are undergoing fast development and that they hold many possibilities for helping policy and decision makers evaluate the complex implications of land use planning decisions. The development and application of the tools also shows that limitations remain to be overcome and that there is scope for much more development and refinement. The beauty of making these tools freely available to anyone who wants to try them out is that they can be tested in different circumstances and used and developed further by other researchers in future.

Of course, models depend on good data and they need to be updated with fresh data. Models are not true reflections of reality but offer somewhat simplified versions, so should not be mistaken for reality. They are certainly not predictive since there are many assumptions and massive uncertainties in all of them. However, in terms of testing the "What if?" questions they can be very helpful decision support tools.

## References

Bateman I, Carson RT, Day B, Hanemann M, Hanley N, Hett T, Jones-Lee M, Loomes G, Mourato S, Ozdemiroglu E, Pearce DW, Sugden R, Swanson J (2002) Economic valuation with stated preference techniques: a manual. Edward Elgar, Cheltenham

Boyd J, Banzhaf S (2007) What are ecosystem services? The need for standardized environmental accounting units. Ecol Econ 63(2–3):616–626

Brécard D, Fougeyrollas A, Le Mouël P, Lemiale L, Zagamé P (2006) Macro-economic consequences of European research policy: prospects of the Nemesis model in the year 2030. Res Policy 35:910–924

Curran SR, Sherbinin A (2004) Completing the picture: the challenges of bringing "Consumption" into the population–environment equation. Popul Environ 26(2):107–131

Daily GC, Alexander S, Ehrlich PR, Goulder L, Lubchenco J, Matson PA, Mooney HA, Postel S, Schneider SH, Tilman D, Woodwell GM (1997) Ecosystem services: benefits supplied to human societies by natural ecosystems. Issues Ecol 1(2):1–18

De Groot RS (1992) Functions of nature: evaluation of nature in environmental planning, management and decision making. Wolters-Noordhoff, Groningen

De Groot RS, Wilson M, Boumans R (2002) A typology for the description, classification and valuation of ecosystem functions goods and services. Ecol Econ 41:393–408

EEA (European Environment Agency) (2006b) The thematic accuracy of corine land cover 2000. Assessment using LUCAS, land-use/cover area frame statistical survey, Technical report no. 7/2006. Retrieved 25 Nov 2008 from http://reports.eea.europa.eu/technical_report_2006_7/en/technical_report_7_2006.pdf

EEA (European Environment Agency) (2006c) The thematic accuracy of corine land cover 2006. Assessment using LUCAS, land-use/cover area frame statistical survey, Technical report no. 17/2007. Retrieved 09 Nov 2010 from http://www.eea.europa.eu/publications/technical_report_2007_17

EEA (European Environmental Agency) (2006a) Urban sprawl in Europe—the ignored challenge. European Environmental Agency, EEA Report 10/2006, Copenhagen

EEA (European Environmental Agency) (2009) Corine Land Cover report. http://www.eea.europa.eu/publications/COR0-landcover. Accessed 12 Jan 2009

Haase D, Bauer A, Reuther I, Schiffers B, Schwarz N, Seppelt R, Weinert J (2009) Spatial development in the Leipzig-Halle region. Brochure at www.plurel.net

Haines-Young R, Potschin M (2010) Proposal for a common international classification of ecosystem goods and services, CICES for integrated environmental and economic accounting. Report to the European environmental agency, Nottingham

Haude W (1955) Zur Bestimmung der Verdunstung auf möglichst einfache Weise. Mitt. d. Dt. Wetterd. 2 (11). Deutscher Wetterdienst, Frankfurt

Helminen V, Ristimäki M, Kontio P, Vuori M (2010) Response function for commuting. PLUREL deliverable report D2.3.7. http://www.plurel.net/images/D237.pdf. Accessed 13 Mar 2012

Helminen V, Hannu R, Ristimäki M, Kontio P (2012) Commuting to the centre in different urban structures. Environ Plann B 39(2):241–261

Johnson MP (2001) Environmental impacts of urban sprawl: a survey of the literature and proposed research agenda. Environ Plann A 33(4):717–735

KC S, Barakat B, Goujon A, Skirbekk V, Sanderson WC, Lutz W (2010) Projection of populations by level of educational attainment, age, and sex for 120 countries for 2005–2050. Demogr Res 22:383–472

Kroll F, Müller F, Bell S, Haase D, Helminen V, Kabisch N, Piorr A, Schwarz N, Strohbach M, Taylor T, Zuin A (2010) Indicator framework for evaluating impacts of land use changes. PLUREL deliverable report D4.3.1. http://www.plurel.net/images/D431.pdf. Accessed 13 Mar 2012

Kroll F, Müller F, Haase D, Fohrer N (2012) Rural–urban gradient analysis of ecosystem services supply and demand dynamics. Land Use Policy 29(3):521–535

Lambin EF, Geist HJ (2006) Land use and land cover change: local processes and global impacts. Springer, Berlin
Lambin EF, Meyfroidt P (2010) Land use transitions: socio-ecological feedback versus socio-economic change. Land Use Policy 27:108–118
Liu JG, Dietz T, Carpenter SR, Alberti M, Folke C, Moran E, Pell AN, Deadman P, Kratz T, Lubchenco J, Ostrom E, Ouyang Z, Provencher W, Redman CL, Schneider SH, Taylor WW (2007) Complexity of coupled human and natural systems. Science 317:1513–1516
MA (Millennium Ecosystem Assessment) (2005) Ecosystems and human wellbeing: synthesis. Island Press, Washington, DC
Markandya A, Harou P, Bellu LG, Cistulli V (2002) Environmental economics for sustainable growth: a handbook for practitioners. Edward Elgar, Cheltenham
Nefs M, Alves S, Zazada I, Haase D (in press) Retirement cities: analysing the opportunities and challenges of a co-existence of ageing and urban shrinkage in Europe. Env and Plan A
Nuissl H, Haase D, Wittmer H, Lanzendorf M (2008) Impact assessment of land use transition in urban areas—an integrated approach from an environmental perspective. Land Use Policy 26:414–424
OECD (Organisation for Economic Co-operation and Development) (1997) Toward sustainable development: environmental indicators. OECD Publication, Paris
Petrov L, Haase D, Lavalle C (2009). MOLAND model results for the Leipzig region. PLUREL Newsletter 5, p 5 at www.plurel.net
Piorr A, Müller K (2009) Making the multifunctionality concepts operable for impact assessment. In: Piorr A, Müller K (eds) Rural landscapes and agricultural policies in Europe. Springer, Berlin, pp 9–23
Ravetz J (2000) City region 2020: integrated planning for a sustainable environment. Earthscan, London
Ravetz J, Rounsevell M (2008) Scenarios—crystal balls for the urban fringe. PLUREL Newsletter 3, at www.plurel.net
Reginster I, Rounsevell M (2006) Scenarios of future urban land use in Europe. Environ Plann B 33:619–636
RIKS (2009) Metronamica—model descriptions. RIKS, Maastricht, The Netherlands. http://www.metronamica.nl/. Accessed 13 Mar 2012
Schwarz N, Bauer A, Haase D (2011) Assessing climate impacts of local and regional planning policies—quantification of impacts for Leipzig (Germany). Environ Impact Assess Rev 31:97–111
Strohbach MW, Haase D (2012) Estimating the carbon stock of a city: a study from Leipzig, Germany. Landsc Urb Plan 104:95–104. doi: 10.1016/j.landurbplan.2011.10.001
UNESA (2007) World population prospects: the 2006 revision. http://www.un.org/esa/population/publications/wpp2006/wpp2006.htm. Accessed 13 Mar 2012
Verburg PH, Eickhout B, van Meijl H (2008) A multi-scale, multi-model approach for analyzing the future dynamics of European land use. Ann Reg Sci 42:57–77
Wessolek G, Trinks S, Duijnisveld WHM (2004) Entwicklung von Boden und nutzungsspezifischen Regressionsgleichungen zur Ermittlung der Sickerwasserrate aus dem Boden: Das TUB-BGR-Verfahren. Bundesanstalt für Geowissenschaften und Rohstoffe/TU Berlin, Geozentrum Hannover/Berlin/Hannover

# Part II
# Case Studies

# Chapter 5
# Introduction to the Research Methodology and Role of Stakeholders

**Carmen B.E.M. Aalbers**

Part Two contains the central part of this book—the knowledge of land use change and peri-urbanisation processes and strategies in selected case study rural-urban regions (Figs. 5.1 and 5.2). While each chapter has been written by a team of local researchers and stakeholders and therefore differs in style, beneath this there are common objectives, a common methodology—the Joint Analytical Framework—and a common focus, making it possible to carry out the comparative analysis to be found in Chap. 13. This section introduces these objectives and methodology so that it does not have to be presented in each case study. In addition to the work carried out by the local research teams, all chapters contain the results of some or all of the spatial and quality of life modelling described in Chap. 4—the RUG Model, MOLAND and QoLSim. In the case of the MOLAND-Light and QoLSim the data for individual case study regions can be tested directly using the software downloadable from the project website (www.plurel.net).

## 5.1 Objectives of the Governance and Spatial Planning Research in the Case Studies

The main aim of the governance and spatial planning research of the PLUREL project was to generate knowledge that contributes to the development of strategies that promote a more sustainable development of rural-urban regions, especially in terms of provision and protection of green open space in the urban fringe areas of these regions. In peri-urban areas the dynamics of land use change and conflicts of interest are varied and intense, as demonstrated in Chap. 2. In many countries new forms of governance are developing, implying a less dominant position for the state

C.B.E.M. Aalbers (✉)
Team Stadsregionale Ontwikkeling, Alterra Green World Research,
Droevendaalsesteeg 3 Wageningen, Netherlands
e-mail: Carmen.Aalbers@wur.nl

K. Nilsson et al. (eds.), *Peri-urban futures: Scenarios and models for land use change in Europe*, DOI 10.1007/978-3-642-30529-0_5,

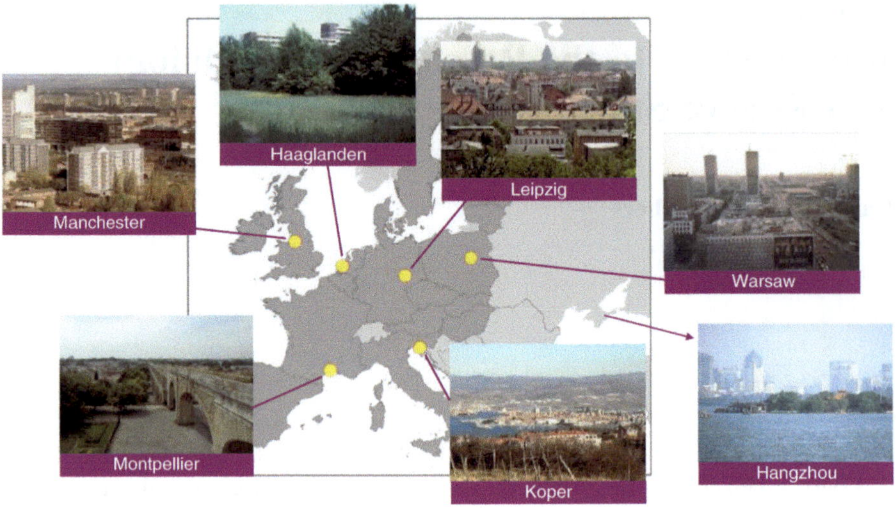


**Fig. 5.1** Case studies in the PLUREL project

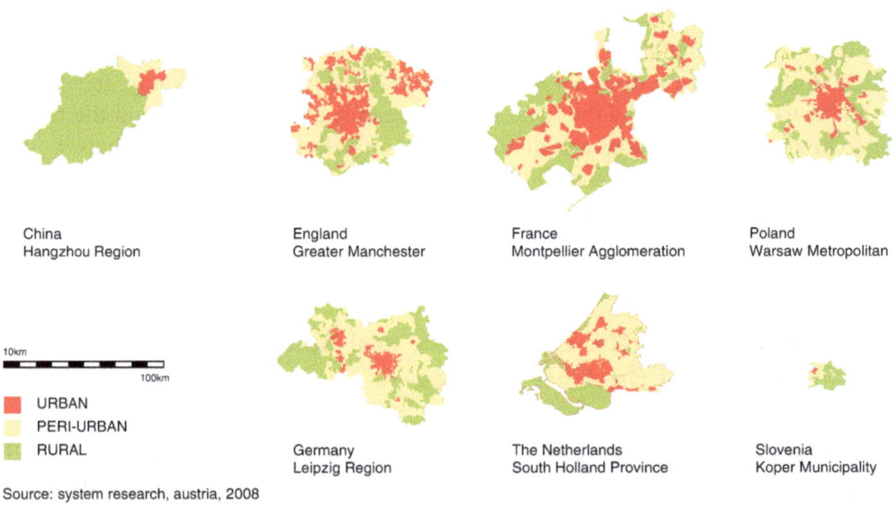


**Fig. 5.2** Urban, peri-urban and rural areas in the case study regions (Prepared by Duzan Doepel)

or national government and the emergence of multi-actor and multi-level networks and expanding interference zones between the state, the market and civil society (Van der Zouwen 2006). Multilevel and multi-actor governance and integrative spatial planning concepts are considered to be useful and more effective for promoting sustainable land use in rural-urban regions. How should a regional practitioner deal with this? Seven rural urban regions were selected at the outset of the project representing different regions, urban structures and dynamics and where there was local interest from stakeholders. The case studies analyzed and assessed the mechanisms of decision making and the use of integrative spatial plans in these rural-urban regions. The cases represented polycentric and mono-centric regions, comprising large and/or small towns. The governance types in the case-study regions range from hierarchic to more cooperative governance and the regions are located in North-West, Eastern and Southern Europe and in China. Among them are examples of post-communist (Warsaw, Koper) and communist regimes (Hangzhou). Also in demographic terms the regions differ. In some the population is growing fast, in other population growth is expected to come to a halt or is locally shrinking. This variety is important for understanding how context specificities influence developments, and for practitioners to decide what knowledge is relevant to their own region. In the case studies the different levels and drivers of decision making and the role of the public and other stakeholders were appraised. This was done in a comprehensive manner, taking into account the given regional specificities and the wider, national and international setting.

## 5.2 Trans-disciplinary Approach

Interdisciplinary research creates new knowledge by integrating experts from different disciplines such as the humanities, social sciences and natural sciences. For sustainability—implying the integrated consideration of society, economy and environment—integrated research is a requisite. Yet the case studies (with the exception of the reference study of Hangzhou) were also aimed at developing practically relevant and state of the art knowledge for EU and regional planners and policy makers. Thus it is important for the research findings to be understandable and usable by practitioners. Therefore we adopted a trans-disciplinary approach. Trans-disciplinary research adds an extra level of integration to that of interdisciplinary research, by involving practitioners and other stakeholders (Tress et al. 2003a, b). Trans-disciplinary research can also provide a societal context to the research, since practitioners usually bring with them information about public demands and other social aspects affecting their area. This trans-disciplinary approach demands significant attention by a research coordinator to create and manage a joint process of knowledge development with the different parties. In our research we therefore cooperated with stakeholders and practitioners from within each case study region (except in Hangzhou where the governmental system and the role of stakeholders is completely different) and a wider group of parties representing possible end-users of the project findings, such as national and European partners. Public authorities, businesses (mostly farmers' organisations), civil society groups (e.g. Birdlife International, Friends of Midden Delfland and hybrid forms of these (e.g. Pennines Prospects) participated. We referred to all these parties as "stakeholders."

**Table 5.1** Prioritization of land use issues for the research according to three priority levels

| Region/issue | Warsaw | Koper | Montpellier | Haaglanden | Manchester | Leipzig | Hangzou |
|---|---|---|---|---|---|---|---|
| 1 Land press. due to housing and business development | 1st | 1st | 1st | 1st | | 1st | 1st |
| 2 Agriculture under pressure | 3d | | 2nd | 2nd | 2nd | | 2nd |
| 3 High value nature at risk | 2nd | 2nd | | | 1st | | |
| 4 Integration of tourism | | | 3rd | 3rd | | 3rd | 3rd |
| 5 Traffic | | | | | 3rd | 2nd | |
| 6 Water management | | 3rd | | | | | |

## 5.3 Questions and Issues Raised by Stakeholders

At the outset of the project the regional-level stakeholders and researchers identified the following issues for the project:

- How to coordinate planning and decision making in a complex situation with a large number of powerful actors with competing interests and a competitive attitude?
- How to keep a balance between the needs of small rural municipalities and pressures from big cities?
- What is the role of identity for development and sustainability of land use relations in the urban fringe?
- What interactive models and tools can be provided to simulate dynamic land development patterns and their environmental impact, for use in discussions during the planning process?
- How can approaches in different urban regions be benchmarked?
- How to disseminate knowledge effectively to the professional communities, NGOs and the public?

The questions clearly reflect the practitioners' concerns and notions of integration, of working between scales, of power balances and the ultimate utility of the expected research results. Their questions were addressed in the formulation of the description of work of PLUREL.

Shortly after the project started, a new get-together with the regional authorities from the seven case study regions was organized in order to identify the most pressing land use issues in the regions. This resulted in a prioritization (Table 5.1):

## 5.4 Research Methodology

The research was multi-method, ranging through background work through literature review, especially of regional- and national-level policy documents and plans, site visits to explore the regions and to see examples of issues on the ground, compiling inventories of actor groups, communication situation mapping, interviews with parties involved in regional planning, interviews of citizens, regional, national and international trans-disciplinary panel discussions on research orientation and outcomes, regional scenario development workshops, expert meetings on scenarios and modelling. The groups of stakeholders or practitioners identified as research partners for each case study region were invited to review the research sessions, to comment on the scientific reports, to participate in the formulation of research questions and to list criteria for the evaluation of regional strategies generally aimed at preserving green open space in the urban fringe. Their help was also solicited in the development of end products (e.g. the scope and contents of this book, the utility of modelling systems) and in facilitating data collection. The practitioners willingly engaged in these activities and became close partners in the project.

The trans-disciplinary approach was far from easy. The involvement of practitioners ensures a strong focus on practical problems and needs, which does not always come naturally to scientists working by themselves. So it was necessary to invite practitioners to discussions and to persuade them or their organization to be members of an advisory board. This then enabled them to have their internal discussions separate from those with researchers, so they could be prepared to speak up—when necessary—against what the researchers were saying or proposing. Separate visits for practitioners were also organized, during which they could single out specific planning situations and problems upon which they wanted the research to focus. In parallel, the scientific quality of the research could be assured by coordination between different research teams while allowing for some of the natural preferences of the individual researchers to emerge.

## 5.5 The Joint Analytical and Assessment Frameworks

In order to focus the research by the regional teams on the afore-mentioned common objectives and issues brought forward by the stakeholders and to allow for the eventual comparison of findings between the regions, the research was guided by two joint research frameworks. These frameworks defined the scale and context to be considered in each case study and the concepts and criteria to be used in the data collection and analysis. In terms of scale the research was focused on rural-urban regions and the regional efforts to control developments. These efforts could be those of a formal body or loose forms of regional cooperation between municipalities or of one authority controlling a whole region

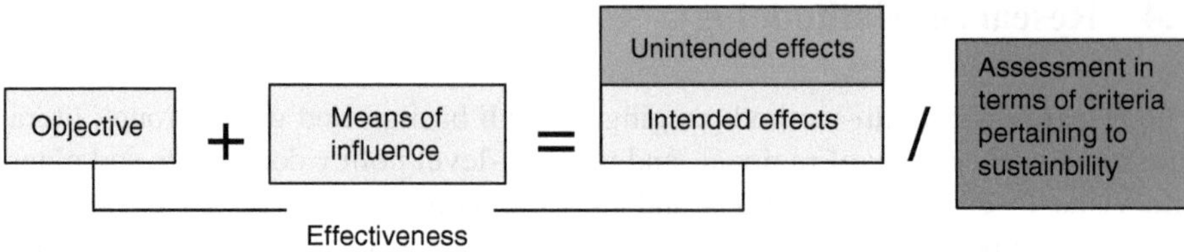


**Fig. 5.3** Schematic image of a strategy, its effects, effectiveness and assessment

(eg. Koper municipality, Hangzhou city respectively). The impact level was that of the single local municipalities or communes. The wider context was to be described in terms of temporal scale (short history), economic, social and physical development and national steering. In order to produce usable knowledge and to include the strategic intentions of actors in planning and governance we allowed for the inclusion of small planning scales. As a central governance unit we used the concept of a strategy. We defined a strategy as a designed succession of decisions and actions that an actor makes in order to achieve his objectives. The interaction between different actors and the regionally-specific physical and social context within which the strategy operates influences its effect. Other concepts and research questions in the joint analytical frameworks considered the actor networks, legal means, alliances, political mandate (votes), discourses, financial resources and knowledge devised by regional and higher authorities to influence local land use decisions.

The comparison—actually the stakeholders asked for a benchmark—between regional governance and spatial planning required a set of criteria with which the results of the preceding analysis of the context in relation to the actors' strategies could be assessed. The modelling and scenario discussions provided additional limited information on the workings and impact of the strategies. The assumptions that need to be made when modelling the complexity of urban developments in the urban fringe and the scenario basis for the models limit the utility for impact assessment of the strategies. The assessments are based rather more on local knowledge. The comparison between the strategies proved helpful to conceive improvements to regional strategies. Figure 5.3 shows the conceptual relationship between a strategy and its effectiveness and its assessment.

Among the research questions that were raised by the case study research were the following:

- Which are useful criteria for assessing and comparing the performance of a selection of–three–main strategies?
- How do the main strategies perform in terms of the developed criteria?
- How can the information on explaining factors for performance and experiences in other countries be used to improve current spatial planning and governance strategies to maintain green land-use in the different case study regions?

**Table 5.2** Criteria for assessing the outcomes of a strategy and the process it comprises

| Concerning outcomes of the strategy |
|---|
| 1) The strategy is resilient—robust and flexible enough to cope with changes in its context and stays effective over the long term (>25 years). |
| 2) It serves multiple objectives—it employs synergy to create maximum effect (PP or PPP[a]) or creates many 'winners'. |
| 3) The strategy is effective—it actually produces the outcomes it was designed for. |
| 4) The strategy pushes land use away, creating new land use conflicts elsewhere or at another level. |
| 5) The strategy pushes land use conflicts away, not causing new land use conflicts elsewhere nor at another level. |
| 6) The strategy strengthens the unique qualities of the area it pertains to. |
| 7) The strategy contributes to a sustainable dispersion of land use at a regional level, with a balance between resource availability and use. |
| 8) The strategy enables existing rural types of land use to stay or to develop. |
| 9) The strategy creates a new or additional urban economy. |
| 10) The strategy leads to accessibility for city people to peri-urban, open landscapes/agricultural land. |
| 11) The strategy protects land with best agricultural production capacity, based on soil quality. |
| **Concerning the process comprised by the strategy** |
| 12) The strategy helps the process of decision-making by clarifying a complex situation. |
| 13) The strategy raises awareness among business, developers in specific sectors, academics, citizens, other sectors of authorities (more choices possible) |
| 14) The strategy involves different actors: individual business, business interest groups, individual citizens, civic society groups, nature NGOs, other authorities (more choices possible) |
| 15) The strategy enables bottom up initiatives by citizens or business, semi-private organizations: individual business, business interest groups, individual citizens, civic society groups, nature NGOs (more choices possible) |
| 16) There is a clear time span for meeting the objectives contained in the strategy. |
| 17) The objectives of the strategy are clearly defined and in a comprehensible manner. |
| 18) There is a legal, statutory, financial or cultural commitment to support the process. |
| 19) The strategy provides for monitoring and evaluation of its internal and external consistency and impacts over time, using existing available data. |
| 20) The strategy empowers producers, developers, citizens, local government, supra local government, national government (more choices possible) |
| 21) The strategy restricts free riding behaviour/costs incurred with those who carry the benefits. |

[a]People, Planet, Profit

The different assessment criteria that the researchers and practitioners enumerated and that were used in the final assessment by them are shown above (Table 5.2) (Aalbers and Van Dijk 2008b). Only the assessments of those outputs or processes which were supported by arguments that could be traced back to the analysis of the strategies have been included in the comparative analysis in Chap. 12. The results of the assessments appear together with the analysis and scenario works in the case study Chaps. 5, 6, 7, 8, 9, 10 and 11. In Chap. 12 the results of the comparison can be found and in Chap. 13 some further comparison of government, planning and financial instruments is presented, also using the findings from the case studies (except for Hangzhou) but setting them in the wider European context.

This ensemble of case study chapters and the two comparison chapters provide the answers to the research questions, embedded in a description of regional context and European trends in governance and planning.

## References

Aalbers C, van Dijk T (2008a) JAF remix: analysis framework. A framework for research of spatial planning and governance strategies in Rural-Urban Regions to promote sustainability of rural-urban linkages. PLUREL. http://www.plurel.net/images/D321.pdf. Accessed 14 Mar 2012

Aalbers C, van Dijk T (2008b) Framework for the Asssessment of regional strategies. D3.2.1. extra. http://www.plurel.net. Accessed 18 September 2012

Tress B, Tress G, van der Valk A, Fry G (eds) (2003a) Transdisciplinary landscape Studies: potential and Limitations, vol 2. DELTA, Wageningen

Tress G, Tress B, Bloemmen M (eds) (2003b) From tacit to explicit knowledge in integrative and participatory research, vol 3. DELTA, Wageningen

Van der Zouwen M (2006) Nature policy between trends and traditions. Eburon Academic publishers, Delft

# Chapter 6
# The Hague Region: Negotiating the Common Ground in Peri-Urban Landscapes

**Judith Westerink and Carmen Aalbers**

## 6.1 Introduction

This chapter deals with the case study region with the highest average urban density, the biggest area below sea level and the largest area of greenhouse horticulture of all PLUREL case studies, The Hague Region, in the Netherlands. The chapter introduces the area and then the planning system in The Netherlands and the position of the city region in peri-urban governance. Three regional strategies are examined in more depth following the analysis of the area carried out using the joint analytical framework presented in the introduction to this section of the book. These are: giving incentives to farmers for landscape management (known here due to the large amount of water in the Dutch landscape as 'green and blue services') to strengthen agriculture in the urban fringe, examples of discourse development for making green space politically more important, and the use of inner city densification to prevent urban sprawl. Two scenarios (based on the SRES scenarios presented and discussed in Chap. 1) were developed for The Hague Region in addition to a Business as Usual scenario: 'Peak Oil' and 'Fragmentation'. The scenarios were modelled using MOLAND (see Chap. 3), with and without strategies, in order to test their robustness in different situations. To conclude, recommendations are given for planning and governance in the peri-urban zone of The Hague Region and for improvement of the strategies which were assessed.

J. Westerink (✉)
Team Nature and Society, Alterra Wageningen UR, Wageningen, The Netherlands
e-mail: Judith.westerink@wur.nl

C. Aalbers
Team Governance, Alterra Wageningen UR, Wageningen, The Netherlands

K. Nilsson et al. (eds.), *Peri-urban futures: Scenarios and models for land use change in Europe*, DOI 10.1007/978-3-642-30529-0_6,

## 6.2 Squeezed, Wet and Diverse

The Hague Region (with one million inhabitants) is classified as an urban polycentric region (see Chap. 2 and Loibl et al. 2008) of 451 $km^2$; it is part of the large continuous urban region known as Randstad Holland located in the west of The Netherlands, situated on the North Sea coast. Its main constituent cities are The Hague (474,000 inhabitants), Delft (95,000) and Zoetermeer (118,000), although the region also comprises a number of other towns and villages, some of which have expanded to become part of a continuous urban fabric with The Hague (such as Voorburg, Leidschendam and Rijswijk). Moreover, it is not an isolated city region: to the South it borders the Rotterdam region and to the north Holland-Rijnland which includes the city of Leiden.

The cities and towns in the region all have a distinct character, due to their location, history and economic activities. The Hague is the largest city of the region. It is the residence of the Dutch government, although Amsterdam is the capital city of The Netherlands. Many embassies and international institutions are located in The Hague, among which is the International Court of Justice. The city likes to present itself as the 'International City of Peace and Justice'. Delft has a technical university, which has attracted a cluster of knowledge institutes and innovative companies. Zoetermeer is a satellite city of The Hague, a new town which grew from a village of 10,000 inhabitants in 1960 to a city of 100,000 by 1990 as a result of national planning policies. Many of the inhabitants of Zoetermeer work in The Hague, although the city also has a developing sports and leisure sector. An important economic sector in the region, with a high impact on the landscape, is greenhouse horticulture. The main greenhouse areas in The Netherlands are found in The Hague Region and they rival the city itself in terms of the area under glass.

Because of the unstable, shrinkable and wet peat soils (Fig. 6.1), towns and cities in this part of The Netherlands developed on sites on higher and more stable ground, such as sand dunes (The Hague, Wassenaar, Rijswijk, Voorburg) and river banks. This led to a clear distinction between 'urban' and 'rural' until relatively recent times. However, most of the suburban expansions and the whole of the city of Zoetermeer were necessarily built on lower ground, owing to the lack of suitable land, and only possible because improved building techniques and water pumping technology allowed this. The dune reserve of Meijendel was spared from construction because of its importance in coastal defence and for drinking water extraction. Bordering Meijendel, the municipality of Wassenaar also managed to maintain a green character. It is known for its estates and expensive villas. It is one of the three 'richest' municipalities in The Netherlands.

The Hague Region has very little space for urban expansion, because it is constrained to the south by Rotterdam and the river mouth of the Rhine, to the west by the North Sea, to the east by the Green Heart peat meadow area in the midst of the ring of Randstad cities, which has a national protection status and to the north by the urban area of Leiden. The peri-urban areas in the region have become

**Fig. 6.1** Peatland as it covered most of The Hague Region in the early Middle Ages

important 'green enclaves' between cities (Aalbers et al. 2009). Traditional meadow landscapes are now considered scarce and policy makers are considering planning instruments for their protection, similar to those used for nature reserves. The city of The Hague is especially 'cornered', being squeezed between the sea (west), greenhouses (south), a nature reserve (north) and urban fabric of other towns (south). For that reason, The Hague has had to expand beyond its borders. Firstly, Zoetermeer was built as noted above (but it remained a separate municipality). More recently, large urban expansion zones were developed near Nootdorp (Fig. 6.2a, b) and Leidschendam. These areas were added to the administrative territory of the municipality of The Hague. Currently, the town of Pijnacker is also growing quickly because of the housing expansions at its southern edge.

Population growth in the region is slow (1 % on average in 2006), mainly through in-migration of various types, but there are differences between municipalities. The city of The Hague has been shrinking in population since 1960 but has shown some growth again since 2000 as a result of the recent expansions to its area as noted above. It is expected that the rate of population growth in the region will continue to slow and will stabilize at a modest annual rate of increase of 0.3 % by 2030 (CBS 2009). Currently, Pijnacker-Nootdorp is the fastest growing municipality. A stagnating population will not mean a halt to building, however. Declining neighbourhoods will need to be renewed, but more importantly, average household size is diminishing, leading to a need for more houses per capita. For the Netherlands as a whole, twice as many houses are needed per capita now as 90 years ago (CBS 2009). In The Hague Region it is expected that the average household size will fall from 2.1 persons in 2010 to 2.05 by 2025 (Haaglanden 2010).

The Hague and Delft are building high-rise office and apartment buildings around their railway stations. In the case of The Hague, this development means

**Fig. 6.2** (**a**, **b**) This recent urban expansion of The Hague has a range of living environments, from highly urban to single houses

the displacement of some of the existing (industrial) activities. Some of these activities are now planned to be relocated to the peri-urban Vlietzone (Fig. 6.3a, b). Vlietzone, currently a rather structureless, messy urban-fringe area, will thus experience major changes in the near future. An integrated development of housing, business, water, nature and recreation is planned here. In the Regional Structure Plan (Haaglanden 2008) it is indicated as a 'multifunctional linking zone'.

Urban fabric in the region has been gradually expanding but in a controlled way. The map of the region looks quite 'organized' as a result (see Fig. 6.4). The government has tried to provide space for recreation and nature in the urban fringe, for instance around Delft, by the means of land purchase, at the expense of continued agricultural land use (e.g. Van Rij et al. 2008). During the twentieth century, recreation areas and peri-urban parks were mainly designed and planted as

**Fig. 6.3** (**a**, **b**) Vlietzone, between Rijswijk and Delft, has a mosaic of peri-urban land uses such as small fields with horses, allotment gardens and golf courses

forests. More recently, however, the recreational value of the traditional meadow landscape has been recognised and ways are being looked at to improve the recreational accessibility of privately-owned farmland. With a subsidy from the central government, scattered greenhouses are gradually being removed from the meadow landscapes. Thus, the physical and visual separation between 'grass' (meadow) and 'glass' (greenhouse) landscapes can be enhanced.

The case study in The Hague Region described in this chapter explores strategies proposed by various actors in the region to strengthen 'green' (or green-blue) land use types in the urban fringe, especially agriculture and recreation. In both forms of land use the cultural landscape is an important issue. The strategies were selected in consultation with stakeholders from the city region and the ministry of Agriculture, Nature and Food Quality. First, an analysis was made of the region, planning and

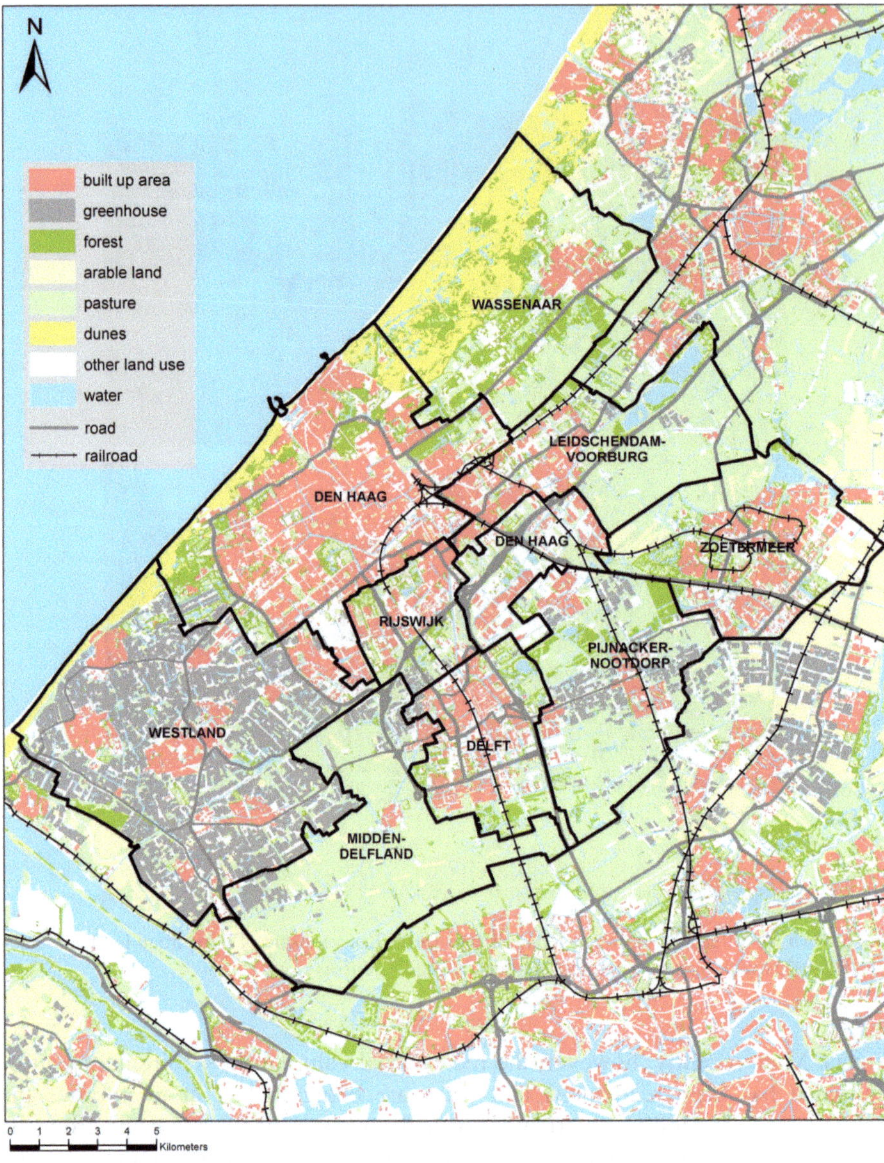


**Fig. 6.4** The nine municipalities of The Hague Region (land use 2004)

governance of its peri-urban areas and the three strategies selected through action research, interviews and literature study as noted in the introduction to this section of the book and as reported in Aalbers et al. (2009). Then, the strategies were assessed using a variety of methods, including scenario development (as reported in Westerink et al. 2010a). In the following section, a short introduction will be given of the Dutch planning system and the role of the city regional authority. Then the three selected strategies will be described as they are currently applied. After this the scenarios will be the focus: the storylines as they were developed together with stakeholders and the modelling results and land use maps based on these storylines (as reported in Mubareka and Lavalle 2010 and Westerink et al. 2010a). The lessons learnt from the scenarios and assessing the strategies lead to policy recommendations in the final section.

## 6.3 Spatial Planning in The Netherlands and the Role of the City Region

Responsibility for land use planning in The Netherlands is shared between municipalities, provinces and the state. Municipalities are the lowest and most important government level with respect to zoning and are responsible for issuing building permits. Zoning plans need to be updated at least every 10 years. The provinces and the state can only influence municipal zoning plans through targeted planning outlines communicated in advance of the zoning process. However, for issues that concern their level, the provinces and the state can also prepare zoning plans. In this case, the provincial or national zoning plan overrules the municipal plan. Under the current Planning Act (in effect since July 2008), city regions no longer have a formal role in planning (Aalbers et al. 2009). Their formal role as described in the Joint Arrangements Act (WGR+) still stands: organizing cooperation between municipalities with respect to a number of tasks, among which is planning. However, the current government, elected in 2010, aims to repeal this act, weakening the position of the city regions even further.

The highest proportion of the municipal budget comes from the state, based mainly on the number of inhabitants and the number of addresses. This may therefore be an incentive for municipalities to strive for population growth (see Tosisc and Gertheis 2010 and Chap. 13). Municipalities have a limited authority to raise taxes themselves, the most important being property taxes. Many municipalities are also active in the land market—selling land to property developers is an important source of income. A non-financial incentive for continued urbanisation is the employment argument, which leads to competition between municipalities to attract inward investment by companies by building business sites. In 2008 this led to a public discussion in The Netherlands, when environmental NGOs claimed that there was a surplus of business sites as a result of this competition.

At the state level, several ministries touch upon peri-urban planning issues. In the past, each ministry used to make its own sectoral plans from a perspective of, for example, housing, agriculture, traffic or economic development, but the previous national spatial plan (Nota Ruimte: VROM 2005) was a joint effort by four ministries. The Ministry of Culture had a great influence on the current discourse with respect to landscape preservation. It launched the 'Belvedere' programme, which promoted 'preservation by development' as an approach to cultural heritage and landscapes, as a reaction to rigid conservation and top-down planning.

**Fig. 6.5** The boundary between the municipalities of Westland and Midden-Delfland is clearly visible in the landscape

'Preservation by development' was believed to be a more practical approach that could allow development as long as the total quality of an object or an area would be improved. The Ministry of Housing sets housing targets which are negotiated with provinces, city regions and cities. Subsidies are available for inner city renewal in the bigger cities. The Ministry of Agriculture delegated much of its responsibilities and funds for nature development, rural development and agri-environment schemes to the Provinces. The Provinces develop their own rural development plans, but have developed a joint agri-environment scheme.

The Hague Region, as regional authority, consists of nine municipalities and is part of South Holland Province. Special municipalities are Midden Delfland and Westland, formed by the merger of several smaller municipalities in 2004. The framing of these municipalities was deliberate: Westland as a 'glass' municipality and Midden-Delfland as a 'grass' municipality (see Fig. 6.5). The objective was to make two non-urban municipalities each with a distinct identity, with the goal of maintaining their character and keeping urbanisation in check.

The spatial plan of The Hague Region is called the Regional Structure Plan (RSP: Haaglanden 2008) and was prepared in consultation with the municipalities. 'Network thinking' is its motto. The green-blue network is one of the main items in the plan. Current green areas are to be physically connected to each other, both as ecological and recreational corridors. Green areas are to be made more accessible by improving recreational routes within them and from the inner city to the urban fringe. The RSP envisions the peri-urban enclaves to be developed into Regional Parks. Network thinking also applies to the transport system, especially public transport. This is to be improved and extended, combined with inner city densification and quality improvement. Quality is aimed both at residential quality of life and at commercial attractiveness. The overall goal of the RSP is to improve the competitiveness of the region.

The Hague Region as regional authority has limited powers and resources for planning (Aalbers et al. 2009). 'Concertation' (networking, negotiation, consultation, promoting cooperation, persuasion and other soft tools) is its main instrument, trying to gain support and raising funds for joint projects. Several of the peri-urban areas (or enclaves) of the region extend into the neighbouring city regions. This means that for policy development in these areas, cooperation is needed with the surrounding regions. 'Concertation platforms' are formed for joint policy development for these areas and other cross-border issues. For example, The Hague Region takes part in the Randstad South Wing administrative platform, where it consults with Rotterdam Region, South Holland Province and other parties.

Acting in concert is not only done with other governments: the region's idea of 'network governance' includes exchange of ideas with commercial private parties and NGOs. A large number of NGOs and lobby organisations are active in the region. Noteworthy among these are the area-based groups, such as the 'Friends of ...' (followed by the name of the area). These civil initiatives can be very influential for a period of time. The Farmers' Union is a must-have partner in any consultation about peri-urban issues. The three Environmental Cooperatives (associations of civilians and farmers focussing on the agri-environment) in the region also have a growing influence. Nature reserves may be managed by organisations ranging from the drinking water company to the state forest service and NGOs.

The main influence of commercial private parties in the peri-urban area is through land acquisition. Investing in land in areas that could be developed in future can be very strategic. However, other large companies and institutions are also of interest to The Hague Region as the regional authority, because of the international competitiveness of the region and the quality of life desired by and for their employees. In terms of the current agricultural land use, the dairy and the horticultural sectors are important. Diversified farms (for instance campsites at farms) and non-agricultural businesses (for instance consultancies) in former farmsteads are also increasing.

In summary, the main task of the city regional authority is organising cooperation between municipalities. With respect to planning, The Hague Region puts this into practice through developing a joint spatial strategy and by initiating and taking part in consultations with other tiers of government, the private sector and NGOs. In the following section, three strategies aimed at land use in the urban fringe that are in use in The Hague Region will be studied in more depth: green and blue services, discourse development and inner city densification.

## 6.4 Strategies for Governance of the Peri-Urban Area

### *6.4.1 Green and Blue Services to Strengthen Agriculture in the Urban Fringe*

Agriculture in the urban fringe is under pressure as a result of several factors. The area of meadowland is diminishing because of urbanisation, glasshouse expansion and the establishment of nature and recreational areas. The remaining farmers face high land prices (DLG 2009), which makes farm enlargement unaffordable. Many farmers do not have successors because their children are not interested in the farming life (Vonk Noordegraaf and Gloudemans 2004). Furthermore, peat shrinkage, as a result of mineralization and drainage is increasingly recognised as a problem by policy makers. The only way to reduce the mineralisation of peat is by increasing water levels, creating less favourable conditions for agriculture. Another 'competing claim' is the need for space for storm water storage (an increased demand as a result of the increase in built areas and sealed surfaces, combined with sea level rising and increasing peak discharge of the rivers), most of which is sought in the peri-urban areas (Delfland 2004).

For the government, in a situation with fewer or no active farmers, other ways would need to be found to manage the landscape. The maintenance of publicly managed nature and recreation areas has been found to be expensive. The recreational demand in the region surpasses the supply, in terms of parks and recreational areas (Fontein et al. 2009). The cultural landscape is therefore needed as additional recreational space. The government prefers to keep farmers in the area managing the landscape, not only because of the high costs of management by government agencies, but also because of the linkage between the cultural landscape and the dairy farming system which created it (Aalbers et al. 2009).

The strategy analysed in this section concerns local agri-environmental subsidy schemes set up to pay for green and blue services delivered by farmers. These services include measures to promote on-farm nature, to improve landscape quality, to supply storm water storage and to provide public access. Green and blue services are seen as a form of multifunctional agriculture, because food production is combined with landscape management. The strategy has two goals: to provide additional sources of income for farmers and to improve the ecological, aesthetic and recreational value of the landscape. Green and blue services can be considered as a governance strategy, because the concept opens up the normally top-down and centralised agri-environmental policy development to bottom-up initiatives and local variation (Westerink et al. 2010a). Local stakeholders, such as farmers and environmental cooperatives may design tailor-made subsidy schemes in cooperation with local, regional and national authorities and civil society groups.

The Hague Region has produced several of the first green and blue services initiatives in the Netherlands, including the Green Fund for Midden Delfland and the Farming for Nature project in the Biesland polder. These two schemes were developed in parallel: the Green Fund in 2003–2006 and Farming for Nature in 2003–2007. In addition to these schemes, which are currently operational, a scheme started in the Land van Wijk en Wouden area, which resulted in the establishment of six walking trails on farmland (around 40 km in total). More measures were envisioned, but insufficient funds were available. In 2008 a new pilot project started in this area, in cooperation with the Rijnland water board, to develop 'blue' services aimed at water quality and aquatic ecology in ditches.

**Table 6.1** Varying arrangements for green and blue services in three areas

| | Green fund Midden-Delfland | Farming for nature in Biesland Polder | Land van Wijk en Wouden |
|---|---|---|---|
| Area | Farms in area of 6,700 ha | 97 ha | 40 km of public footpaths |
| Number of farmers participating | 76 | 1 | Unknown |
| Annual expenditure payments | Approx. €225,000 | Approx. €100,000 | On av. approx. €40.000 |
| Fund | 9–12 M€ | 1.9 M€ | €200.000 |
| Duration | 6-year contracts, paid from interest of fund | Contract 30 years, funding for 20 years, additional funding is sought | Unknown |
| Approach | Measures on farm: meadow bird protection, landscape elements, heritage elements. Measures are selected and applied for by the farmer | Farming system: self-sufficient for nutrients, adjusted water levels, and landscape elements: creating conditions for biodiversity and water quality (whole-farm) | Public footpaths. Other measures could not be funded |
| Main actors (apart from the farmers) | Midden Delfland Green Fund, Environmental Cooperative Vockestaert | South Holland Province, Friends of Biesland, Alterra | Area Committee, Environmental Cooperative Wijk en Wouden |
| Funding | The Hague, Delft, Midden Delfland | Ministry of Agriculture, South Holland Province, The Hague Region, Delft, Pijnacker-Nootdorp, Delfland water board, The Hague | South Holland, Zoetermeer, Leiden, Zoeterwoude, Leidschendam-Voorburg, Leiderdorp, Alphen ad Rijn and private donors |

All these initiatives are characterised by an 'area fund' from which payments are made to farmers for their 'services'. The fund is provided with contributions from the surrounding municipalities, the province and sometimes the water board, the city region and the state. In the case of Land van Wijk en Wouden, the idea included a contribution per inhabitant from the municipalities (€0.75/head). The contributions from the governments to the funds are voluntary. The green and blue services initiatives were therefore preceded by painstaking concertation processes. Farmers, environmental cooperatives, 'friends of' groups and researchers took part in these activities. They led to different designs for measures and arrangements in the three areas (see Table 6.1). State aid procedures with the European Commission added to the preparation time.

When assessing the strategy, a distinction should be made between the general concept and the specific initiatives in the region. The selected initiatives differ with respect to establishment, funding and the impact of the measures. In general, far-reaching measures can be expected to have a larger effect on biodiversity and landscape quality. However, such measures imply higher payments and will need longer-term contracts. The participation level of farmers is expected to be higher when the measures are less rigorous. Far-reaching measures also have a larger impact on farm management: many farmers prefer to incorporate green and blue services into their current way of farming without too many adjustments (Ruto and Garrod 2007). In Midden-Delfland and Land van Wijk en Wouden, a lack of funds limits the number of measures carried out and hence both the public effect and the effect on farm income.

To improve the value to the public of the peri-urban enclaves, far-reaching environmental measures and increased public access on farmland should be aimed at. Conflict resolution is also necessary. A balance should be found between protecting meadow birds and welcoming more people onto farmland, which might disturb the birds, especially in the nesting season. However, higher water levels and meadow birds can be a good combination with fewer conflicts. Higher water levels may be locally desirable for water management and for countering subsidence of peat and could be arranged as a 'blue service'.

The strategy is aimed at the main actors for sustaining agricultural land use: the farmers. It serves multiple goals (sustaining agriculture, promoting biodiversity) and leads to multifunctional land use (food production, biodiversity, public access). The strategy leaves, in theory, much space for locally developed and tailor-made solutions. However, the bureaucratic European context of the Common Agricultural Policy (CAP) and state aid regulations may hinder bottom-up initiatives (PLUREL 2008). It is hard to assess the extent to which goals are clear and effects are measurable. Participation by farmers in local agri-environmental schemes is voluntary and the evaluation of ecological effects is complex. In other words, the strategy has a large amount of uncertainty built into it. However, positive incentives may be more effective than (restrictive) rules when it comes to landscape management by farmers (Westerink et al. 2009).

### 6.4.2 Discourse Development to Raise Political Support for Green Space

The description of the formal planning system made it clear that the city regions in The Netherlands have very limited formal power with respect to land use planning. Power and financial resources are with the municipalities, the province and the state. The main task and instrument of the city region is therefore the soft method of bring people together in an informal way (in concert) (Aalbers et al. 2009), striving for cooperation between the municipalities and gaining support for joint objectives and projects. Discourse development is an important strategy of The Hague Region in these processes. Discourses are lines of thinking, perceptions, perspectives, narratives and ideas. They are used for persuasion, for gaining support for certain policies, for lobbying and sometimes for branding. Discourses developed or adopted by The Hague Region often connect green space to other issues, in order to make green space more important in the policy arena (Westerink et al. 2010a). Linking themes is also used to obtain access to funds at a provincial or national level.

As an example, the public administration of The Hague Region forged links between green space and culture and heritage. In this way, green space is given a place in the 'regional sense of identity', especially concerning traditional agricultural landscapes or estates. Cultural history adds a valuable dimension to green space. Addressing landscape from this angle, an attempt can be made to raise interest from new target groups, in this case one which is culturally sensitive. The idea is to gain broader support both with the public and with the municipalities for preservation of and investments in green open space. In the course of the PLUREL project, the use of landscape paintings to stimulate public debate about changes in the landscape was investigated (Westerink et al. 2010b), using the historical painting of the same scene as a trigger for discussions about its current state.

The Hague Region also links green open space in a discursive manner to preferences of the expatriate community ('expats'). This community is important for The Hague Region because The Hague strives to be attractive for international institutes and companies to establish offices there and there are many embassies and other organisations with foreign staff. Linking green space to this 'international competitiveness' puts green space into the economic policy arena. According to this discourse, expats prefer to live in cities with abundant green space. Therefore the availability of green space is a factor for international organisations to come or to stay, because of the interests of their employees. In this context, green space is referred to as the 'green gold of the region' by the politicians responsible for it. Research among expats in three Dutch cities, including The Hague, confirms that both the availability and the quality of green space are important factors in expat living preferences (Veer and Luttik 2010, see also Florida 2002). Businesses also take into consideration the preferences of their employees when deciding on locations for setting up new operations (Love and Crompton 1999).

The use of spatial concepts fits into the use of discourses. In the Regional Structure plan the term 'connection' is much used, for instance with respect to green areas and the distances between them, or from the city core to the peri-urban zone. The connection concept illustrates both an attempt to improve the accessibility of green space and to combine or integrate themes (in this case green space and infrastructure). Another interesting spatial concept is that of the envisioned Regional Park Duin, Horst en Weide (Aalbers et al. 2009). This concept includes a landscape sequence, offering a diverse and interesting landscape to the urbanite visitor. The spatial concept (landscape sequence) is used to support the discourse ('the peri-urban area is a consumption landscape' Westerink et al. in press a).

To determine the significance of the effects of discourses with respect to land use change is a major challenge. In the course of PLUREL, we did not manage to do this effectively, so discourse development as a strategy could not be introduced into

**Fig. 6.6** Cycling is one of the main recreational activities in the peri-urban areas of The Hague Region. Cycling below the water level in Midden Delfland, skyline of Rotterdam

the MOLAND modelling processes (see below). Although it is generally believed that discourses precede behaviour and decision making, they are rarely formulated in a measurable way. They are designed to inspire, not to monitor.

The discourse 'peri-urban areas are important as recreational space for the city dweller' includes a risk of excluding those groups of city dwellers for whom the peri-urban areas are not (yet) important from policy development. In peri-urban planning, attention should be given to the variety of recreation preferences and future changes in recreational behaviour (Fig. 6.6). The needs of immigrants from non-European cultures have, for instance, not yet been sufficiently recognized (Aalbers et al. 2009).

Discourses can be an effective instrument in communication about planning and area development. Recognizing the value and legitimacy of emotions, place attachment and identity may greatly improve communication processes. In addition, art forms can be considered as alternative communication tools to gather information about people's perception of an area and to start a discussion about the future that is not limited to the rational domain.

### *6.4.3 Urban Densification*

Inner city densification is a strategy at the Randstad South Wing level. Randstad South Wing is an administrative concertation platform in which The Hague Region is one of the participating regions and The Hague is one of the participating cities. The parties agreed to strive for realising urbanisation 'as much as possible' within the existing urban fabric. The Hague Region adopted an 80 % objective of new building to be within the inner city, in line with the South Wing aim, which is more than the national government's objective of 40 % inner city building for the Randstad as a whole (VROM 2008). This densification in Randstad South Wing (both commercial and housing) is to be concentrated around public transport nodes. The public transport system is to be expanded and improved at the same time.

The strategy has a number of goals. First and foremost, it aims at improving the international competitiveness of Randstad Holland (and South Wing in particular). Accessibility in Randstad South Wing is considered a major limitation on the current competitive position. Also, the quality of the landscape, as noted above, is seen as important for the attractiveness of the region. Space is very scarce and the peri-urban enclaves are treasured. Quality of life/environmental quality are acknowledged in the discussion platform as 'lagging behind' the economic development as policy themes thus far (Zuidvleugel 2006). However, quality of life *in* the city is a bottleneck for inner city densification. Inner city densification at the cost of, for instance, parks or other public spaces is a risk. Also, dense urban areas will have little opportunity for private green space—also seen as valuable by many people. The RSP acknowledges the need for private green space especially for families with children. In the Regional Structure Plan of The Hague Region, mixing functions, layered land use and high quality public space are identified as strategies to maintain quality of life in the dense areas, combined with improvement of the public transport network. Revitalisation of residential and industrial areas also needs to contribute to densification.

Building 80 % inside the city still implies that 20 % of construction will take place in peri-urban areas. The strategy for building in these areas is very different from building in the city: in the peri-urban the RSP envisions high-quality houses in a green environment: villas, housing estates and mixtures of housing and green space.

The strategy was derived from the compact city paradigm but applied to a polycentric region and worked out as a spatial network concept Westerink et al. (in press b). As a compact city strategy, its aims are geared towards sustainability in all three domains (environmental, social, and economic). However, work reported in the international literature has pointed to risks in compact city thinking, especially with respect to environmental scaling (environment at regional scale may improve, but may deteriorate locally in the city) and to social justice (can people live where they want to live?) (see for instance Schweitzer and Zhou 2010 and Burton 2001). The inner city densification strategy as described in the Regional Structure Plan acknowledges these risks in the sense that quality solutions are sought, such as mixed (including green) land use and underground parking (Fig. 6.7). However, part of the problem (the car as space consumer, noise producer and fine dust emitter), is not tackled, while car use in The Hague is much higher than in the other main Randstad cities (CBS, figures 2002–2003). The strategy assumes that urban residents will use the peri-urban area in addition to or in exchange with urban parks. Indeed, the peri-urban areas in The Hague Region are intensively used for recreation but not by all population groups. A survey in Delft showed that immigrants do not (yet) visit peri-urban areas: they depend on urban (neighbourhood) parks for leisure (Aalbers et al. 2009).

**Fig. 6.7** These trees can only grow between high-rise buildings and on top of a car park because of an underground structure

The Hague Region has the highest urban density of all case study regions in PLUREL with on average 1,259 households and 2,202 inhabitants per km$^2$ (CBS 2006). There has to be a limit to densification. As a concept it may be too much of a quantitative approach that may sooner or later damage quality of life and the competitiveness of the region. In this way, economic and population growth as a policy objective may bite its own tail in the long term. A more qualitative approach to 'growth' that incorporates quality of life, environment and social cohesion (see Box 6.1), may be more sustainable (Westerink et al. 2010a).

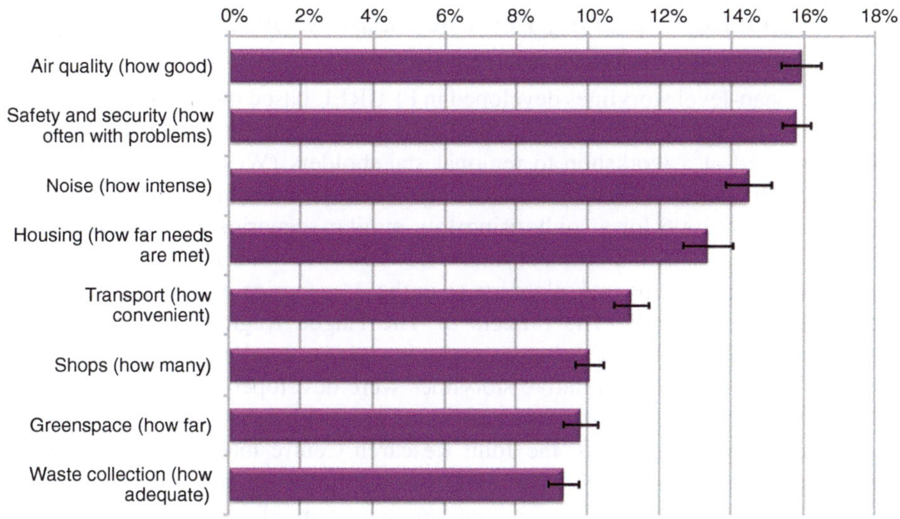


**Fig. 6.8** A graph showing the ranking and percentages of importance of different indicator values for The Hague (Source: OPENspace research centre, Edinburgh College of Art)

**Box 6.1 Quality of Life in The Hague Region: Indicators of Land Use Change and Its Effect**

As part of the quality of life indicator work (see Chap. 3), a limited sample of people were asked about their perceptions of different factors of the environment in relation to their perceived quality of life. Although the results are not statistically valid owing to the restricted sample size, they offer some insights into the factors which people view as important (Fig. 6.8). The most important issue as an indicator of environmental quality of life is air quality, possibly because people are concerned about pollution and dust from cars, for example. This is followed by a feeling of safety and security which might change if areas become more densely populated. Noise is the next most important factor—traffic noise, aircraft and noisy neighbours may all have an impact. Housing suitability comes next followed by the convenience of transport—if land use change results in slower and less convenient transport or commuting times then people are less happy with their living environment. Availability of shops comes next with accessibility to green space coming second bottom before adequacy of waste collection. Thus, if land use change as a result of urban densification starts to affect the higher level indicators negatively people may try to move to somewhere better. More samples would have enabled a more sophisticated analysis of the variability amongst the population and their current environment—such as whether they lived in a leafy suburb or inner city. The Quality of Life Simulator enables some of these factors to be tested along with predicted land use changes.

## 6.5 Scenarios Exploring Possible Futures for the Hague Region

Four European-level storylines developed in PLUREL based on the SRES scenarios (see Chaps. 1, and 3, and Box 6.2 'Regional Urban Growth (RUG) modelling') were presented at a workshop to regional stakeholders (Westerink et al. 2010a). Unfortunately, not all four scenarios could be worked out for the regional level (see Bouwman et al. 2006 for a study using four similar scenarios for the same area). The stakeholders selected two storylines, B1 'Peak Oil' and B2 'Fragmentation', because a situation with weak planning as in the A scenarios was deemed unrealistic for The Netherlands. The officers of The Hague Region were particularly interested in B2, a regionalisation scenario, in order to explore the position of the regional authority. The qualitative storylines were developed further in the workshop for the specific situation of the case study area with reference to the PLUREL list of indicators. Afterwards, the Joint Research Centre interpreted the regional storylines for modelling with MOLAND (see Chap. 3). In addition to these two scenarios, a 'Business as Usual' scenario was worked out based on general projections by the Dutch central statistical bureau (CBS). All three scenarios were run with and without the strategies of inner city densification and green and blue services. Modelling the strategy of discourse development with MOLAND was not feasible because of its qualitative nature, as explained above. The storylines were mainly developed for The Hague Region with some reference to Rotterdam Region. The MOLAND modelling, however, was done for the whole of South Holland Province to comply with the RUR typology (see Chap. 2).

**Box 6.2 Regional Urban Growth (RUG) Modelling for The Hague Region**
All four scenarios were modelled with RUG for 2025, with less detail than MOLAND and without stakeholder consultation. The scenarios show a clear 'economy versus environment' divide and the two environment-oriented scenarios give very similar results. As there are very few rural zones in The Hague Region, the *Hyper-tech* scenario does not result in counter-urbanisation despite the technological change. High population and economic growth and few planning constraints lead to high urban growth throughout the peri-urban and urban areas. In the *Extreme water* scenario, the storm surge risk makes the coast less attractive. However, here this is partly compensated by the fact that the coast lies higher than the peri-urban areas further inland (below sea level) and the improved flood defences put in place since the 1953 flood. The *Peak oil* and *Fragmentation* scenarios show very similar results, as environmentally-aware planners forbid new construction in the large flood risk zone between the cities (centre of map). In the *Peak oil* scenario, high fuel costs and reliance on public transport further contribute to concentrating populations in city centres. In the *Fragmentation* scenario, clustered communities live closely side by side due to space constraints. There are no areas in the region where enclaves might form, so older local people either

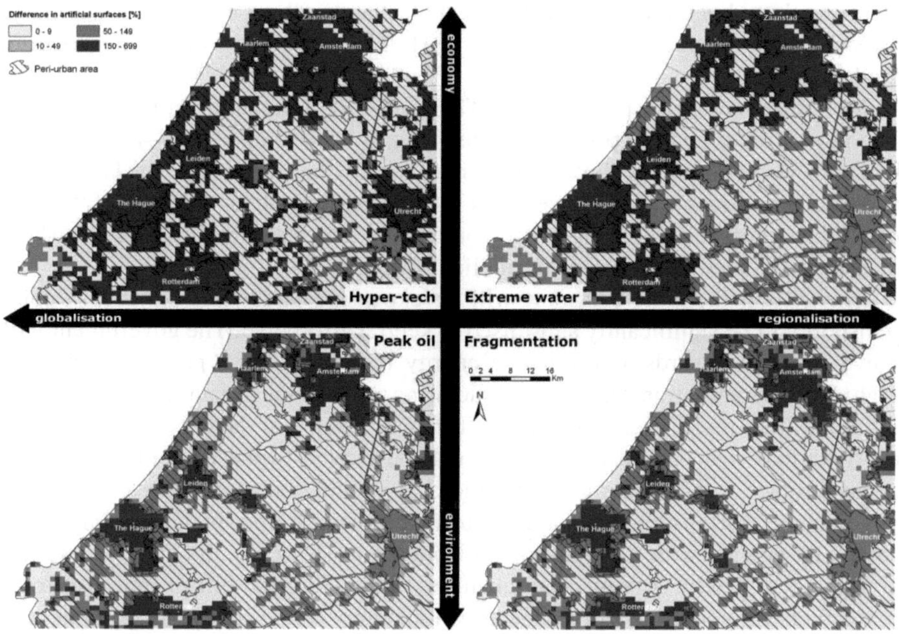


Fig. 6.9 (a–d) RUG modelling

form communities within the city or move to rural regions further afield (Fig. 6.9).

Source: University of Edinburgh

## 6.5.1 B1: Globalization, Strong Planning and Peak Oil – Regional Storyline

In the case of B1, the stakeholders expect The Hague Region to have a relatively favourable competitive position because of the international institutes in The Hague and the technology cluster in Delft. Traditionally there has been little industry in The Hague Region and the trend of globalisation and inter-governmental cooperation will lead to a growing importance of the institutes in The Hague and a strengthening of the economy related to this sector. There will also be a growing demand for the technological knowledge of the institutes in Delft and therefore they will also continue to grow. However, Rotterdam will shrink economically because of the declining harbour economy as a result of the energy crisis. The greenhouse sector in Westland and around Pijnacker will suffer from the rising transportation cost, but will have developed into a net producer of energy and a $CO_2$ consumer, due to advanced and efficient technology. From the current export-driven strategy, the emphasis will shift towards closer markets, but the Greenport Westland/Oostland will not collapse. In summary, the stakeholders expect that the economy of The Hague Region will continue to grow moderately.

More than half of the population in The Hague Region will be of foreign origin. The proportion of elderly people continues to grow, up to maybe 30–40 % over 65 by 2040. Many of them will want to return to the city because of the levels of health and social services there. However, a strong movement to the cities from the peri-urban parts of The Hague Region is not expected, because of the short distances involved and the ease of public transport.

Energy and therefore transport will be expensive under this scenario. Public transport will be significantly subsidised to keep it affordable. The government will move strongly towards a sustainable energy transition. Land prices will remain high. This combination leads to restructuring of existing urban areas into more intensive and multifunctional, energy-efficient forms of land use. Urban land use functions that need a lot of space will be moved to the Rotterdam harbour, because space will become available there. In a situation of strongly cooperating government, it seems logical that the housing problem of The Hague Region will be solved by dealing with it together with Rotterdam. However, the harbour will not collapse completely. Among other changes, there will be a shift of transport from fossil fuels to biomass energy. New developments in the Rotterdam harbour will take place at the side of Maasvlakte, extending into the sea. The coastline will be broadened with extra sand deposition. This new land will not be used for housing (due to the risk of flooding), but for nature development and recreation.

The production costs of land-based agriculture will increase (land, energy, livestock feed). After a difficult period, soil-bound agriculture will recover, because B1 also offers opportunities after some time. Dairy farmers will produce energy through the fermentation of manure. The competitive disadvantage of local production will diminish. Alternative supply chains will emerge but the world market will remain influential.

In spite of the high land pressure, strong planning ensures that the green open areas, including the meadow areas remain more or less intact. Conservation and development of the green-blue network will remain an explicit policy goal. Land use will be planned ‘up to the last $mm^2$’. The Regional Structure Plan (RSP) will be implemented, including intensification/densification and high-rise development around public transport nodes. The public transport network will improve, including its image. Densifying the city makes it urgent to take measures to maintain quality of life and public space, among others by creating green meeting places and using green roofs, walls and balconies in the densest areas. Multifunctional and multi-layered land use will be growing in importance. Building 80 % of new houses and offices in the current urban fabric (a goal in RSP) still means that 20 % will have to be built outside. Further urbanisation will, however, be small-scale, in the urban fringe next to the current built-up area, across the region. Even in the situation of an energy crisis, construction in existing urban space will be more expensive than in the fringe. It is uncertain whether the economy will be strong enough for that. High-rise apartment buildings will be constructed at the edge of

Westland, with a view over Midden-Delfland. Rising house prices lead to the necessity of subsidising housing for lower income groups. 'A house with a garden is for the rich'.

In B1 a strong government is needed for effective policy implementation, but The Hague Region as a government tier is designed to work through consultation. In B1 there will be cooperation between governments at higher tiers and it will be more top-down. It seems logical in B1 to upscale The Hague Region to the level of Randstad South Wing in order to become a more influential actor in the policy arena ('South Wing Authority').

### 6.5.2 B2: Regionalisation, Strong Planning and Social Fragmentation – Regional Storyline

According to the stakeholders, a B2 scenario would be disastrous for The Hague Region in economic terms because the regional economy depends on globalisation. The diminishing cooperation between countries and governments and the growing distrust among social groups in B2 will lead to the termination or shrinkage of international organisations and cooperation bodies. Many of the current institutes in The Hague will be closed down. The international community in The Hague will be decimated. This will have a large impact on the economy in the city. The EU will be dismantled or will at least become less important. European countries will again start to protect their own markets. The export from Westland/Oostland horticulture will collapse. Many companies will go bankrupt: half of the greenhouse area will fall out of use. At first, dairy farming will diminish, but it will recover due to a re-orientation to the regional market. Delft, however, will flourish because of the growth in demand for clean and efficient technology, as a result of the efforts of the government to achieve energy independence.

There will also be less cooperation, less concertation, less understanding, more intolerance and more differences of opinion between countries. This may lead to conflict situations and even war. In cities, criminal behaviour will increase. City centres will become unattractive for the higher social classes and for tourists. The population of the region will shrink because of the economic downturn and the limits to immigration ('our own people first'). Because of a lack of international cooperation, Randstad South Wing will face higher peak flows in the rivers, and floods will occur more often. This will limit the possibilities for land use.

The Hague Region has a strong international orientation and the Dutch trading spirit will not disappear in B2. People will try to benefit from two sides (hoping for the international market to recover or remain active, while protecting the home market), but it is questionable whether this strategy will work. The international market will not disappear completely, but exports will shrink. Also in this scenario, the Rotterdam harbour will reduce in size. The share of greenhouse horticulture that survives will aim more for the national and regional market. Land prices will not drop dramatically, but will be lower than in B1 (Peak Oil) because of the shrinking economy and the available space in the greenhouse areas and the Rotterdam harbour. The surviving dairy farmers will aim for the national and the regional market. Some of the dairy farmers will change their farming strategy to produce meat. There will also be an increase in part-time farmers and care farmers aimed at the elderly. There will be a growing need for allotment gardens.

Tourism and recreation will remain but the emphasis will shift. People will remain interested in recreation, but international tourism will fall. The Dutch will prefer to spend their holidays in their own country because in B2 they cherish the Dutch identity. Whether recreation and tourism can develop into a strong economic sector, is hard to say. There may be a growing demand for cheaper forms of recreation, such as small-scale (exclusive) private campsites. In Midden-Delfland, horse-keeping will change the landscape. Well-off people from Delft will buy old farmhouses and convert them into private houses. These developments can lead to a 'messy' landscape. The landscape is, however, appreciated as part of the regional identity. It remains to be seen if the government will still be willing to pay for the maintenance of green space. However, the preservation of agricultural land use will be less disputed, because food production will again be important (self-sufficiency at regional or national level).

Delft will become the most economically important city in the region. From the Delft elite and what is left of the Hague elite, there will be a demand for 'gated communities' across the region, but with concentrations in Rijswijk and Vlietzone (the area between Rijswijk, Delft and Nootdorp) and along the coast of Westland. This could be new enclosed neighbourhoods, or reconstruction of former embassy buildings. The international zone of The Hague will be transformed into luxury dwellings for the elderly as well as educational and religious institutes—at least, if there will be sufficient capital for this. Degradation of this zone is also possible. The real need, in contrast, will be for cheap houses. However, government resources will be low. Because the poor will have to live in impoverished and degraded neighbourhoods, and only the well-off will be able to afford new housing or renovation, there will be greater differences in quality of life, in spite of the aim of the government for social sustainability.

Half of the greenhouse floor space will fall out of use. There will be no money, however, to dismantle them: this leads to a degraded landscape with all kinds of ad-hoc use of empty greenhouses for other purposes. Because the empty greenhouses are spread across the area, the government will try to reconstruct its land use pattern. This calls for strong planning and it is an expensive process. The reconstruction will concentrate the glass in the centre of Westland. Local initiatives will emerge, for instance for energy production with empty greenhouses and small-scale vegetable growing by civilians. Empty greenhouses will be reconstructed into energy plants. Other greenhouses will be covered with solar panels. Also, wind turbines will be erected. There will be an innovation subsidy for the reorientation of horticulture. At the Midden-Delfland side and along the coast there will be space for housing. Most likely, the development of luxury houses will be allowed in order to finance the reconstruction.

**Table 6.2** Model characteristics

| | |
|---|---|
| Resolution | 100 m |
| No. of regions | 1 |
| Land use classes | No. of vacant states = 3, no. of functional states = 8, no. of feature states = 7 |
| Input | Land use map 1995 and 2004, zoning map, suitability maps, transportation network |
| Calibration | Kappa (corr.) 0.191345 |

The Hague Region as an authority will achieve a central position in representing the regional interests. At a supra-regional level, there will be less cooperation between governments. However, at the local level, governance will remain important. Nevertheless, the position of The Hague Region will need constant attention, even in a world of 'Fragmentation'.

### *6.5.3 MOLAND Results and Comparison of Scenarios*

The qualitative storylines were translated into quantitative assumptions for input in the MOLAND model (Table 6.2). One of the most important of these was population growth. For business as usual (BaU), the projections from CBS were used as noted earlier. Other trends were assumed for B1 and B2 based on the storylines. After 2015 a declining population trend is expected for South Holland Province. In B1 (Peak Oil), the population could still grow because of the relative good competitive position of The Hague Region in the storyline, but B2 (Fragmentation) would mean a substantial population shrinkage.

All three scenario's show a limited growth of the urbanized area. In BaU the expected population shrinkage prevents strong growth of the urban area; in B1 the energy crisis is expected to lead to more efficient and therefore compact development in spite of the population growth; in B2 population shrinkage and economic crisis will even lead to massive vacancy. The land use changes and the differences between the scenarios only become clear when taking a closer look at the maps and when comparing the statistical results (Fig. 6.10a–d; Table 6.3).

In BaU, residential development takes place at the expense of greenhouses, business sites and the Eastern part of Rotterdam harbour. The harbour grows further out into the sea. Delft grows a little to the Northwest, there is infill in The Hague, and some middle-sized towns expand. The model predicts densification of Wassenaar and a growth of business sites southeast of Zoetermeer and Leiden. In BaU, all vacant land is filled in towards 2040. There is little loss of pasture and arable land.

The projected land use map for the B1 Peak Oil scenario shows limited urbanisation, but slightly more than for BaU, mainly because of the growth of work locations (see Table 6.3). More than half of the vacant urban land is used by 2040. Residential development mainly takes place in the form of urban infill, but

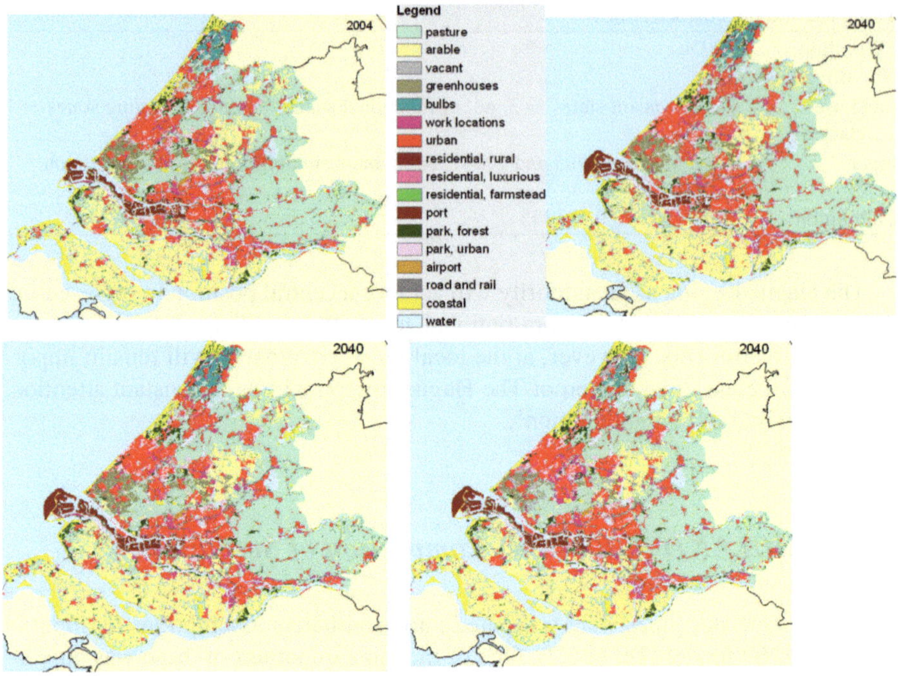


**Fig. 6.10** (**a–d**) Land use projections for 2040 based on the scenarios Business as Usual (*top right*), B1 Peak Oil (*bottom left*) and B2 Fragmentation (*bottom right*) compared to the land use situation in 2004 (*top left*)

**Table 6.3** The scenarios compared by their resulting land use changes (Source: JRC)

| | | Projected change 2004–2040 | | | | | |
|---|---|---|---|---|---|---|---|
| | 2004 | BAU | | B1 'Peak oil' | | B2 'Fragmentation' | |
| Land use class | ha | ha | % | ha | % | ha | % |
| Pasture | 104,167 | −1,408 | −1.35 | −793 | −0.55 | −799 | −0.77 |
| Arable | 60,048 | −294 | −0.49 | −1,528 | −2.45 | −1,462 | −2.43 |
| Vacant | 874 | −872 | −99.77 | −513 | −90.62 | 6,477 | 741.08 |
| Greenhouses | 8,950 | 152 | 1.70 | 154 | 1.70 | −4,472 | −49.97 |
| Bulbs | 4,028 | 70 | 1.74 | 68 | 1.74 | −2,017 | −50.07 |
| Work locations | 7,432 | 344 | 4.63 | 587 | 7.90 | 143 | 1.92 |
| Urban | 45,309 | 1,455 | 3.21 | 1,442 | 3.18 | 1,319 | 2.91 |
| Residential rural | 600 | 21 | 3.50 | 26 | 4.33 | 8 | 1.33 |
| Residential luxury | 1,555 | 48 | 3.09 | 73 | 4.69 | 43 | 2.77 |
| Residential farmsteads | 2,105 | 0 | 0.00 | 0 | 0.00 | 276 | 13.11 |
| Port | 5,834 | 566 | 9.70 | 566 | 9.70 | 566 | 9.70 |
| Park forest | 7,513 | 0 | 0.00 | 0 | 0.00 | 0 | 0.00 |
| Park urban | 17,237 | −1 | −0.01 | −1 | −0.01 | −1 | −0.01 |
| Airport | 270 | 0 | 0.00 | 0 | 0.00 | 0 | 0.00 |
| Roads/rail | 10,078 | 0 | 0.00 | 0 | 0.00 | 0 | 0.00 |
| Nature (reeds/sand/marsh) | 14,688 | −75 | −0.51 | −75 | −0.51 | −75 | −0.51 |
| Water | 50,087 | −6 | −0.01 | −6 | −0.01 | −6 | −0.01 |

also in work locations and greenhouses. There is also some growth in the greenhouse area, mainly through infill. The Rotterdam harbour extends further into the sea and derelict land in the harbour area is filled in. There is some housing development in the Eastern part of the harbour, near the city centre, but less than in BaU. Some work locations in the province grow because of their good accessibility by highway. The model predicts vacancy in other work locations, to be filled by luxury residences towards the end of the period. In Wassenaar, densification takes place in the upmarket neighbourhood. There is little loss of pasture and arable land: compared to BaU there is less loss of pasture and more loss of arable land.

The most striking land use changes in B2 Fragmentation occur in the greenhouse areas (the stakeholder assumption of 50 % decline in the greenhouse sector was inserted in the model). Much vacancy occurs here, which is only occasionally taken over by housing or work locations. Not only do the greenhouses suffer; also parts of the flower bulb cultivation area falls out of use. Other major contributions to vacancy are from work locations (932 ha), luxury residences (413 ha) and the port (171 ha). Vacancy occurs in some of the commercial areas in Rotterdam and The Hague. These are taken over by housing, or again by work locations. The economic crisis in B2 is illustrated by vacancy in the upmarket residential areas of Wassenaar. Some vacancy appears along the edges of the Rotterdam harbour in the city centre. Some housing appears here, but in most of the area the harbour functions return. Housing and working land use classes grow by much less than in the other scenarios. Urban housing takes place mainly in the form of urban infill, in The Hague replacing former work locations. Larger extensions of work locations are expected by the model in the vicinity of Delft, because of its economic attraction effect. Peri-urban Vlietzone will be filled up mainly with work locations. Also notable is the development of farmstead residences at the edges of the case study area and especially the South Holland islands. There is little loss of pasture and arable land: compared to BaU there is less loss of pasture and more loss of arable land. However, around Delft much pasture is converted to work locations. The low net loss of pasture is due to conversion of arable land to pasture elsewhere in the province.

After developing the different storylines, the performance of the strategies in these different 'worlds' was explored. The strategies 'inner city densification' (Strategy 1) and 'green and blue services' (Strategy 2) were added to the baseline scenarios for extra runs with the MOLAND model. The effect of the strategies on land use change in the three scenarios is presented in Table 6.4.

The modelling results for the strategies are sensitive to the assumptions made in the modelling process. The formulas developed for inner city densification (Strategy 1), for instance, prevent the model from re-urbanising vacant land, which in the B2 Fragmentation scenario leads to high vacancy levels in the city in addition to the massive amount of vacancy in greenhouses and flower bulb cultivation areas. Another important assumption was the improved chance of farm survival as a result of the green and blue Services strategy (Strategy 2). However, with the assumptions made (which are elaborated in Mubareka and Lavalle 2010), it is interesting to see that the strategies perform differently in the three scenarios.

**Table 6.4** Comparison of baseline scenarios and two strategy alternatives on land use change in South Holland 2004–2040

| | | BAU 2040 | | | B1 2040 | | | B2 2040 | | |
|---|---|---|---|---|---|---|---|---|---|---|
| | | Baseline | Strategy 1 | Strategy 3 | Baseline | Strategy 1 | Strategy 3 | Baseline | Strategy 1 | Strategy 3 |
| Land use class | 2004 (ha) | Change (ha) | Change (ha) | Change (ha) | Change (ha) | Change (ha) | Change (ha) | Change (ha) | Change (ha) | Change (ha) |
| Pasture | 104,167 | −1,408 | −613 | −573 | −793 | −227 | −691 | −799 | −499 | −1,887 |
| Arable | 60,048 | −294 | −112 | −114 | −1,528 | −1,111 | −150 | −1,462 | −1,125 | −398 |
| Vacant urban | 874 | −872 | −516 | −218 | −513 | −752 | −474 | 6,477 | 17,091 | 6,500 |
| Greenhouses | 8,950 | 152 | 152 | 12 | 154 | 14 | 14 | −4,472 | −4,472 | −4,472 |
| Bulbs | 4,028 | 70 | 70 | 0 | 68 | −2 | −2 | −2,017 | −2,017 | −2,017 |
| Work locations | 7,432 | 344 | 200 | 230 | 587 | 97 | 204 | 143 | −1,284 | 144 |
| Urban | 45,309 | 1,455 | 322 | 138 | 1,442 | 1,452 | 565 | 1,319 | −8,053 | 1,320 |
| Resid rural | 600 | 21 | −2 | 2 | 26 | 14 | 6 | 8 | −107 | 8 |
| Resid luxury | 1,555 | 48 | 15 | 15 | 73 | 30 | 20 | 43 | −267 | 42 |
| Resid farmsteads | 2,105 | 0 | 0 | 24 | 0 | 1 | 24 | 276 | 249 | 276 |
| Port | 5,834 | 566 | 566 | 566 | 566 | 566 | 566 | 566 | 566 | 566 |
| Park forest | 7,513 | 0 | 0 | 0 | 0 | 0 | 0 | 0 | 0 | 0 |
| Park urban | 17,237 | −1 | −1 | −1 | −1 | −1 | −1 | −1 | −1 | −1 |
| Airport | 270 | 0 | 0 | 0 | 0 | 0 | 0 | 0 | 0 | 0 |
| Roads/rail | 10,078 | 0 | 0 | 0 | 0 | 0 | 0 | 0 | 0 | 0 |
| Reeds/sand/marsh | 14,688 | −75 | −75 | −75 | −75 | −75 | −75 | −75 | −75 | −75 |
| Water | 50,087 | −6 | −6 | −6 | −6 | −6 | −6 | −6 | −6 | −6 |

**Table 6.5** Land use change 2004–2040 for agricultural classes and reduction of loss, compared to land use change for urban classes and reduction of growth, as result of green and blue services strategy (Source: JRC)

| 2004–2040 | BaU | | B1 | | B2 | |
|---|---|---|---|---|---|---|
| Land use change (ha) | Baseline | Strategy 3 | Baseline | Strategy 3 | Baseline | Strategy 3 |
| Pasture | −1,408 | −573 | −793 | −691 | −799 | −1,887 |
| Arable | −294 | −114 | −1,528 | −150 | −1,462 | −398 |
| Total | −1,702 | −687 | −2,321 | −841 | −2,261 | −2,285 |
| Reduction of loss agricultural classes | | 60 % | | 64 % | | −1 % |
| Work locations | 344 | 230 | 587 | 204 | 143 | 144 |
| Urban | 1,455 | 138 | 1,442 | 565 | 1,319 | 1,320 |
| Total | 1,799 | 368 | 2,029 | 769 | 1,462 | 1,464 |
| Reduction of growth urban classes | | 80 % | | 62 % | | 0 % |

Inner city densification has varying impacts for the different scenarios, but was in general quite effective in containing urban growth, according to the modelling results:

- In BaU, inner city densification has no influence on greenhouses and flower bulb cultivation, but leads to lower growth in the urban and luxury residential class and even a decrease in the rural residential class.
- In B1, inner city densification leads to a lower loss of farmland than in the scenario without strategies. Work locations grow much less as a result of urban densification. Rural residences and luxury residences are also constrained, which leads to a slight growth of the urban class compared to a B1 baseline. There is no growth in greenhouses and flower bulb cultivation as a result of the urban densification strategy.
- In B2, inner city densification has no influence on greenhouses and flower bulb cultivation, but the strategy leads to shrinkage in work locations, urban residential and rural and luxurious residential land use classes. Implementation of urban densification in B2 results in a total of 17.091 ha vacant land in 2040, or 10,614 ha extra compared to the scenario without this strategy.

Both in BaU and in B1, green and blue Services have quite a significant effect on land use change and are fairly effective in protecting farmland from urbanisation and the growth of greenhouses and flower bulb cultivation, in spite of the difference in subsidy level (see Table 6.5). In B2 Fragmentation, the effect of green and blue services is negligible: the reduced loss of work locations is more than compensated by extra residential farmsteads.

## 6.6 Conclusions for Improved Strategies

Peri-urban areas represent a problem of governance and scaling (Padt and Westerink in press). They are the ‘back side’ of cities and are intersected by municipal and regional boundaries. The peri-urban enclaves of The Hague Region are no exception to this. Some of these extend into neighbouring regions. Coherent planning in these areas therefore requires cooperation between a large number of local and regional authorities. The second section of this chapter explained how The Hague Region tries to operate in this governance context, with limited formal competences and budget (Aalbers et al. 2009). The most important government tier in planning is the municipality, which has a directly elected democratic representation. Democratic representation within the board of city regions is indirect. For guiding developments in its peri-urban landscapes, the regional authority may have insufficient power and resources and the area of jurisdiction of The Hague Region may be too small. The Hague Region is smaller in size than the rural–urban region of Rotterdam-The Hague (Loibl et al. 2010). For a more coherent and powerful directing of land use change in the rural urban region, the Province could take a more active role. The Province is the only tier of government that oversees the peri-urban areas of The Hague Region as a whole. The provincial tier, between the municipalities and the central government, has more power and resources than the city region and has direct democratic representation.

So far, planning efforts have been quite effective in preserving peri-urban landscapes and in containing the cities of the region (see e.g. Van Rij et al. 2008). The Hague Region has the highest urban density of all case study regions in PLUREL. Further densification is a policy goal to spare the peri-urban landscapes, but densification may have unwanted consequential effects, such as traffic congestion, loss of green space in the city and an urban flight of those who can afford to move to a house out of town. The combination of urban densification with improved public transport, multifunctional and intensive land use and urban renewal are among the strategies in the region to improve the quality of life in the cities. However, the availability of green space in the city is a point of concern. For the growing group of immigrants, urban parks are more important for daily leisure than the peri-urban areas. In city regional planning, attention should be given to the variation of recreational preferences and future changes in recreational behaviour.

Urban containment can contribute towards keeping options open for the future. Storm water management, local food production, renewable energy production and public and private green space should be among the issues addressed by urban planning. The urban fabric should be flexible enough for conversion and re-use in the future if necessary. Mixing functions, creating flexible spaces and buildings, establishing gardens on roofs and providing 'quality of place' seems to call for a shift of tasks from planners to designers. The 'human scale' in urban design should be aimed at, and possibilities for high density without high-rise should be explored (see e.g. Uytenhaak and Mensink 2008). Densification is typically an issue that could lead to much resistance from citizen groups. Urban renewal drastically changes the area where people have sometimes lived for decades. Citizen participation is therefore necessary to take account of these emotions and to create renewed place attachment. Recognizing the value and legitimacy of emotions, place attachment and identity may greatly improve communication processes.

In communicating about planning and area development, discourses can be powerful instruments. They are very important in the consensus-seeking governance style common in The Netherlands. Discourses can be valuable when they are combined with or lead to other dimensions of policy arrangements, such as coalitions, legislation and financial means (see Chap. 13).

The peri-urban agricultural landscapes in The Hague Region are a clear example of public goods delivered by the private sector (in this case agriculture): they are appreciated by urban dwellers but not paid for directly by those who enjoy them. Local agri-environment schemes are arrangements trying to correct this 'market failure', by using tax funds to pay for them. Green and blue services will not solve all the problems of agriculture in the urban fringe of The Hague Region. However, the strategy can contribute more to strengthening agriculture if:

- European state aid regulations would allow payments akin to adequate levels of profit to farmers for public goods and services;
- Agri-environment payments are classified in the World Trade Organisation 'Green Box', to avoid lengthy state aid procedures and to foster local initiative;
- Farmers' groups are encouraged to develop new ideas for green and blue services;
- More budget is reserved for agri-environmental schemes, for instance through the Common Agricultural Policy (CAP);
- Green and blue services are combined with other strategies, such as land banking, zoning (which has been quite effective in keeping the land price down) and the development of commercial urban–rural relationships.

The modelling results show rather modest land use changes (with the exception of the high greenhouse vacancy in B2 that was an input rather than a modelling result) based on explorative storylines. Other modelling exercises (Kuijpers-Linde et al. 2007; VROM 2008) show a heavier urbanisation of the region based on scenarios with higher projections of demographic and economic growth than used in this research. The large differences illustrate that scenarios should be used with care, because policy implications can be significant. While the national government is still aiming for growth in Randstad Holland (VROM 2008), South Holland Province is starting to anticipate a slight population decrease in the near future, leading to modest projections for land use change (Provincie Zuid Holland 2010). Re-thinking growth seems a major challenge for governance and land use planning in the case study area for the near future.

## References

Aalbers CBEM, van Dijk, T, van der Jagt PD, Westerink J (2009) Analysis of regional spatial planning and decision making strategies and their impact on land use in the urban fringe. The Hague Region. PLUREL report D 3.3.1. Alterra Wageningen UR, unpublished report

Bouwman AA, Kuiper R, Tijbosch HW (2006) Ruimtelijke beelden voor Zuid-Holland. Rapport 500074002/2006. Milieu- en Natuurplanbureau, MNP, Bilthoven

Burton E (2001) The compact city and social justice. In: A paper presented to the housing studies association spring conference. Housing, environment and sustainability, University of York, 18–19 Apr 2001

CBS (2006/2009) Data accessed through statine, online database of Centraal Bureau voor de Statistiek, www.cbs.nl

Delfland (2004) Waterkansenkaart. Hoogheemraadschap van Delfland, Delft

DLG (2009) Grondprijsmonitor 2008. Recente ontwikkelingen in de agrarische grondmarkt. Dienst Landelijk Gebied, Den Haag

Fontein RJ, de Boer TA, Breman BC, Goossen CM, Henkens RJHG, Luttik J, de Vries S (2009) Relatie recreatie en Natuur. Achtergronddocument bij Natuurbalans, Werkdocument 160. WOt Natuur & Milieu, Wageningen

Haaglanden (2008) Regionaal structuurplan. Stadsgewest Haaglanden, Den Haag

Haaglanden (2010) Woningmarktmonitor 2009. Stadsgewest Haaglanden, Den Haag

Kuijpers-Linde MAJ, Geurs KT, Knoop JM, Kuiper R, Lagas P, Ligtvoet W, de Niet R, van Oostenbrugge R, Westhoek HJ (2007) Nederland Later. Tweede Duurzaamheidsverkenning, deel Fysieke leefomgeving Nederland. Milieu- en Natuurplanbureau, MNP, Bilthoven, MNP-publicatienummer 500127001/2007

Loibl W, Köstl M (2010) Report on a methodology to delineate RUR sub-regions. PLUREL report D 2.1.4, unpublished report

Loibl W, Köstl M, Steinnocher K (2008) List of generic rural–urban region types; quantitative classification. PLUREL report D2.1.2/D2.1.3. Version V3 15 Apr 2008, unpublished report

Love LL, Crompton JL (1999) The role of quality of life in business, relocation decisions. J Bus Res 44(3):211–222

Mubareka S, Lavalle C (2010) Land use projections based on MOLAND output. PLUREL report D 2.4.2. Joint Research Centre, Ispra, unpublished report

Padt FJG, Westerink J (in press) Addressing scale in open space preservation. Learning from the Hague region in The Netherlands. Tijdschrift voor economische en sociale geografie

PLUREL (2008) Final minutes parallel workshop green and blue services. 1 and 2 Oct 2008, The Hague. Part of 3rd General meeting of the PLUREL project. http://www.boerenvoornatuur.nl/uploads/downloadables/publicaties/2009%20Parallel%20workshop%20Green%20and%20Blue%20Services%20Final%20Minutes%20webversion.pdf. Accessed 14 Mar 2012

Provincie Zuid Holland (2010) Visie op Zuid-Holland. Ontwikkelen met schaarse ruimte. Provinciale Structuurvisie. Vastgesteld op 2 juli 2010. Provincie Zuid-Holland, Den Haag

Ruto E, Garrod G (2007) Farmers' preferences for agro-environmental contract design. ITAES WP7 final report. University of Newcastle upon Tyne

Schweitzer L, Zhou J (2010) Neighborhood air quality, respiratory health, and vulnerable populations in compact and sprawled regions. J Am Plann Assoc 76:363–371

Tosisc I, Gertheis A (2010) Review of further instruments with indirect effect on regulation of peri-urban land uses. PLUREL report D2.2.2, unpublished report

Uytenhaak R, Mensink J (2008) Steden vol ruimte: kwaliteiten van dichtheid. Rotterdam, 010

Van Rij E, Dekkers J, Koomen E (2008) Analysing the success of open space preservation in the Netherlands: the Midden-Delfland case. Tijdschrift voor Economische en Sociale Geografie 99 (1):115–124

Veer PM, Luttik J (2010) Naar buiten! To the Great Outdoors. Groen- en recreatiewensen van internationale kenniswerkers in Nederland. Alterra-rapport 1940, 114pp. Alterra Wageningen UR

Vonk Noordegraaf A, Gloudemans SAH (2004) Komt het Instrument van de Grond? Inventarisatie van het draagvlak voor een grondinstrument in Midden-Delfland, Stichting Kavelruil Zuid-Holland, WLTO-afdeling Delflands Groen

VROM (2005) Nota Ruimte. Ministerie van Volkshuisvesting. Ruimtelijke Ordening en Milieu, Den Haag

VROM (2008) Structuurvisie Randstad 2040. Naar een duurzame en concurrerende Europese topregio. Ministerie van Volkshuisvesting. Ruimtelijke Ordening en Milieu, Den Haag

Westerink J, Breman BC, Smits MJ, van Alebeek FAN, Migchels G, Bakker G, van Amersfoort MC, Schrijver RAM (2009) Maatschappelijke prestaties. Meer publieke diensten en minder publieke lasten door de landbouw met behulp van het Gemeenschappelijk Landbouwbeleid. Alterra-rapport 1961. Alterra Wageningen UR

Westerink J, Kiers M, Kok K, Aalbers CBEM (2010a) Scenarios and enhanced strategies. Case study The Hague Region, The Netherlands. PLUREL report D.3.4.1, Alterra Wageningen UR, unpublished report

Westerink J, van der Jagt PD, Aalbers CBEM (2010b) Assessment of strategies. Case study of The Hague Region. Unpublished PLUREL Milestone report

Westerink J, Lagendijk A, Dühr S, Jagt Pvd, Kempenaar A (in press a) Contested spaces? The use of place concepts to communicate visions for peri-urban areas. European Planning Studies

Westerink J, Haase D, Bauer A, Jarrige F, Ravetz J, Aalbers CBEM (in press b) Dealing with sustainability trade-offs of the compact city in peri-urban planning across European city regions. European Planning Studies

Zuidvleugel (2006) De Zuidvleugel: Motor in de delta. Visie op de ontwikkeling van de Zuidvleugel. Bestuurlijk Platform Zuidvleugel, Den Haag

# Chapter 7
# Warsaw: Spatial Growth with Limited Control

**Mirosław Grochowski, Piotr Korcelli, Elzbieta Kozubek, Tomasz Sławiński, and Piotr Werner**

## 7.1 Introduction

In this chapter the case of the Warsaw Metropolitan Area (WMA) is presented as an example of how systemic political and economic change resulting from decentralization and democratization influence urban development processes. The shift from the system of central planning and rigid control to a decentralised system promoting local autonomy and to market-based rules of development, as well as a re-birth of local democracy and the empowerment of local government has created strong incentives for economic development over the last 20 or so years. However, dynamic development processes also often assume a spontaneous character which may lead to conflicts, impede development processes and make efficient use of the development potential of a region impossible.

The chapter starts by presenting some basic information about the Warsaw Metropolitan Area, the main factors that have impact on development processes in the area and the main development challenges, in order to provide a context. The next section describes the impact of democratization processes on approaches and practices of local and regional development. The spatial planning system and

M. Grochowski (✉)
University of Warsaw, Warsaw, Poland
e-mail: mgrochow@uw.edu.pl

P. Korcelli
Institute of Geography and Spatial Organization, Polish Academy of Sciences, Warsaw, Poland

E. Kozubek
Institute of Geodesy and Cartography, Warsaw, Poland

T. Sławiński
Mazowian Office for Regional Planning, ul. Solec 22, 00-410 Warsaw, Poland
e-mail: t-slawinski@tlen.pl

P. Werner
University of Warsaw, Warsaw, Poland

K. Nilsson et al. (eds.), *Peri-urban futures: Scenarios and models for land use change in Europe*, DOI 10.1007/978-3-642-30529-0_7,

**Table 7.1** Basic information on Mazowieckie voivodship, Warsaw metropolitan area and the City of Warsaw (Rocznik Statystyczny Województwa Mazowieckiego/Mazowieckie Voivodship Statistical Yearbook/2010, US Warszawa, 2010)

| | Mazowieckie voivodship | Warsaw metropolitan area | Warsaw |
|---|---|---|---|
| Area (sq. km) | 35,598 | 6,205 | 517 |
| Number of population (millions) | 5 | 3 | 1.7 |
| Population density (per sq. km) | 140 | 474 | 3,291 |
| Administrative status | Region, (voivodship), NUTS 2 level, unit of territorial subdivision, self government and government level of public administration | Area delineated for planning purposes, composed of 72 municipalities, no legal status and management structures | Urban municipality (gmina), NUTS 3, 4 and 5 level, unit of territorial subdivision governed by local self government |

the system of self government in Poland are examined in the third section. Following this is a description of peri-urban areas in the WMA and the results of in-depth studies conducted in selected municipalities. Next is a description of development scenarios which were developed following a workshop and discussions with stakeholders. The results of land use modelling performed for two scenarios are then presented, followed by an evaluation of the main strategies employed to guide development of the WMA. The last section of the chapter summarises the main finding of the studies and draws some conclusions.

## 7.2 Warsaw Metropolitan Area: Factors of Change

The Warsaw Metropolitan Area is located in the Mazowieckie voivodship (a voivodship is the Polish regional tier of government). This is the largest in Poland in terms of both area and population (11.4 % of Polish territory; 13.1 % of total population of Poland) (see Table 7.1). The City of Warsaw is the biggest metropolitan area in Poland and is experiencing dynamic development. It is also the wealthiest Polish municipality and the growth pole of the Warsaw Metropolitan Area. However, the WMA is very diverse in terms conditions and level of development and one may also find relatively poor rural municipalities within the WMA.

The municipalities that are under greatest urbanisation pressure tend to be those situated close to major transport routes. Land use changes in the metropolitan area triggered by urbanisation pressures rapidly accelerated from the early 1990s onwards. Rising demand for land for development results in the conversion of agricultural land into housing and services. Proactive behaviour by developers, together with the lack of firm spatial planning policies at both local and regional level, results in a situation where more and more farmers decide to sell their land. In addition, due to this ad hoc process, there are no infrastructural investments preparing the land for development, while the size and shape of plots put on the market is frequently not suitable for anything other than agricultural purposes (the typical narrow strips of land forming the Polish agricultural pattern).

Mazowieckie voivodship is the region attracting migrants, mainly from elsewhere in Poland and these are concentrated in the area of Warsaw and its surroundings. While people generally migrate to urban municipalities in Poland, this is not the case of the WMA, where many immigrants select municipalities located next to Warsaw which are not so urbanised. The social and economic profiles of local communities are thus shaped to large extent by migration. Between 1995 and 2005 the average annual increase in population through in-migration in some municipalities of the WMA exceeded 4 % per annum. This represents the combined effect of migration inflow from other regions and of residential mobility within the metropolitan area, in particular out from the city of Warsaw to suburban and exurban (urbanized areas isolated from any continuous urban fabric which have arisen as a result of ad hoc development processes) areas situated within the metropolitan ring. The suburbanisation movement, however, is more than offset by in-migration from outside the WMA. According to survey-based estimates, the resident population of Warsaw exceeds its official number by some 250 thousand (Bijak et al. 2007). Also, the number of inhabitants of the metropolitan ring (its rural and peri-urban zones) is underestimated in public statistical sources. Hence, Warsaw, and the metropolitan area as a whole, account for an increasing share of the total population of Poland, the size of which has been stagnating since around 2000.

Behind this unbalanced population growth lie economic driving forces which were unleashed by the systemic transformation of Poland as a whole between 1989 and 1990. Firstly, Warsaw and its metropolitan area benefited from the change from an industry-dominated into a service-dominated economy, a shift that occurred quite rapidly during the 1990s. The costs of this included an economic downturn which negatively affected a number of cities and regions across Poland. By 2000, the employment profile of Warsaw, when measured by shares of individual SIC (Standard Industrial Classification) sections came close to those of Vienna (Gawryszewski et al. 2000) and Berlin (Korcelli-Olejniczak 2004). The founding of the Warsaw Stock Exchange (now the biggest in East-Central Europe) had a snowball effect by attracting headquarters of nationally important firms to the city. Their number, among Poland's 500 largest enterprises, increased from 125 to 192 between 1993 and 2004 (Lijewski 2003; Śleszyński 2007). Especially important in this respect was the process of consolidation within the banking sector which gained momentum in the late 1990s.

At the same time as the developments described above, Warsaw became the prime destination for international investors, drawn to the city and its metropolitan area by the large capacity of the local and regional consumer and investment goods markets, as well as a highly diversified and educated labour market. By the year

2000 Warsaw accounted for 40 % of Poland's foreign direct investments (Domański 2001). In the section of foreign trade the share was as high as 90 % t (Komornicki 2003). The WMA was gradually assuming the position of Poland's economic heartland, the role that during previous decades was held by Upper Silesia. By 2005 the WMA contributed over 13 % of national GDP, a proportion twice as high as its share of the total population of Poland. Internationally, Warsaw moved up the rankings of European metropolitan areas, owing, among others, to expansion of its producer business services sector (Taylor and Ciechański 2006).

The economic, social and political changes over the last two decades have had a far-reaching impact upon land use patterns, particularly felt in the peri-urban and rural zones of the WMA. Over the last 20 years car ownership has expanded from less than 50 to over 400 cars per thousand inhabitants, the level typical of other large European cities (Strategia 2003). Suburban living, a long time aspiration of many people, has become affordable to an increasing segment of the population. Land prices and land rent are once more determining factors in the allocation of economic activities, with a resulting relocation and spatial de-concentration in sectors such as industry and wholesaling. In spite of progressive population ageing (which in the case of the WMA is slower due to substantial in-migration of younger people), there is a considerable, unrealised potential for further residential mobility, with the prevailing direction of movement from city to peri-urban areas. In summary, there are many arguments suggesting that the present trends will continue. This implies further expansion of the peri-urban zone of the metropolitan area and further urbanisation pressure resulting in continuing land use transition from primarily agricultural or vacant land to residential, commercial, transport and recreational uses.

As noted earlier, the attractiveness of the Mazowieckie voivodship for migrants results from the fact that it comprises the largest regional labour market in Poland, with over two million employees (16 % of the total number of employed people in Poland) (Rocznik Statystyczny Województwa Mazowieckiego/Mazowieckie Voivodship Statistical Yearbook/2010, US Warszawa, 2010). The labour market has a relatively modern structure. Employment in industry within the WMA has been gradually decreasing. There are, however, some exceptions to the rule, these being municipalities located along transport routes. The growth of employment in industry takes place in municipalities located 50–100 km from Warsaw.

Problems related to the WMA development have become more serious over the last two decades. Warsaw has acquired new metropolitan functions and the metropolitan area has been reshaped. Firms and residents are more mobile and proactive, looking for the best location which meets their needs and expectations. The WMA offers an attractive environment for businesses as well as relatively good living conditions for employees compared with other Polish regions and metropolitan areas. Ongoing suburbanisation and urban sprawl have an impact on the landscape, on land use patterns and new functional relationships among municipalities situated within the metropolitan area. A need for coordinated development of metropolitan areas in Poland has been expressed many times by representatives of public authorities from different administrative tiers (Grochowski 2010). However, there is no evidence of any change in approaches to the practice of metropolitan area development in terms of legal regulations or any other activities initiated by the central government.

## 7.3 Inadequate Planning Regulations to Face the Challenges Ahead

The Spatial Planning and Land Management Act of 2003 calls for the preparation of spatial development plans for metropolitan areas. It is the regional government which is obliged by law to prepare such plans. This responsibility implies which governmental level should be in charge of the management of metropolitan area development. However, the regulatory responsibilities of different tiers of local government are not clearly stated in the legislation. Regional governments are also obliged to prepare a strategy for regional development and a regional spatial development plan for each voivodship. All these plans must be mutually coherent, reviewed and evaluated periodically, and amended as necessary. Regional development programmes must result from the strategy for regional development and be coordinated with the guidelines incorporated into the regional spatial development plan.

However, since 2003, no spatial development plan for any metropolitan area in Poland has been enacted. Legislative inconsistencies impeded preparation and enactment of these plans. This also concerns the plan for the WMA. Nor is there political agreement on by whom and how the development of metropolitan areas in Poland should be managed. A bill on "metropolitan law" is still under preparation at the time of writing, and it is doubtful whether disputes on relevant management mechanisms and instruments will be concluded over the next months or years.

Among the main development challenges facing the WMA is polarisation of development: the city of Warsaw is the growth pole of the Mazowieckie voivodship and the WMA. Polarisation increases urbanisation pressures in those municipalities with the best environmental conditions (open space, forests, valuable land for agricultural production) and which are attractive for potential residents; thus the WMA is the stage for land use conflicts. The WMA is experiencing "dual–mode" development, leading to rising inequalities in the level of economic development and quality of life in urbanised as opposed to rural areas, and triggering residential relocation to the Warsaw city centre and to the surrounding municipalities. Other development problems are related to poorly developed transport infrastructure and a lack of economic, social, and territorial cohesion. Spontaneous and dynamic development results in urban sprawl that contributes to spatial chaos. Problems also result from the lack of mechanism for the coordination of development efforts designed and implemented at the level of individual municipalities. In the situation of an increasingly complex functional pattern and a lack of any sub-regional development strategy or cohesive spatial policy the Warsaw Metropolitan Area cannot meet criteria for sustainable development (Grochowski 2009; Strzelecki and Kucińska 2005).

## 7.4 Explosion of Local Democracy and Lack of Spatial Coordination

Systemic transformation started in Poland with processes of decentralization in 1989–1990. This was a response to the economic and social crisis of the time and a way of moving from a centrally planned to a market economy through radical changes in political decision-making, policy-making and administrative control systems. The paths of changes within these three systems determined the currently evolving responsibilities and powers of local government tiers. Decentralization started with the establishment of self government at the local municipality level in 1990. The second phase of the administrative reform took place in 1999 when self-governing counties (powiaty) and regions (voivodships) were set up.

Since 1990 municipalities have exercised planning control through deciding on ways of economic and spatial development. Local dimensions of development were re-discovered and appreciated. The fact that the second phase of administrative reform, which was part of the decentralization process, was postponed until 1999 had a big impact on the way the self-government structures currently operate. Autonomy of municipalities is deeply rooted in the thinking about the different tiers of the system of public administration and their tasks. This autonomy is protected by the law as well as by the activities of self-government associations that take care of the corporate interests of different categories of municipality. It is a politically very difficult and risky task to attempt to limit the powers of municipalities, including those of spatial planning. The autonomous character often results in narrow approaches to development from a single municipality perspective. Responsibilities and powers are fragmented, which prevents the attainment of sustainable development of complex, functionally integrated areas. In policy making the idea of a "spatial neighbourhood" is only rarely incorporated into a set of assumptions and factors that determine development processes.

Decentralization trends can also be observed in the institutional framework of implementing programmes funded from EU sources. When Poland joined the EU in 2004 the Integrated Regional Operational Programme (IROP) was prepared to implement the 2004–2006 National Development Plan/Community Support Framework (NDP/CSF). The programme developed NDP goals by setting priorities, trends and the amount of funds for implementing the State's regional development policy. The IROP was prepared by the Ministry of the Economy, Labour and Social Policy (MELSP) in cooperation with regional self-government in all voivodships. In 2007 the IROP was replaced by 16 Regional Operational Programmes prepared by the voivodships. Ongoing decentralisation of the management of regional development (including additional competences and management of financial resources) has not been accompanied by the introduction of measures allowing the coordination and preparation of development plans or the coordination of efforts focused on rational spatial development.

The paradigm of competitiveness shapes the model that the municipalities use. Polish legislation provides the legal basis for cooperation between municipalities: they may establish common goals, unions and associations in order to deal with problems that are important for more than one municipality. However, examples of such common undertakings are very rare. There are few incentives for cooperation or for the promotion of cooperation among municipalities as well as among other levels of governance.

The group of actors that have most impact on spatial structures functions are the different tiers of regional and local government, domestic and foreign private investors, the current and former owners of real estate, residents and other groups. The paradigm of competitiveness places local governments in a relatively weak position *vis-a-vis* some of these actors, especially investors, who bring jobs and/or taxes to the municipalities. Spatial development concepts and plans are sometimes modified to meet the needs of investors at the expense of the natural environment or rationality and functionality of spatial plans.

The development of local democracy has not been followed by appropriate changes in regulations concerning spatial development management. The planning code enacted in 2003 increased liberalism by abolishing the previous hierarchical planning system and providing for the possibility of instant, investor-driven development. When development is based on the private acquisition of undeveloped agricultural land, this may result in a totally dysfunctional configuration of built-up areas, including huge surplus costs for construction and for operation in particular. It may also lead to further "colonization" of the countryside by people building houses in un-regulated ways following land purchase.

## 7.5 Overview of the Spatial Planning and System of Self Government

There are four tiers of public administration in Poland: national, regional, sub-regional and local. Poland is divided into 16 regions (*wojewodztwo—voivodship*), 379 counties (sub-regional unit—*powiat*), and 2,478 municipalities (local unit—*gmina*).

At the national level central government agencies are responsible for preparation of a strategic document called the Concept of National Spatial Development (*Koncepcja Przestrzennego Zagospodarowania Kraju*). This is an instrument to guide structural changes in the country, prepared through consultation with regional and local governments and other sectoral interest groups.

At the regional level voivodships must prepare a Strategy for Regional Development, a Regional Spatial Development Plan and a Spatial Plan for Metropolitan Areas (Fig. 7.1). These documents must be coherent and should be reviewed periodically and amended as necessary. Regional development programmes must be based on the strategy for regional development and be coordinated with the guidelines incorporated in the regional spatial development plan. Regional spatial

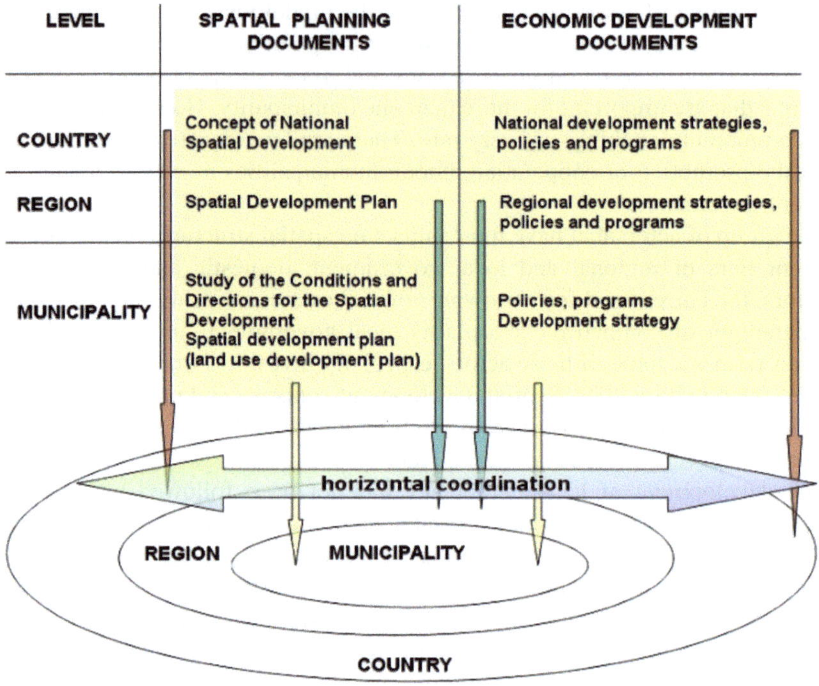


**Fig. 7.1** System of spatial and economic development planning

development plans should thus take into consideration the national and regional tasks of a broader scope.

Municipalities at the county and gmina level prepare spatial development plans and decide on the location of functions, the intensity of land use and the scale and forms of construction. Municipalities prepare two spatial planning documents: a Study of the Conditions and Directions for the Spatial Development and a Land Use Development Plan. In practice municipalities also prepare (although it's not obligatory) a third document related to economic and social development the Local Development Strategy, which is perceived to be useful tool for development management.

The Study of the Conditions and Directions for the Spatial Development is a planning document which determines spatial policy in each municipality and sets the local rules of spatial management. The Land Use Development Plan should ideally be prepared in order to implement a comprehensive, long-term spatial development policy. However, the municipality can function without such a plan. It is only obligatory to prepare the land use development plan for areas indicated in the study. In the absence of the land use development plan local authorities can manage spatial development through two kinds of administrative decisions: a Decision on Development Conditions and a Decision on Investments Serving

Public Purposes. If the decision is to prepare the land use development plan then it must be coordinated with the study (no contradictions are allowed). The plan might be prepared for a part or for the whole municipality area. It is a legal base for detailed spatial management and it is adopted as a local by-law.

## 7.6 Managing Development of Peri-Urban Areas Within the Warsaw Metropolitan Area: Case Studies in Three Municipalities

Mazowieckie voivodship is diverse in terms of functions, their location, and resulting land use patterns. The same applies to the Warsaw Metropolitan Area that encompasses not only intensely urbanised areas, but also agricultural, open, and green areas including the Kampinoski National Park. Within the borders of the voivodship and the WMA there is a mosaic pattern of such areas that can be broadly defined as "peri-urban" (Fig. 7.2). Analysis of land use pattern in the Warsaw Metropolitan Area by municipalities enabled six categories of municipality to be identified: urbanised (dominated by built-up areas), urbanising (experiencing urbanization pressure reflected in changes of land use pattern), traditional agricultural (dominated by arable land and a traditional rural economy), diversified agricultural (mixture of land use pattern and activities related to agricultural production), modern agricultural (dominated by orchards), and forest (dominated by forests and green areas).

Taking into account these categories of municipalities, their functions, and their location within the WMA, the following types of municipalities facing problems of peri-urban development were identified: Type A, suburban zone nucleus, Type B, suburban inner zone, and Type C, suburban outer zone. For each of these types, one case study was selected for further investigation (Fig. 7.3). They are characterised by different transport accessibility: Błonie has very good connections to Warsaw and good access to other towns from the western part of the WMA. In the case of Halinów the access via the road network is relatively poor. However, Halinów has a railway which improves transport links with Warsaw. Leoncin has very poor access to transport networks as a municipality with peripheral location adjacent to the extensive forests of the Kampinowski National Park. All three municipalities are located about 30 km from Warsaw:

| | |
|---|---|
| Type A | Suburban zone nucleus: municipalities located in the suburban zone of Warsaw; dominated by urban functions; with local urban centre—providing jobs opportunities and services for neighbouring rural areas, surrounded by open space; well equipped with technical infrastructure, with clear urban edge understood as the boundary between built-up areas and open space; represents a transitional form—from urban to suburban in terms of size, intensity, functions, functional relationships with surrounding areas. |
| | The municipality of Błonie is an example of the suburban zone nucleus type (Fig. 7.4a, b): a municipality with compact town in the centre, surrounded by a high quality natural environment with strong urban pressure, dynamic development processes and level of development above average in the whole Mazowieckie voivodship and the WMA, operating as a local service centre, attractive for investors. Błonie has |

(continued)

| | |
|---|---|
| | very good connections to Warsaw and good access to other towns from the western part of the WMA. |
| Type B | Suburban inner zone: municipalities located in the suburban zone of Warsaw; mixed functions, open spaces, more rural in terms of landscape, size and functions of settlement units; not necessarily bordering the central zone of the metropolitan area; represents transitional form—from suburban to peri-urban in terms of size, intensity, functions, functional relationships with surrounding areas. |
| | The municipality of Halinów is an example of a suburban inner zone type (Fig. 7.5a, b): a municipality with an average level of economic development, mixed functions, under urban pressure since the municipality offers high quality natural environment. Halinów is a municipality with a significant number of migrants—people who moved to Halinów from Warsaw looking for good location to build buy a house. Halinów is an example of a suburban inner zone municipality under the strong influence of Warsaw. In the case of Halinów the access via the road network is relatively poor. However, Halinów has a railway which improves transport links with Warsaw. |
| Type C | Suburban outer zone: municipalities located in the suburban zone of Warsaw; agricultural functions; rural landscape; moderately affected by urban pressure; represents transitional form—from peri-urban to rural in terms of size, intensity, functions, functional relationships with surrounding areas. |
| | Leoncin is an example of a suburban outer zone (Fig. 7.6a, b): a rural municipality with weak urbanisation pressure, high value open space including the protected area of the Kampinoski National Park. Leoncin has very poor access to transport networks as a municipality with peripheral location adjacent to the extensive forests of the Kampinowski National Park. Poor transport accessibility results in relative spatial and functional isolation of the municipality. |

## 7.7 Quality of Life Preferences and Perception in the Three Peri-Urban Municipalities

A questionnaire survey of 1,200 respondees was administered in order to identify inhabitants' preferences concerning their place of residence and the factors that impact their attractiveness; to identify the perception of respondents of peri-urban areas, their attractiveness and usefulness for different purposes; to canvass inhabitants' opinions on sub-urbanisation in the WMA—specifically about its consequences; and to learn how they evaluate the WMA functioning as the area where they live and work.

The results of the research demonstrate that the most important components of the living conditions for inhabitants from all municipalities are: "quality of the natural environment" and open green areas (Fig. 7.7). The residential location should also ensure a feeling of "safety", and of "being at home". This factor emerged more often in Halinów—the municipality with many people who moved there (mainly from Warsaw). In the rural municipality of Leoncin, inhabited mostly by people who have lived there for several generations "feeling of being at home" was mentioned rather rarely. They are just at home as a part of traditional local community.

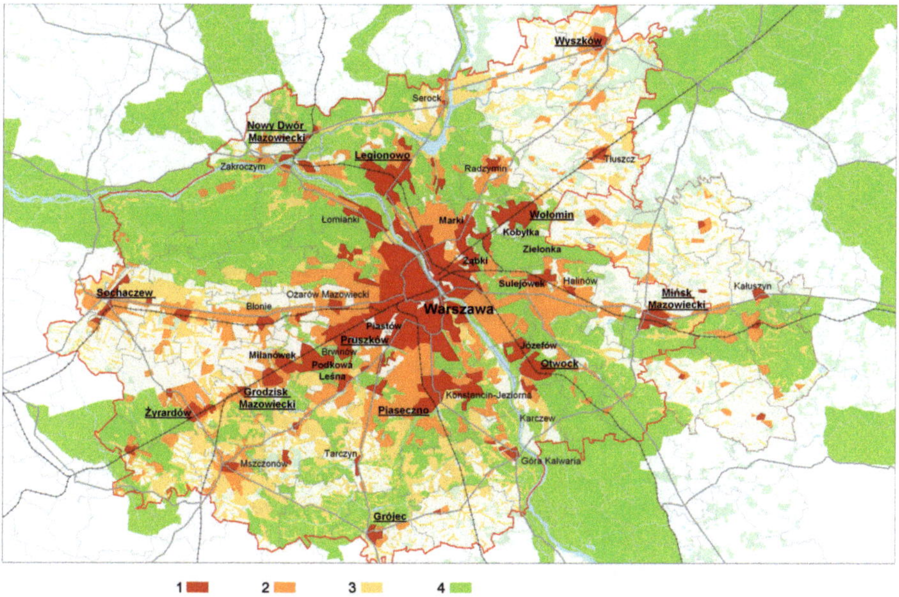


**Fig. 7.2** Peri-urban areas within the WMA intensive urbanization zone (categories of municipalities according to official planning documents such as the 'Plan of Spatial Development of the Warsaw Metropolitan Area': urbanized). *1*. Intensive urbanization zone (categories of municipalities: urbanized); *2*. Supplementary development zone (categories of municipalities: urbanized and under urbanization); *3*. Adaptation zone (categories of municipalities: traditional agricultural, diversified agricultural, modern agricultural); *4*. Open zone (categories of municipalities: traditional agricultural, modern agricultural, forest type)

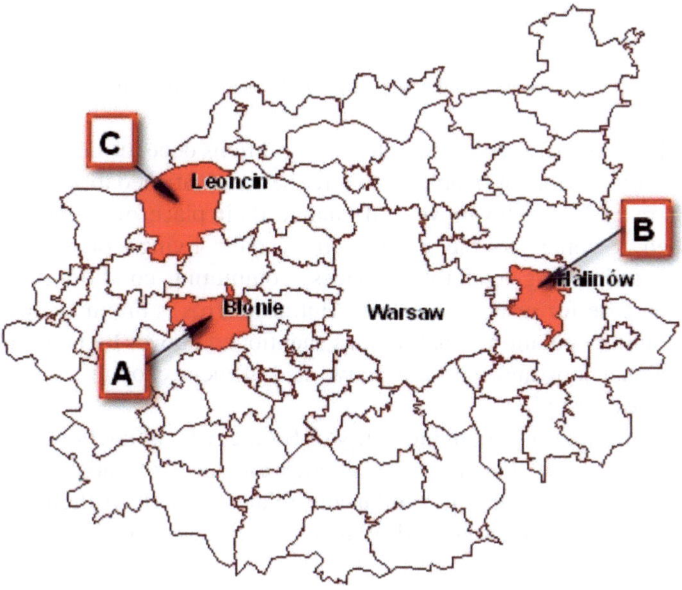


**Fig. 7.3** Municipalities selected for in-depth studies

**Fig. 7.4** (**a**) Błonie—central part of the city, (**b**) Błonie—commercial area (logistic centre)

The level of satisfaction of current living conditions depends on the time spent in the municipality: the longer people live there the more satisfied they are. Inhabitants who are more mobile, who move and change their place of residence are more critical when it comes to the evaluation of their living conditions and their expectations concerning available services, commuting conditions, etc. This is especially the case for respondents from Halinów. This is the municipality with a relatively young population, inhabited, as mentioned above, by many newcomers with professional jobs and who are active and affluent enough to move to more attractive places for living.

The interesting case is the municipality of Błonie, which functions as a suburban nucleus. Regardless of the time of residence the level of satisfaction of living conditions among inhabitants is similar and relatively high. Although Błonie has good transportation connections with Warsaw it's citizens usually only travel to the

**Fig. 7.5** (**a**) Halinów—new housing developments, (**b**) green areas of Halinów

capital to use specialized services (15 % commute to Warsaw on regular basis/ everyday/or often). The rest of their needs are fulfilled locally. Thus the spatial proximity of Warsaw does not necessarily trigger daily commuting. However, regardless of the type of municipality and its offer for inhabitants they perceive Warsaw as a "nearby" city, where they can find everything that is lacking in their home municipality. Warsaw is seen as an important, available asset: although respondents do not live there they can use Warsaw's resources. The results of the study show that respondents appreciate what Warsaw has to offer but still enjoy a suburban lifestyle more (Figs. 7.8 and 7.9). Respondents from Halinów serve as an example: almost 40 % of them travel every day to Warsaw to work or to attend school.

A majority of respondents from all three municipalities describe Warsaw as a crowded place, with traffic and noise. Respondents from Halinów also pointed out

**Fig. 7.6** (**a**) Leoncin—border with the Kampinowski National Park, (**b**) Leoncin—rural housing

the high costs of living. For respondents from Leoncin Warsaw is also an unsafe place. They see the city as a living environment which contrasts with their municipality. More positive opinions about Warsaw were expressed by people (usually younger) who look for specialised services (culture, leisure, recreation, education) or attractive jobs.

Although the three selected for case studies municipalities are not equally equipped with services (transport, shops, etc.) access to them was not the key factor in evaluating their living conditions. Costs of renting or buying an apartment or house are important for younger respondents who are usually more mobile. The hierarchy of criteria considered while looking for an apartment or house are similar regardless of the type of municipality.

The notion of *peri-urban* is very new to the public. Respondents had problems in defining what peri-urban areas are, what features they have, how to delineate them and how they can benefit from them. Respondents were much more familiar with

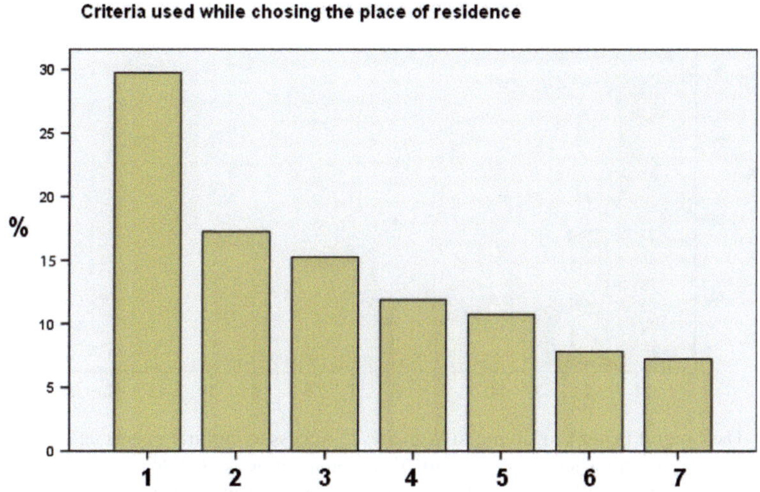


**Fig. 7.7** Criteria used while choosing the place of residence: *1*. Quality of natural environment; *2*. Safety; *3*. Opportunity to find a job; *4*. All factors are important; *5*. Cost of buying/renting house/apartment; *6*. Difficult to say; *7*. Good transportation to place of work

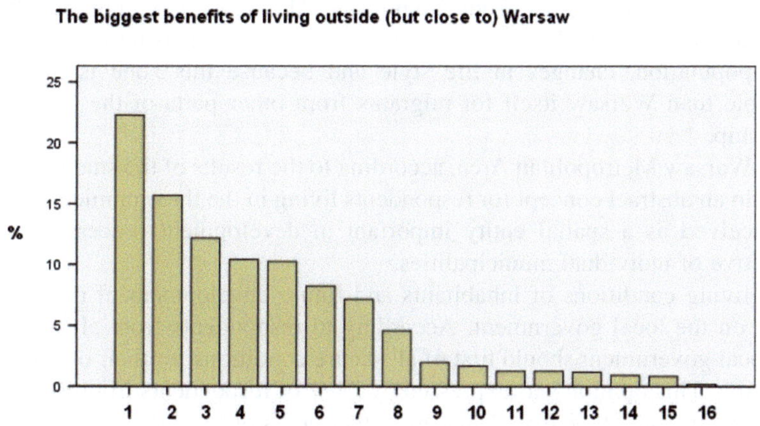


**Fig. 7.8** The biggest benefits of living outside (but close to) Warsaw: *1*. Nice surroundings, landscape; *2*. Living in the open outdoors; *3*. Clean environment; *4*. Feeling of "being at home"; *5*. Quietness/stillness; *6*. Living in a house not flat/apartment; *7*. Open spaces; *8*. Lower costs of living compared to Warsaw; *9*. Possibility to run small business at home; *10*. Low density housing developments; *11*. Possibility of having a garden; *12*. Safe environment; *13*. Possibility of having bigger flat/house for the same money; *14*. Possibility to breed small animals; *15*. Others; *16*. Difficult to say

the notion of a *suburban zone*. Perceptions of this were shaped by its functions not by its values related to landscape or natural resources. Suburbanization in the WMA was perceived as a very dynamic process; as messy, uncontrolled, and chaotic but at the same time as something natural that cannot be stopped or directed. The remedy

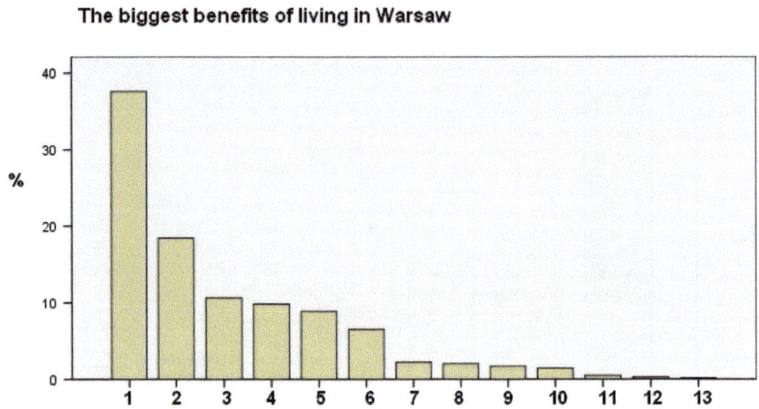


**Fig. 7.9** The biggest benefits of living in Warsaw: *1*. Access to cultural events and entertainment services; *2*. Easy to find a job; *3*. Access to educational services; *4*. Interesting opportunities to spend free time; *5*. Easy to find a well paid job; *6*. Access to different services; *7*. others; *8*. Good opportunities for professional development; *9*. Good housing conditions; *10*. Anonymity; *11*. Social life/social contacts; *12*. Difficult to say; *13*. Luxurious living conditions

for rapid development of suburban areas was seen as development of technical infrastructure. Respondents from Błonie and Halinów also suggested that the suburban zone will still be developing because of the increasing economic status of the population, changes in life style and because this zone is more easily accessible than Warsaw itself for migrants from other parts of the Mazowieckie voivodship.

The Warsaw Metropolitan Area, according to the results of the study, seems still to remain an abstract concept for respondents living in the three municipalities or is not perceived as a spatial entity important in development processes from the perspective of individual municipalities.

The living conditions of inhabitants and future development of municipalities depend on the local government. According to respondents from all the samples areas local government should first of all secure conditions for their own economic well-being. This opinion was expressed by 73 % of respondents from Błonie, 65 % from Leoncin, and 59 % from Halinów. Other goals of local development policies listed by respondents were: in Błonie—new housing constructions; in Halinów and Leoncin—improvement of living conditions through protection of the natural environment.

Development strategies and plans are supposed to be prepared through participatory means. Citizens' expectations, needs, preferences, and priorities should be addressed by these documents. However, the level of public participation in all three municipalities is very low. A very passive situation was presented by the more rural community of Leoncin, while the most active was Halinów. The younger and more educated the community the more involvement they seem to have in public life, including planning processes.

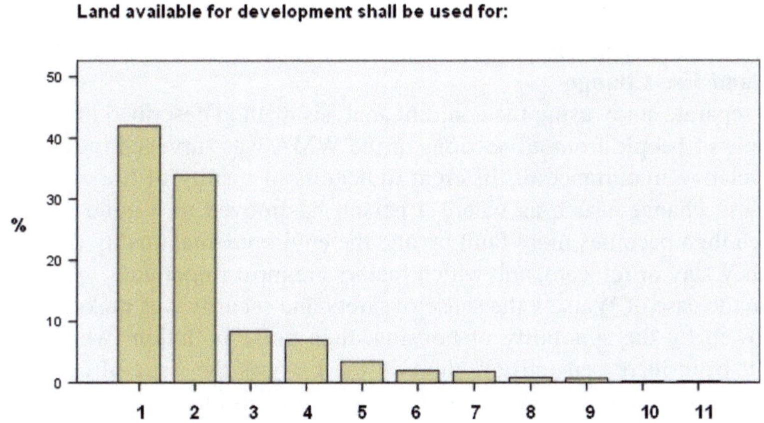


**Fig. 7.10** Usage of available land. *1*. Recreation and sport; *2*. Single houses construction; *3*. Location of blocks with apartments; *4*. Location of industry; *5*. Open spaces shall be left as they are to maintain their natural character; *6*. Others; *7*. Construction of big shopping centers; *8*. Construction of building with office space; *9*. R&D centers; *10*. Logistic centers, warehouses; *11*. Difficult to say

Public participation in planning processes seems to be especially important in the context of the future of open areas and land currently used for agricultural production (Fig. 7.10). For most respondents future development was associated with economic development while expectations concerning open land were that it would be converted into recreational areas. In order to avoid conflicts, to integrate and mobilise local communities as well as to meet citizens expectations in rational way, a more proactive attitude among citizens is needed.

The results of the survey briefly described above suggest that, surprisingly, respondents from three different types of municipalities share to large extent the same values, have similar expectations and preferences, and evaluate suburbanisation processes in a similar way. In no case was the problem of peri-urban development addressed from the perspective of their unique values and assets. They were seen as areas of possible further expansion of urban functions.

Thus the anticipated trajectory of these areas is that development plans will continue to be focused on further economic development; issues of the natural environment were only mentioned in the context of a need for environment protection. Thus for most people growth has top priority and it might be anticipated that development plans will generate increasing urban pressure since they contain programmes focused on attracting new inhabitants and businesses. Natural environment protection plans have a "sectoral" character; there is a lack of a comprehensive approach to protection and utilization of different resources, agricultural land, open spaces, and green areas. Diagnostic parts of development strategies include information on existing or potential urban pressure; however, there is a lack of a clear statement on responses of planning and management character (except restrictions concerning areas protected by law).

**Box 7.1 Quality of Life Across the Whole WMA: Indicators in Relation to Land Use Change**

In a separate study using the conjoint analysis method described in Chap. 4 a sample of people from all sections of the WMA was surveyed in relation to the relative importance of different indicators of quality of life affected by and use change—such as where a person has moved to a quite rural area which then becomes more built up and the environmental quality declines—do they stay or relocate, and which factors are most important.

In the case of Warsaw the sense of safety and security was ranked highest, followed by the suitability of housing, then noise pollution (which would result from increased traffic). Next on the list was the issue of commuting time (which would increase with greater urbanisation), then air quality (reduced if more transport leads to greater pollution) followed by the adequacy of waste collection (which may cause problems if municipalities cannot maintain service levels when development is fast). The convenience of green space is surprisingly low on the list, coming second bottom to the number of shops available.

This study can be used to test scenarios to see how residents would be likely to react to land use changes which affect these indicators. However, some people react differently to others and their tolerance to change is a combination of where they live now, their lifestage and their lifestyle, so that a sophisticated tool is needed, provided by the Quality of Life Simulator (QoLSim) (Fig. 7.11).

## 7.8 Discourse on Metropolitan Areas Development

This section considers the various factors affecting the development of the WMA and sets the context for the scenarios to follow.

The problems of development of the WMA are not unique and can be found in many other cases of functionally integrated areas that experience dynamic development both in terms of growth of area and of population as well as increases of the complexity of spatial and functional structures. The lack of legal regulations concerning metropolitan area development in Poland has triggered discussions on how to eliminate the negative consequences of growth and how to take advantages of the development potential of the central city and its surrounding municipalities. From the very beginning of realising the problems, inter-municipal cooperation has been seen as an attractive solution. As noted earlier, local democracy and autonomy of municipalities (including planning power) are deeply rooted in the system of public administration. No wonder, therefore, that inter-municipal cooperation, the mechanisms of which leave the policy domain of local government intact (because there is no permanent transfer of local competencies) is such an attractive solution to many representatives of local governments from metropolitan areas.

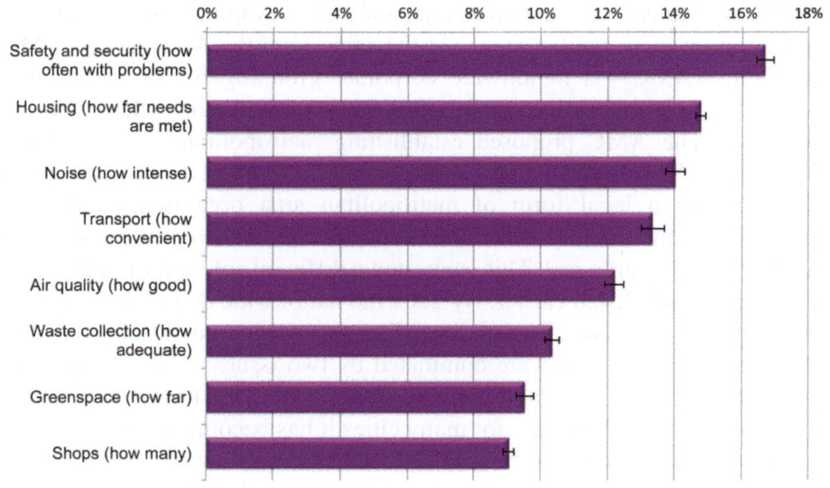


**Fig. 7.11** A graph showing the ranking and percentages of importance of different indicator values for Warsaw (Source: OPENspace Research Centre, Edinburgh College of Art)

Discussions on metropolitan area development in Poland started at the end of the 1990s. Different categories of municipalities, metropolitan city governments, NGOs, central government agencies, regional governments and experts have all contributed to these discussions. In the case of the WMA one very proactive participant of the discourse has been the Metropolitan Warsaw Association (Statut 2009). The Metropolitan Warsaw Association (established in 2000 and legally registered in April 2004) is the result of a grass-root initiative by local leaders from the municipalities of the WMA. At the time the Association was established the City of Warsaw had the legal form of a union of 11 municipalities but the largest central municipality of Warsaw did not join the Association, only 10 municipalities neighbouring the municipalities surrounding central Warsaw. In 2002 the legal status of the City of Warsaw changed to become one municipality divided into 18 districts with limited competencies and powers. Warsaw, as one municipality, joined the Association in 2006. Four working groups dealing with issues crucial for the development of the member municipalities of the Association were set up: spatial planning and administrative organization, communal services, transport development and programmatic issues of the Association's activities. With Association support the bill on metropolitan areas was drafted in 2008.

However, it turned out that the perception of the "metropolitan issue" as understood by Warsaw and the other member municipalities was different. The joint work slowed down and results were few. Some conferences and working meetings were organized by the Association to lobby for proposed solutions but the outcomes of these events were not converted into tangible results. The only exception was an initiative to introduce the "common travel ticket" in the WMA, which was introduced in the so-called *Warsaw Agglomeration* in 2009.

In 2008 the Ministry of Internal Affairs and Public Administration started work on legislation on metropolitan areas. At the same time the Association of Metropolitan Cities (AMC)—a nationwide corporate grouping of the biggest Polish cities—started to lobby for its own proposal for a management scheme for metropolitan areas. The AMC proposed establishing metropolitan counties. The bill prepared by the Metropolitan Warsaw Association proposed a voluntary union of municipalities as a legal form of metropolitan area organisation. These two proposals were contradictory so in 2009 representatives of regional governments prepared yet another proposal. This suggested a different solution from those of the other three proposals listed earlier. By 2009 the nationwide debate narrowed down to focus on one proposal, that of the Ministry of Internal Affairs and Public Administration. Discussions were dominated by two issues: how to divide power (who will be the head of the "metropolitan government") and how many metropolitan areas should be established—for many cities it has become a political ambition to be recognized as a metropolitan city. The debate has become deeply politicized which could be seen on reading new versions of the draft legislation as it was debated. The number of metropolitan areas rose from 2 to 7 and then to 16 and subsequently dropped to 3 and then back again to 2 metropolitan areas (Grochowski 2009).

Warsaw did not take part in this debate. This fact placed the Association in an awkward position. Eventually, at the beginning of 2010 the debate went into hibernation. It also seems that after 10 years of existence the Association also became moribund—it did not manage to realize its main goals. Warsaw as a member of the Association did not strengthen the Association. It did not become a lobbyist supporting new approaches to metropolitan area development. It also seems that Warsaw still retains very little interest in cooperation with municipalities from the WMA.

This discourse on metropolitan area development has never been institutionalised. Information exchange has been chaotic. No platform for structured dialogue has been offered by central government or others involved in the discourse. No attempts have been made to compare and evaluate different proposals systematically. Discourse via lobbying for specific solutions among their supporters has led nowhere.

## 7.9 Scenarios for Warsaw Metropolitan Area

The debate on the future of the WMA was brought back to life during implementation of the PLUREL Project. Representatives of the Association, the City of Warsaw government, representatives of other municipalities from the WMA and other stakeholders participated in different activities under PLUREL including discussions and a workshop on development scenarios based on the SRES scenarios (see Chaps. 2, 4, and Box 7.1 'Regional Urban Growth (RUG) modelling'). Four development scenarios presented to workshop's participants triggered discussion on the future, the most probable paths of the WMA development and factors that would determine them. The conclusions of a critical examination of the four scenarios and the results of discussion on current and future driving forces that would impact development paths led workshop's participants to point out that in the specific conditions of the WMA two other scenarios should be developed: a limits-to-growth scenario and a spatial containment scenario. Although both scenarios had unique features the limits-to-grow scenario was related to the *Hyper-tech* scenario, and the spatial containment scenario was linked with the *Peak oil* scenario. These two scenarios are outlined below. For each some assumptions are presented, then a storyline' followed by the potential consequences if this scenario happens (Box 7.2).

**Box 7.2 Regional Urban Growth (RUG) Modelling for the Warsaw Region**

The four scenarios differ in the intensity and patterns of change in artificial surfaces.

In the *Hyper-tech* scenario, rapid technological change, new transport technologies and few planning constraints lead to peri- and counter-urbanisation. The heart of the city also attracts a share of the population, possibly young urbanites benefiting from the high economic growth.

The *Extreme water* scenario shows moderate growth in all zones. Alternating extreme events such as flooding from the Vistula river and drought may affect various parts of the case study area.

In the *Peak oil* scenario, high fuel costs and strict planning policies concentrate growth in the urban core of the city, though here there is also a fair amount of growth in the peri-urban area, especially where these are well-connected to the centre. This may be explained by the high increase in projected GDP, compared to other areas under the same scenario, which encourages peri-urbanisation despite the oil price shock.

In the *Fragmentation* scenario, urban growth spread within all zones. However, it is less homogenous than in the *Extreme water* scenario, as communities of different age groups, ethnicities, etc. cluster together. Green enclaves appear in rural areas as older native people move out of the socially and ethnically diverse cities (Fig. 7.12).

Source: University of Edinburgh

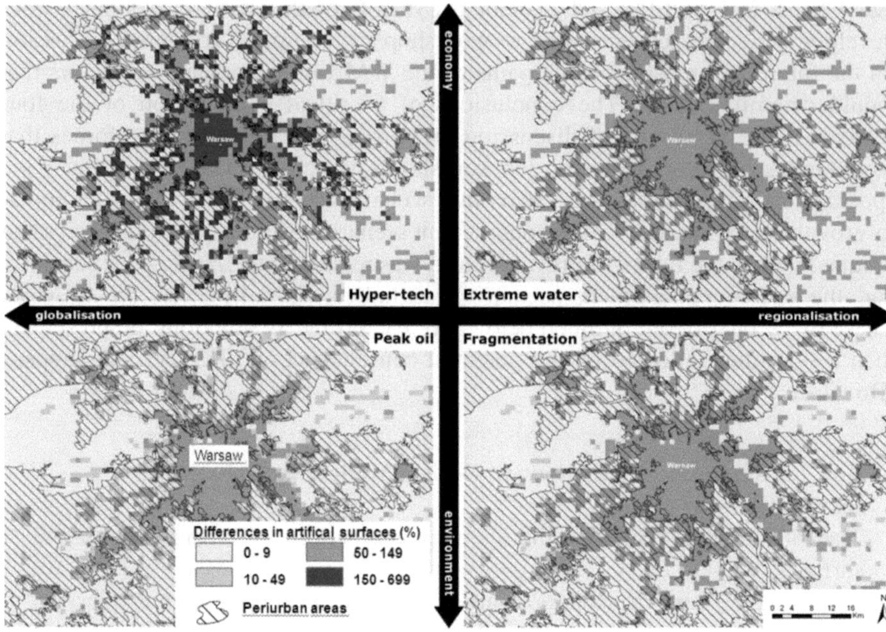


**Fig. 7.12** RUG modelling

## 7.10 Scenario 1: Weak Planning, Strong Growth, Eventually Self-Limiting Growth

### *7.10.1 Basic Assumptions*

Dynamic economic development of Warsaw and the WMA, based on new technologies, under stable macro-economic conditions and with the continuation of current practices of spatial planning and the management of development.

### *7.10.2 Local Storyline*

Warsaw takes advantages of its development potential. The R&D sector leads the regional economy. Trends of development remain the same as they have since 1990. Warsaw is a growth pole of the WMA and Mazowieckie voivodship. Further polarisation of development processes takes place. The majority of firms are concentrated in Warsaw city.

The spatial pattern of development of Warsaw takes the shape of a growing, aggressive octopus with tentacles of business and housing development spreading outwards along transport corridors. Uncontrolled urbanisation leads to chaotic urban colonisation of areas surrounding the city. Although there are no investments in transport infrastructure, new investments are still attracted to Warsaw and its surroundings. The area is not territorially cohesive and its newly developed areas are not easily accessible. Private cars become the main mean of transport. Despite these problems, and because of the lack of planning intervention, urbanisation pressure is still rising. More migrants are arriving in Warsaw and the surrounding municipalities, which offer lower costs of living. These municipalities develop a predominantly housing function while service functions remain underdeveloped. The inhabitants of Warsaw become more affluent. This leads to spontaneous suburbanisation since many of them decide to leave the central parts of Warsaw and move to the suburbs. This is partly because Warsaw is losing its attractiveness as a place to live. Migration brings significant changes to the social and demographic structure of the Warsaw population, which is ageing. The population in the WMA is rising. Demand for housing brings a reduction of agricultural areas and open space.

There is no reaction by the public authorities responsible for development management. Planning regulations and procedures remain unchanged. There is no cooperation among municipalities concerning the preparation and implementation of development plans. Very dynamic but uncoordinated development of Warsaw and the WMA leads to functional paralysis of the system, which happens around the year 2020. After that the number of inhabitants starts to decrease slightly.

### *7.10.3 Consequences*

Warsaw fails to become more attractive at national and European scales because the city does not offer good living conditions. The city and the WMA are poorly served by transport infrastructure and lose their attractiveness as a place of doing business. Social disparities increase bringing serious social conflicts. Underdevelopment of services in suburbanised areas forces their inhabitants to commute to Warsaw to fulfil their needs. The WMA acquires features of a large monocentric type of rural–urban region. The agricultural sector goes through a restructuring process that results in a decrease of areas used for agricultural purposes and also important for their ecological functions. Warsaw is surrounded by highly urbanised areas and suffers from a lack of land for leisure and recreation. The concept of a "Green Ring" surrounding Warsaw has no chance of being realized.

## 7.11 Scenario 2: Strong Planning, Sustained Strong Growth: Spatial Containment Scenario

### *7.11.1 Basic Assumptions*

Dynamic economic development of Warsaw and the WMA is based on new technologies. There are stable macro-economic conditions. There is a new legislation concerning development of metropolitan areas in Poland the Warsaw Metropolitan Area which enables an integrated approach to spatial and economic development of functionally integrated areas to be developed. Strict planning regulations ensure the rational use of assets in accordance with the vision and strategy for the WMA development.

### *7.11.2 Local Storyline*

Warsaw takes advantage of its development potential. The R&D sector leads the knowledge-based economy. Development incentives appear across the metropolitan area thanks to institutionalised cooperation between the regional government, the government of Warsaw and the municipalities of the WMA. Coordinated efforts help to create and offer proper conditions for economic activities in different parts of the WMA. Although Warsaw still dominates in the spatial pattern of the WMA this pattern develops features of polycentricity. Functions become de-concentrated.

Warsaw still attracts migrants. New housing developments are located in Warsaw and the surrounding municipalities according to and following the development of social and technical infrastructure for the WMA. Development of the public transport system is a priority in strategies and plans. Pragmatic and well prepared investments lead to rational use of space. The condition of the natural environment improves. Services for tourists become integral to local economies.

### *7.11.3 Consequences*

The flow of migrants is significantly lower as well as pressure to commute because services are more evenly distributed. The WMA becomes more attractive as a place to live and work. The level and quality of life of the inhabitants increases. Warsaw becomes an attractive European metropolitan city. The creative classes find Warsaw the right place to settle, which triggers development of new forms of economic activity including services. Agricultural areas and open spaces are protected and the concept of the "Green Ring" is implemented. Traditional agriculture remains part of the local economy and contributes to the vitality of peri-urban areas. The spatial system is cohesive and functional and the costs of the system are lower.

## 7.12 Modelling Land Use Changes

To visualize the potential spatial consequences of realising the two scenarios, projections of the situation in the period from 1995 to 2025 were prepared using the MOLAND model as described in Chap. 4. The experiment with modelling was successful and met expectations i.e. they indicated the nature and spatial extent of changes. The results of simulation based on the two scenarios differ considerably.

### 7.12.1 Scenario 1: Limits to Growth

According to this scenario the area covered by urban fabric will increase over the period 1995–2020 from 55,160 ha to 61,430 ha (11 % increase) and will then decrease to 54,450 ha (Table 7.2). Agricultural areas will decrease by 8 % in the same period. This scenario enhances processes of suburbanisation and urban sprawl that may result in an expansion of green urban areas (i.e. areas formerly occupied by agricultural functions where housing and services functions are located), since the city may lose inhabitants moving outside its borders. Thus a limits-to-growth scenario can trigger processes leading to "shrinking" of the city (Fig. 7.13). The area of urban fabric will increase by 11.4 % to 2020; then will shrink by almost the same percentage, as might be anticipated based on projections of demographic change (Fig. 7.14). The crisis envisaged in 2020 will have serious implications for the future development of Warsaw and the WMA. The resolution of spatial conflicts will require additional financial resources and available land. Development opportunities may not return for several years since degradation of physical, economic, and social space will be so serious.

### 7.12.2 Scenario 2: Spatial Containment

Following the second scenario the increase of urban fabric will continue and reach the level of 62,950 ha in 2025 (increase of 12 %) (Table 7.2). Green urban areas will slightly shrink. The same concerns agricultural areas. However, it is not just the number of cells which is important in this case but the degree of concentration of areas of different types.

According to this scenario the process of urban sprawl will be restricted. Rigid zoning will reduce the number of functional conflicts. A mosaic pattern of function locations will be replaced by well-organised structures following already developed and appropriate technical infrastructure. This will bring additional incentives for economic and social development.

Initial assumptions and parameters characterising the situation in the WMA and development trends for both scenarios produced interesting pictures of the potential future situation (Figs. 7.13 and 7.15). Although by the year 2020 no difference in the total cover of urban fabric can be seen there were significant differences in its spatial distribution. As anticipated, the urban fabric as well as that of industrial and commercial areas was characterised by a much higher spatial concentration in Scenario 2.

**Table 7.2** Proportionate area (%) of land use classes in WMA region 1990–2025 in two different scenarios (Metronamica ML)

| Land use | Limits to growth | | Spatial containment | |
|---|---|---|---|---|
| Scenario/year | 1990 | 2025 | 1990 | 2025 |
| Urban fabric | 8.9 | 8.8 | 8.4 | 10.2 |
| Industrial or commercial units | 1.0 | 1.0 | 0.9 | 1.1 |
| Transport areas | 0.5 | 0.5 | 0.5 | 0.5 |
| Mineral extraction or dump or construction sites | 0.2 | 0.2 | 0.2 | 0.2 |
| Agricultural areas | 57.5 | 53.7 | 59.7 | 52.5 |
| Green urban areas | 6.3 | 9.5 | 5.8 | 9.1 |
| Natural and semi-natural vegetation | 0.2 | 0.2 | 0.2 | 0.2 |
| Standing forests | 23.9 | 24.6 | 22.8 | 24.6 |
| Water courses and bodies | 1.5 | 1.5 | 1.4 | 1.5 |
| Total area (ha) | 619,000 | | | |

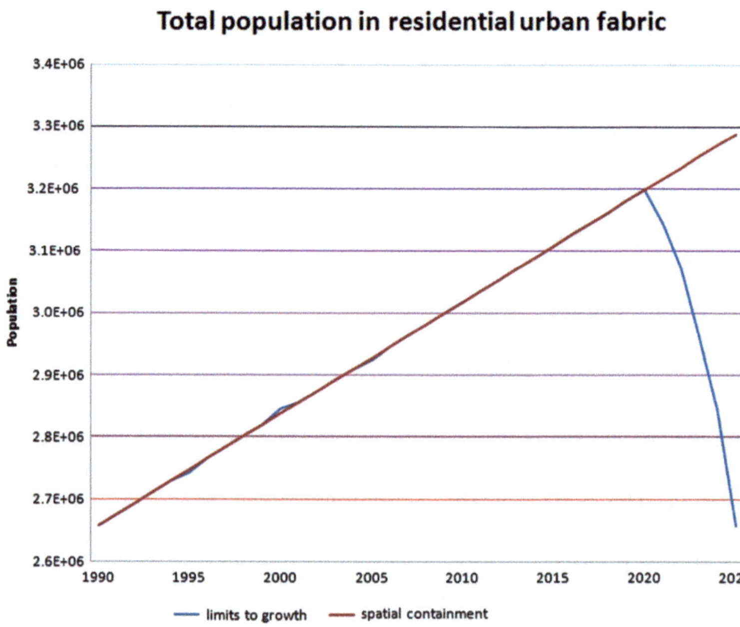


**Fig. 7.13** Total population in residential urban fabric in WMA region 1990–2025 (limits to growth and containment scenarios)

The results of the simulations should be considered carefully by policy makers and planners. Supply and demand for land differ over time and are determined by the needs of inhabitants and investors and by the quality of space. Extensive use of space may lead to its degradation, as it was shown in the self-limiting growth scenario. Intensity of land use is as important a factor as spatial arrangement in

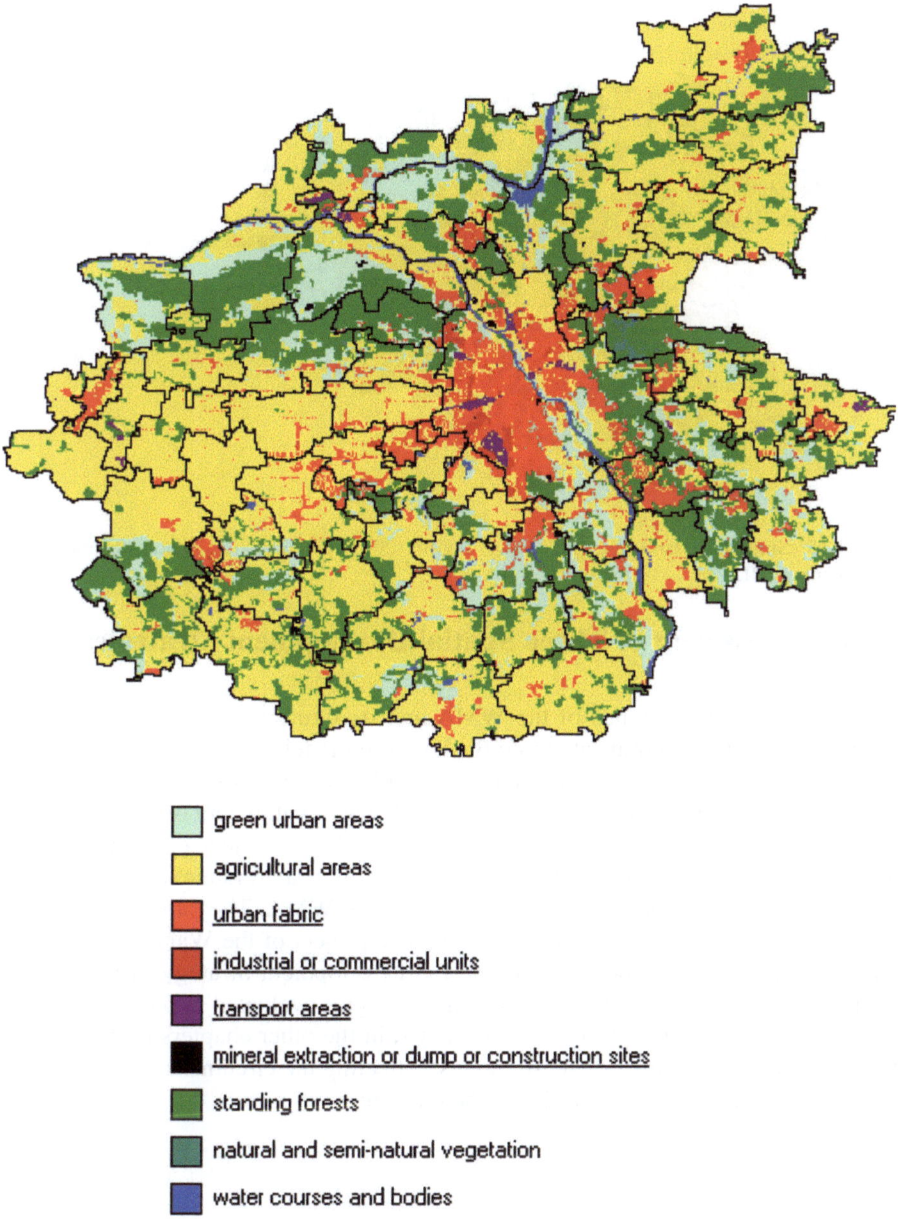


**Fig. 7.14** Simulated land use in WMA region 2025 (Scenario 1: self-limiting growth, Metronamica ML)

terms of location of functions. Unequal distribution and concentration of functions may lead to the forced mobility of some sectors of the population. It may also trigger changes of land use in neighbouring areas.

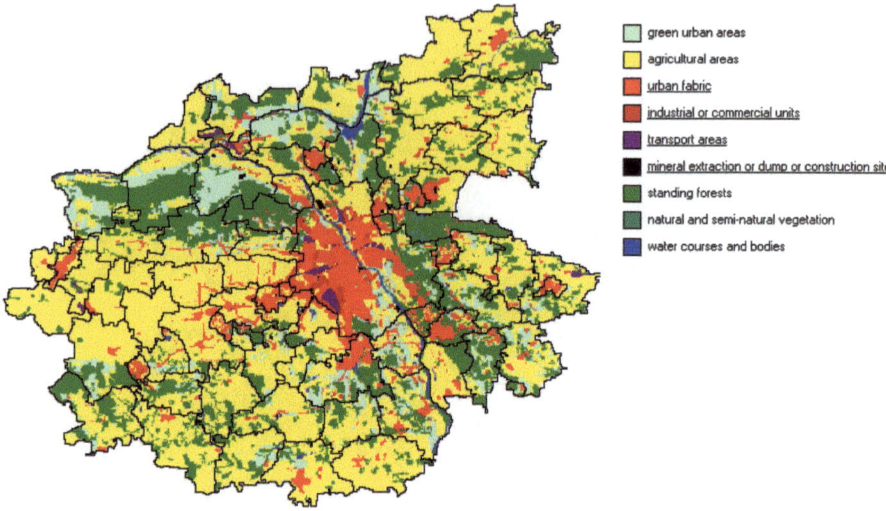


**Fig. 7.15** Simulated land use in WMA region 2025 (Scenario 2: spatial containment, Metronamica ML)

## 7.13 Assessment of Three Spatial Development Strategies

Current development trends of the WMA and its future situation outlined in the scenarios, which were translated into the land use models, have to be addressed by strategies that allow it to meet development challenges. In this section a selection of strategies employed by the main actors to guide the development of the WMA will presented and assessed. The strategies were selected from a range of those available and documented according to the following criteria: spatial range, thematic range, main objectives and role in the development of the WMA. They are: The Strategy of Mazowieckie Voivodship Development, The project of the Warsaw Metropolitan Area Spatial Development Plan and the Development Strategy of the City of Warsaw. Clearly, when compared to some strategies evaluated in other case study regions of the PLUREL project and presented in the other chapters in this section these are high level, "top down" strategies, reflecting the circumstances of the case study and the aspects noted earlier in the chapter.

### *7.13.1 Strategy 1: The Strategy of Mazowieckie Voivodship Development*

This strategy provides basic guidelines concerning strategic development goals and the measures to be used to achieve them. The main assumption of the strategy is that the general economic situation of the region (despite existing differences and polarisation of development as described earlier), the application of cohesion policies and principles of sustainable development will facilitate development processes and allow strategic goals (formulated as follows) to be reached: development of rural areas; cohesive development of Warsaw and its environment; harmonious development of the WMA metropolitan functions; dynamic development of technical and social infrastructure; elimination of areas of poverty; improvement of the health and the quality of life of the inhabitants.

The following actions to reach a detailed list of objectives are foreseen to be necessary in order to achieve these goals:

#### 7.13.1.1 Strengthening Connections Between Warsaw and Regional, Domestic and International Environment

Under the new geo-political conditions resulting from the EU borders moving to the East, the significance of transport corridors which set up a cohesive transport system increases. In the new political reality Warsaw will function as "gateway city", being an "entrance gate" from the east to the EU. Insufficient capacity of the road and rail network and the poor technical condition of the system result in a weak accessibility of Warsaw. Therefore, the strategy aims to improve the quality and safety of transport connections between Warsaw and European metropolises, major economic centres in Poland and other voivodships by the development of air transport infrastructure, adjustment of road and rail network of international importance to European standards and improvement of road and rail connections with the sub-regional centres.

#### 7.13.1.2 Counteracting Degradation of Landscape and WMA's Natural Environment

The tasks to be undertaken in the future will be focused on two groups of actions: counteracting degradation of the environment in the areas of strong urbanisation and the preservation of the protective and preventative character of environmentally valuable areas. Priority is given to improvement of sewage management, increase of water supply reliability quality and the creation of a cohesive system of waste management. A green belt around the capital city will be created and the NATURA 2000 network will be established in order to protect the most important values of the natural environment and to preserve natural connections among ecologically active areas in the city and in the region.

#### 7.13.1.3 Restraint of Chaos in the Spatial Arrangement of the Capital City and Its Surroundings

Development of the capital metropolis under the circumstances of a lack of effective instruments for coordination and control brings many threats for sustainable development. As far as the Warsaw's space and its surroundings are concerned actions in favour of preventing unfavourable phenomena will include the modernisation and revitalisation of city centre areas, strengthening the cooperation between local government units in the filed of spatial planning and programming economic development in the WMA; and concentration of urban development in corridors of rail transport. Creating a cohesive system of legal regulations which protect environmentally valuable areas is a further goal.

#### 7.13.1.4 Protection and Restoration of the Natural Environment Aimed at Providing Sustainable Development

One of the most essential determinants of the improvement of quality of life in the voivodship is the state of the natural environment and its protection, aimed at the preservation of natural environments important for economic and social development in accordance with the principle of sustainable development. Under this objective expected actions will cover, among others, improving the environment pollution monitoring system, adjusted to EU standards, Mazovia's cooperation within the framework of the "Green Lungs of Poland" agreement, an increase of the region's total forest area and forest protection activities by means of planned afforestation, revitalization of degraded post-military areas as well as fostering ecological awareness among inhabitants as a factor to increase the effectiveness of the environment protection actions.

## 7.14 Strategy 2: Project of the Warsaw Metropolitan Area Spatial Development Plan

The project examined here covers much a smaller area and is more specific, as it is required by the Law on Spatial Planning. There are five strategic goals in the Plan.

### *7.14.1 To Secure Sustainable Development of the WMA and Its Surroundings*

This goal will be achieved by the improvement of territorial cohesion of the WMA and the voivodship. It will allow an increased access to metropolitan functions.

### *7.14.2 To Increase the Attractiveness of WMA as a Regional Growth Pole*

It is planned that the regional government will support efforts to build appropriate technical and institutional infrastructure to foster development of an innovative economy within a framework of a polycentric settlement pattern. It should reduce the scale of suburbanisation and urban sprawl and contribute to the protection of natural environment resources.

### *7.14.3 WMA as an Area of High Quality of Life*

Under this goal efforts will be focused on improvement of housing conditions and living conditions generally. The concept of the "Green Belt" around Warsaw is a crucial element in reaching this strategic goal.

### *7.14.4 Harmonious Spatial Development of the WMA*

Under this goal activities will be focused on improvement of the quality of the natural environment including such unique assets as the Wisła Valley and Kampinoski National Park. It is also envisaged that a processes of revitalization of towns and cities in the region and in the WMA will be intensified. Improvement of living conditions in these areas should significantly reduce migration flows to Warsaw and surrounding municipalities.

### *7.14.5 WMA as an Area with an Identity*

This goal was formulated to increase the level of social integration and to contribute to building and mobilising social capital as an important endogenous asset.

## 7.15 Strategy 3: Development Strategy of the City of Warsaw

The following goals and objectives that may have impact on the situation in the WMA are present in the third strategy:

### 7.15.1 Development of Metropolitan Functions, Strengthening Warsaw's Position at the Regional, National and European Level

Warsaw is the most important regional and national cultural centre, the home of international institutions and the largest transport hub. Consolidation of metropolitan functions will increase the international status of the Polish capital and will allow it to compete successfully with other European metropolises. Warsaw's development will be harmonised with the development of its surrounding area (neighbouring municipalities, the region and other cities). Potential competition must give way to a network of cooperative connections and collaboration of respective centres, whose specialisation will allow all to participate proportionally in the growth of the metropolitan area of Warsaw.

To ensure efficient internal and external transport systems for the WMA the following programmes are being developed: (a) improvement of external road links; (b) improvement of public transport in the metropolitan area. In order to reinforce Warsaw's position as an important European economic, financial, scientific and political centre the following programmes are designed: (a) attracting European economic and financial institutions to Warsaw; (b) construction of new congress, trade and exhibition centres; (c) participation of Warsaw in international networks of cities.

To establish institutions necessary for the efficient functioning of the metropolitan area of Warsaw the following programmes are designed: (a) organisation of institutions concerned with the development of the metropolitan area of Warsaw; (b) institutional integration of strategic planning on a regional scale.

In cooperation with the Marshal of the Mazovia Voivodship, presidents and mayors of the towns and municipalities of the metropolitan area, goals and appropriate measures will be specified to ensure speedy development of the whole area. The strategy also envisages the establishment of institutions that will deal specifically with the development of the WMA.

### 7.15.2 Organization of the Spatial Structure of the City that will Protect Green Areas Forming the Basic Ecological System

Zoning will be carried out consistent with the requirements of the natural environment systems and the functions of urbanized zones. Protection of the areas constituting the city's natural system, as well as green belts and borderlines, is a prerequisite for a good quality natural environment throughout the city and in recreational facilities for its inhabitants. In the strategy the suburb zone was highlighted. The most important tasks for this zone include the protection of environmentally valuable areas and the organisation of housing development, especially along main roads.

To secure smooth implementation of the development plans, the city government aims to develop a multi-year plan of property sales and acquisitions which will take into account the requirements resulting from execution of the city's development strategy. The City will determine which properties are required for the execution of planned investments and the scope of its property purchases for the purpose of such investments. Properties unnecessary for the execution of public tasks will be sold or provided under perpetual common ownership, rented, leased or made available in some other way. The city will formulate uniform, city-wide rules and procedures for the sale and acquisition of property. This applies to, among others, procedures for attaching neighbouring properties or their parts to shared properties, changes of shares in shared properties and settlement and refund of investments made by purchasers in properties. The City will prepare an offer of sale for selected properties, directed mostly to outside investors. It will also prepare an offer of multi-year lease of property and equipment to entities which, after the requisite investment, will conduct specified public services at those locations.

## 7.16 Results of the Analysis of the Strategies

The results of the analysis of the strategies should determine whether the development goals formulated in these documents correctly refer to the challenges of future development of the three areas for which the strategies were prepared. In all strategies the WMA ought to be present as a key element of the functional structure of the whole region (voivodship). However, the practice of development of the Warsaw Metropolitan Area as well as specific areas—"transition zones" and "zones under transformation" (peri-urban areas are examples of the two types) proves that these documents do not properly fulfil their tasks.

- The first reason is that every strategy addresses the WMA development from a different perspective determined by the particular competencies, responsibilities and power of the authority that was responsible for the preparation of the strategy.
- The second reason is that every strategy plays a different role as an instrument of spatial development management.
- The third reason is that the spatial planning system (described in the third section of this chapter) is set so that strategies and plans prepared at the regional (voivodship) and sub-regional (in this case metropolitan) levels can be reduced merely to blue prints (structural plans), that provide guidelines which are not obligatory or binding on those who are responsible for local development plans. Even if the strategies from these levels are prepared properly, according to a strict methodology, they may not be translated into real action on the ground. Goals, objectives, rules and guidelines incorporated in strategies from regional and sub-regional levels are formulated in a very general way. Thus it is quite simple to prove that goals from the strategies and plans prepared at the

municipality level are not in conflict with them. It is just a question of wording and interpretation.

- Strategies do not work because the system of planning is not designed to coordinate development undertakings. Under current spatial planning legislation there is no hierarchy of plans. Regions and municipalities are obliged to agree upon and incorporate into their plans guidelines from the national level. This especially concerns so-called "investment of public benefit". At the same time the system is set so that municipalities can choose not to prepare a development strategy or spatial development plan. In this case they govern spatial development through the mechanism of "administrative decisions"—permissions to locate investment. Even if a municipality has a spatial development plan and development strategy, vertical coordination can merely secure growth—there is no guarantee that vertically coordinated efforts will result in growth and sustainable spatial development.

The number of goals and objectives of every strategy and the way they are formulated make it difficult to assess their impact on relations between the voivodship, the WMA, and the city of Warsaw, to assess their contribution to sustainable development of the WMA, or to assess their contribution to realisation of one of the scenarios described earlier. However, analysis of detailed records from strategies justifies the statement that: Strategies 1 and 2 are relatively cohesive and that Strategy 3 has both positive and negative potential (Table 7.3).

## 7.17 Suggestions for Enhancement of Strategies

Interviews, discussions, and results of workshops organised for researchers and stakeholders allow also the formulation of opinion, so that at least three informal strategies are under consideration for use in coping with problems of spatial development:

### *7.17.1 Strategy: Building a Framework for Effective Development Planning in the Region and Metropolitan Area*

This is a "strategy of multi-level governance"—that is the key issue for development of the WMA. Under this concept long-term (strategic) planning is supposed to be integrated with medium-term and short-term (operational) planning. A leading role is played by the regional authorities. This strategy aims to form a development coalition that will be instrumental in searching for consensus among conflicting interests. Under this strategy, programmes of protection of agricultural land, organic farming, eco-tourism and improvement of agricultural production can be formulated and potentially implemented through programmes and projects

**Table 7.3** Contribution of strategies to sustainable development of the WMA

| | | Strategy 1 | Strategy 2 | Strategy 3 |
|---|---|---|---|---|
| Direct impact | Positive | X | X | X |
| | Negative | | | X |
| | Neutral | | | X |
| Indirect impact | Positive | X | X | |
| | Negative | | | X |
| | Neutral | | | X |
| Level of cohesion with other strategies | High | | | |
| | Moderate | X | X | X |
| | Low | | | |
| Potential contribution to limits to growth scenario | High | | | X |
| | Moderate | | X | |
| | Low | X | | |
| Potential contribution to spatial containment scenario | High | X | X | |
| | Moderate | | | |
| | Low | | | X |

prepared both at regional and local level. Players from the regional level are involved in discussions on formal regulations concerning the legal framework for spatial planning and management of functional urban areas. A system of monitoring of development processes, forecasting methods, and development models are specific tools to be created under this strategy.

## 7.17.2 *Strategy: Harmonization of Development of the Warsaw Metropolitan Area*

The strategy's goal is to create a spatial concept for sustainable development of a functionally integrated Warsaw Metropolitan Area. Under this strategy rules and parameters of land use are being formulated (in the Plan of Spatial Development of Warsaw Metropolitan Area that is ready but not enacted because of a lack of legal instruments). Planners from the regional level are responsible for the concept preparation. The strategy will allow the following goals to be reached: establishment of a set of regulations to be followed by entities subordinated to the regional authority and coordination of sub-regional development programmes and projects. The spatial development concept will serve as a platform for the negotiation of goals and development ideas among different municipalities and counties. An important goal of the strategy is to promote the idea of sustainable development and to educate the public about opportunities and threats resulting from specific development paths.

### 7.17.3 Strategy: Sectoral Programmes as a Tool to Facilitate Sustainable Development of Peri-Urban Areas (Suburban Zone Nucleus, Suburban Inner Zone, Suburban Outer Zone)

Sectoral programmes lack a spatial dimension. This strategy results from the "Strategy of Development of Mazowieckie Voivodship" and its concepts are incorporated into those of the WMA spatial development plan and will enable sectoral programmes to be tailored to specific local conditions. The strategy is being seen as a tool to be used to achieve goals of spatial character like the creation of the Green Belt around Warsaw.

## 7.18 Concluding Remarks

The case of the Warsaw Metropolitan Area shows how decentralization and democratisation in Poland have changed the conditions for planning and managing social, economic, and spatial development. The case of the WMA proves that sometimes a shift from an oppressive regime to democracy can be both a blessing and a problem simultaneously.

The self-limiting growth scenario, together with the simulation of its spatial consequences, provides convincing arguments for a change of current practices of planning and managing development processes in the WMA. Efficient means of control over further "colonization" of municipalities from the WMA are very much needed. It is unacceptable that new housing developments are constructed on land not equipped with basic technical infrastructure. However, lack of technical infrastructure is possibly a less important problem than a lack of understanding by decision makers of the potential consequences of ongoing "colonization" which may result in conflicts and problems of an economic, social, and spatial nature.

Suburbanisation, urban sprawl and the issue of peri-urban development are not being recognised by an individual municipality as long as they do not create problems for this municipality. This is the result of a specific "culture of cooperation" among municipalities: usually they only cooperate when they face common problems. From the perspective of building strategies and designing spatial development plans the horizontal aspect of coordination is a key issue. The most "urbanisation vulnerable" municipalities are these which posses the most attractive natural environment. This is the most powerful driving force that pushes people to locate their homes in suburban municipalities. Uncontrolled urban pressure may destroy resources of a specific municipality and destroy an important component of an ecological system that goes beyond the borders of one municipality.

# References

Bijak J, Kicinger A, Kupiszewski M, Śleszyński P (2007) Studium metodologiczne oszacowania rzeczywistej liczby ludności Warszawy (Feasibility study for the estimation of the actual population size of the city of Warsaw). Środkowoeuropejskie Forum Badań Migracyjnych i Ludnościowych (Central European forum For migration and population research CEFMR), Working paper 2/2007, Warsaw

Domański B (2001). Kapitał zagraniczny w przemyśle Polski. Prawidłowości rozmieszczenia, uwarunkowania, skutki (Foreign capital in Polish industry. Regularities of distribution, determinants, and results). IGiGP, Jagiellonian University, Kraków

Gawryszewski A, Korcelli P, Nowosielska E (2000) Funkcje metropolitalne Warszawy (metropolitan functions of Warsaw). IGiPZ PAN, Warsaw

Grochowski M (2009) Cooperation and networking in metropolitan areas—from community of problems to community of common interests in the Warsaw metropolitan area. In: European urban system: metropolization and networking, Europa XXI, vol 19. IGiPZ PAN, Polish Geographical Society, Warsaw

Grochowski M (2010) Samorząd na obszarach metropolitalnych. Raport z badań (Self government in metropolita areas. Report), Samorząd Terytorialny, Nr 6/2010, Wolters Kluwers, Warsaw

Komornicki T (2003) Export competitiveness of Polish poviats and big towns on the market of an enlarged European Union. Europa XXI, No. 10, IGiPZ PAN, Warsaw

Korcelli-Olejniczak E (2004) Funkcje metropolitalne Berlina i Warszawy w latach 1990–2002. Współzależność pozycji w systemie miast Europy Środkowej (Metropolitan functions of Berlin and Warsaw. Interdependence of positioning in the system of cities in Central Europe). IGiPZ PAN, Warsaw

Lijewski T (2003) Koncentracja ośrodków aktywności gospodarczej w Polsce w świetle list 500 firm (Concentration of economic activities center based on the list of 500 enterprises). Przegląd Geograficzny, nr 53, 3, IGiPZ PAN, Warsaw

Ministerstwo Gospodarki, Pracy i Polityki Społecznej (Ministry of Economy, Labour, and Social Policy) (2003) Strategia Rozwoju Województwa Mazowieckiego (Strategy of Mozowieckie Voivodship Development), Mazowieckie Biuro Planowania Regionalnego, Warszawa

Śleszyński P (2007) Gospodarcze funkcje kontrolne w przestrzeni Polski (Economic control functions in the Polish space). Prace Geograficzne, 213, IGiPZ PAN, Warsaw

Statut Stowarzyszenia Metropolia (Charter of the Association Metropolia Warszawa) (2009) Warsaw

Strategia Rozwoju Województwa Mazowieckiego (Strategy of Mozowieckie Voivodship Development). Mazowieckie Biuro Planowania Regionalnego, Warszawa, 2003

Strzelecki Z, Kucińska M (2005) Żywiołowe rozprzestrzenianie sie metropolii warszawskiej (Sponteneous spreading of Warsaw metropolis). In: Kozłowski S (ed) Żywiołowe rozprzestrzenianie się miast—narastający problem aglomeracji miejskich w Polsce (Spontenenous spreading of cities in Poland—growing problem of urban agglomerations in Poland), KUL-PAN, Białystok, Lublin, Warsaw

Taylor Z, Ciechański A (2006) Deregulation in Polish rail transport. Transp Rev 26(3):305–324

## References

# Chapter 8
# Manchester: Re-Inventing the Local–Global in the Peri-Urban City-Region

**Joe Ravetz and Pam Warhurst CBE**

## 8.1 Introduction

Some of the most populous parts of England are neither urban nor rural, but somewhere in between: a new kind of peri-urban landscape emerging in the fringes and hinterlands of cities and city-regions. Such peri-urban areas reflect both a more networked, mobile, globalised society, and also one which increasingly values local character and quality of life. The Manchester city region is one of these areas, a poly-centric agglomeration of several large towns and cities, between which an extensive peri-urban zone flows, linking them together like a form of connective tissue.

Manchester was one of the world's first industrialized and global trading cities, creating unheard of levels of pollution and sprawl as it developed. Over 180 years later, the peri-urban city-region of Manchester and satellite towns such as Salford, Bolton and Wigan, is developing a new kind of structure and purpose. However, this is in many ways problematic and divided: there is restored green infrastructure side-by-side with post-industrial wasteland; expensive twenty-first century housing and high value business located next to hollow and shrunken nineteenth century towns. In each situation, local agendas have developed to try to respond to the national and global economic and cultural forces, in order to re-invent some kind of role and identity for themselves and for the future.

The methods used in the Manchester case study were, as for all of them, based on the PLUREL Joint Analytic Framework (as described in the introduction to Part 2 of this book) adjusted to take account of the UK situation. This centred on a series of 25 interviews and 5 workshops with policy-makers and stakeholders, including

J. Ravetz (✉)
Centre for Urban & Regional Ecology, University of Manchester, UK
e-mail: joe.ravetz@manchester.ac.uk

P. Warhurst CBE
FRSAM, Chair of Pennine Prospects Ltd, Chair of the Forestry Commission Great Britain

K. Nilsson et al. (eds.), *Peri-urban futures: Scenarios and models for land use change in Europe*, DOI 10.1007/978-3-642-30529-0_8,

spatial planners, economic development and environmental officers, utilities, property, house building and landowners associations, together with representatives in each of the two focus areas within the case study. This was combined with a desk study of policy analysis and assessment: and a parallel 'futures' strand of participative scenarios, with land use modelling and economic analysis, in a 'policy-scenario-testing' method.

In this chapter we aim to disentangle the complexity of this historic conurbation. We look at the general pattern and history of spatial development across the Manchester City-region. Then we explore the structural dynamics of change, using the multi-step framework outlined in Chap. 1, which looks at problems and opportunities in three types of transition:

- 'Metro-politization': an urbanising transition, occurring across wider peri-urban and rural areas, drawing on the work of Soja (2000);
- 'Cultural capitalism': a globalising transition—new patterns of networked economic and social structures and activities, as outlined by Scott (2000, 2006);
- 'Spatial ecology': a localising and green infrastructure transition, with new patterns and identities in places and communities (Douglas and Ravetz 2011).

After this, we focus on two selected sub-regional areas with examples of peri-urban partnerships, in the South Pennines and the Mersey Belt areas. Then we look at selected spatial, environmental and economic planning strategies and policies: Green Belt policy, green-blue infrastructure, and local development. Following this we turn towards the future with alternative scenarios, including land use modelling and participative debate on the future 'problem space' and 'opportunity space'. Finally, we explore new concepts for sustainable peri-urban development in the light of major changes in spatial planning and development policy, as proposed by the UK Coalition government in 2010–2011, (which started to be implemented during the final stages of the research). The conclusion and recommendations section reviews the lessons from this, and asks—where next?

## 8.2 The Manchester City-Region

### *8.2.1 From Industrial Decline to Post-Industrial Re-Invention*

There is no simple definition of the Manchester city-region (MCR), or of its peri-urban hinterland; in fact the term 'peri-urban' is not often used in the UK, and other terms are more common, such as 'urban fringe', 'rural–urban fringe', or 'countryside around towns'. The many layers of the MCR can be summarised (Fig. 8.1) as:

- The City of Manchester, the municipality at the centre of the conurbation (population 437,000).
- Greater Manchester is the name of the former County, including the 10 Districts (i.e. municipalities) of Manchester, Trafford, Salford, Wigan, Bolton, Bury,

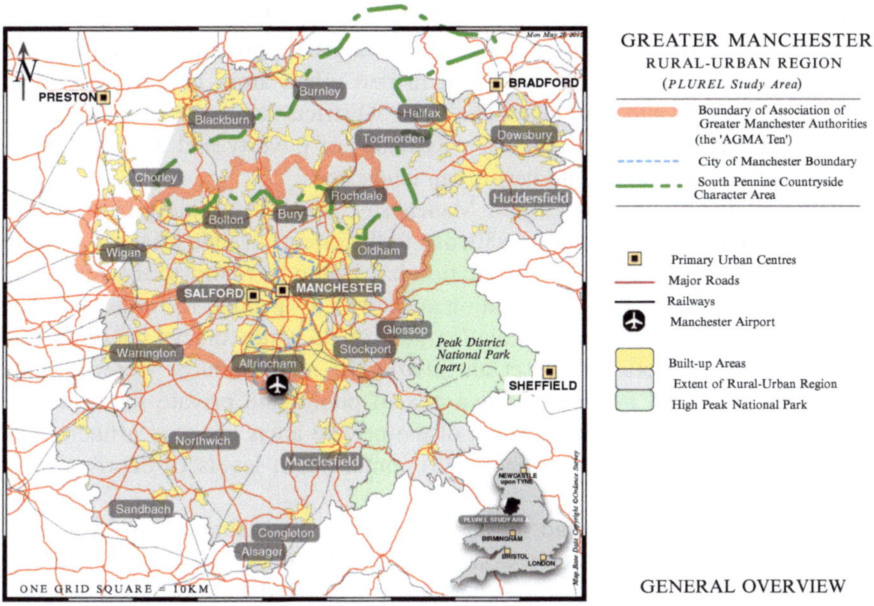


**Fig. 8.1** Manchester city-region; general overview

Rochdale, Oldham, Tameside and Stockport. With a total population of 2.5 million, these cooperate through the Association of Greater Manchester Authorities (AGMA), and from 2011 through its freestanding 'Combined Authority', and also through an independent Local Enterprise Partnership.

- Greater Manchester Economic Development area, an informal grouping which includes six more Districts in the southern commuter hinterland;
- The PLUREL case study boundary extended the above by a further six Districts to the east and north, to include a wider range of peri-urban types: this covers roughly a 1-h commuting radius, with a total population of over 4 million.
- At larger scales there is the sub-regional agglomeration of Greater Manchester/ Merseyside. A looser inter-regional agglomeration of the Trans-Pennine 'super-city' is also sometimes used, which until 2010 overlapped with the 'Northern Way' policy concept which linked all the urban areas from east to west coasts.
- Meanwhile, the North-west region of England (NUTS 2 level) was abolished as a governance and planning unit by the Coalition government of 2010, with impacts yet to be fully realized on the capacity for strategic planning and regional investment.

Greater Manchester grew rapidly as the world centre of the cotton textile industry during the eighteenth and nineteenth centuries at the height of the British Empire, helped by the construction in 1830 of the world's first commercial railway between Liverpool and Manchester, and in 1894 the Manchester ship canal, which enabled a global trade in textiles and other products. The MCR now has a diverse economy of over £35 billion GDP, with the largest regional cluster of finance, law, media, creative industries and higher education in the UK. It also contains some of the worst unemployment, pollution, crime, social deprivation and inadequate housing, both in the central areas, and in peri-urban areas alongside sprawling suburbs of wealth and privilege. The outer Districts are each based on former industrial towns which likewise show extremes of poverty and affluence, with a complex mixture of urban and peri-urban landscapes.

At the region's core, the City of Manchester grew very rapidly between 1750 and1900 and then declined after 1950, due to the effects of industrial restructuring. More recently, since 1990, the population has begun to return to the regenerated city centre and some other regeneration areas which surround it; further out, some neighbourhoods have stabilised and/or gentrified, while others continue to be fragmented and chaotic (Ward et al. 2010). The outer suburbs were developed mainly in the period 1920–1980, as a mix of public and private sector housing, generally with lower densities. Some of the private areas are very affluent, while many 'peripheral' public housing estates contain high levels of deprivation and social exclusion.

Urban expansion was rapid throughout the twentieth century. Even when the urban economy and population was shrinking, many people chose to relocate to the suburbs or peri-urban communities, leaving a perforated and obsolete inner urban structure in many areas, similar to what has been seen in many other cities. Many peri-urban settlements also experienced industrial decline, and a rapid transformation from being productive working towns to service-based and/or commuting towns.

The main urban areas are surrounded to the north and east by the hills of the South Pennines, the site of former industrial activity, and to the south by the arable farmland of the Cheshire Plain. To the west, the 'Mersey Belt' between Manchester and Liverpool contains a complex mosaic of peri-urban villages, small towns, new commuting settlements, peripheral public housing estates, scattered small settlements interspersed with farmland, transport infrastructure, waste tips, river valleys, new business or shopping parks, and former industrial areas (Handley and Wood 1998). Generally, much of the legacy of industrial pollution has been cleaned up, natural areas have been (mostly) conserved, and the heavily contaminated river valleys have been reclaimed and rehabilitated as country parks (Nicholson-Lord 1987). However there are many areas which have been recently affected by urban infrastructure and commercial development. For example, the M60 orbital motorway, completed in 2003, cuts through many of the remaining green areas and river valleys which surround the central conurbation (Hyde et al. 2004). In terms of land use, livestock farming in the South Pennines has declined, while arable farming in Cheshire is booming, within a wide range of landscape types (Blair 1987) such as

- Disturbed landscapes:—various impacts from minerals, waste and industry: as seen in much of the MCR peri-urban area;
- Neglected landscapes: low intensity marginal farming, much of it on the lower parts of the South Pennine hills;
- Industrial agriculture: high intensity mono-cultures, in the Mersey Belt and lowland Cheshire farming landscapes;

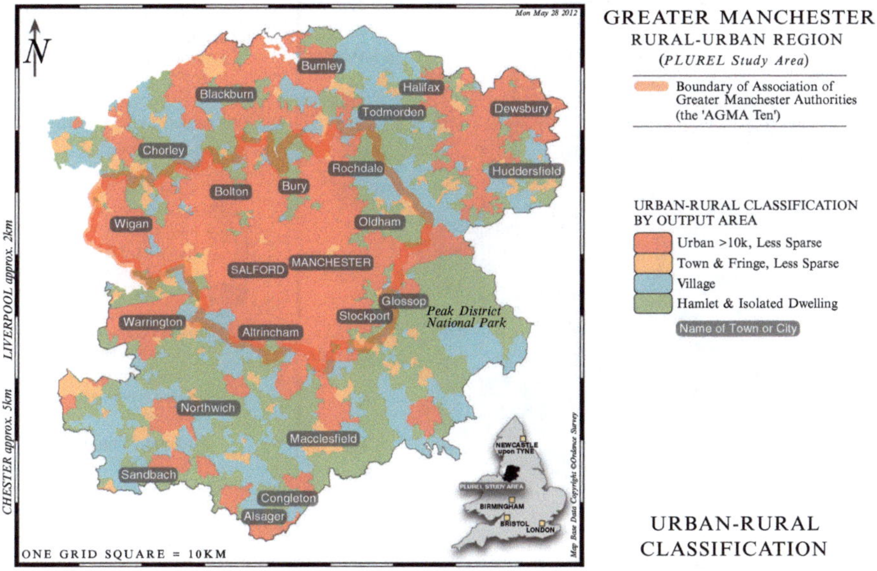


**Fig. 8.2** Manchester city-region: urban—rural area classification

- Traditional agriculture: mixed farming with diverse landscapes, seen in the more hilly parts of Cheshire;
- Amenity landscapes: woodland, country parks, large estates, open moorland; a large proportion of the peri-urban area is in some kind of amenity use but this often overlaps with other land use types.

The landscape type probably most directly at risk of peri-urbanisation is 'traditional agriculture', which is covered by 'Less Favoured Area' status (mostly in the South Pennines), 'Farm Stewardship schemes', and other support mechanisms. The concept of disturbed landscape also applies to visual and psychological qualities: 'tranquil' areas are those which are more than 3 km from major roads or urban areas and other sources of noise or light pollution, and they cover less than 3% of the Greater Manchester area (CPRE 2007). The 'urban-rural area' is the official classification of small area types, which straddle the urban boundary (Bibby and Shepherd 2004). The mapping of the MCR shows large areas of 'less sparse', i.e. peri-urban mixed settlement patterns (Fig. 8.2).

### *8.2.2 Peri-Urban Economic and Social Structure*

The results of the dynamic processes of the peri-urban seem to concentrate relatively affluent and relatively deprived areas into different localities around the

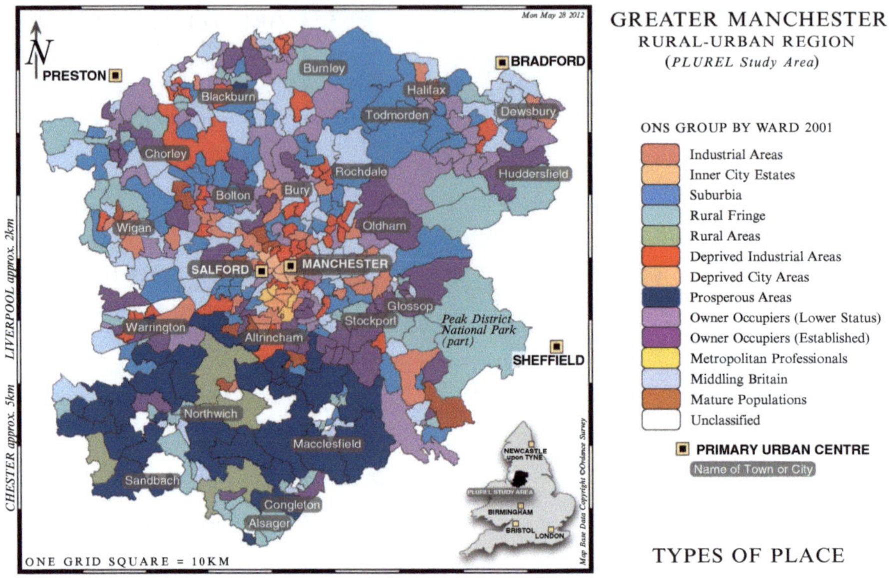


**Fig. 8.3** Manchester city-region: social area types

MCR: some are highly self-contained while others are dominated by commuting. We can observe several processes, involving the mobility of incomers and existing residents (GMCVO 2007):

- Existing wealthy residents of peri-urban/rural areas are often descendants of originally landed or semi-landed aristocracy, merchants or industrialists. Over time their assets in land, buildings and capital have consolidated. Some nouveau riche incomers have supplanted the original landowning families.
- Incoming affluent households are attracted to high quality landscapes and can afford to commute, or else have high mobility, knowledge-based occupations.
- Existing low income households may find changes in the rural economy and more limited employment opportunities destructive to their livelihoods; insecure housing tenure or high housing costs adds to economic pressure.
- Low income households were in many cases relocated, as part of spatial planning policy, from the inner city slums to peripheral public estates. They often found themselves isolated from employment and services, and unable to afford private transport.

The resulting pattern is shown in the mapping of 'social area types' (Fig. 8.3). This can be overlaid with the analysis of commuting distance, with a clear commuting flow visible particularly on the wealthier southern side of the peri-urban MCR (where more than 20 % of residents commute long distances, with serious congestion on local roads as a result) (RTPI 2006).

### 8.2.3 Dynamics of Change and Transition

The dynamics of peri-urban change work at several degrees or levels of complexity. Using the framework set out in Chap. 1 we can identify the direct patterns and driving forces of urban expansion alongside the underlying structural effects and system-wide transitions.

- Firstly, urban expansion is a direct result of population and economic growth: in the MCR, population growth took place in phases starting in the mid-eighteenth century, and then levelled off around 1950, with decline in many inner urban areas. Economic development then resulted in further expansion, in the shape of increased floor space per person (for housing, workspace and services).
- Secondly there is the regional agglomeration effect: here, the merging of Manchester, Liverpool and many satellite towns into an extended conurbation began in the nineteenth century and has been shaping the peri-urban area ever since, with the most recent stage being the completion of the orbital motorway. Since the 1960s, the Green Belt and related policies have generally succeeded in physically containing urban development. However the economic and social agglomeration process has continued, aided by expansion of transport systems, labour markets, and retail and services catchment areas.
- The third aspect is the structural dynamics of power and political ideology, which can be seen shaping the peri-urban space; and the fourth aspect of system-wide transitions and resilience effects operate across the MCR. Both these are set out below. A fifth aspect of policy and strategy is explored in the next section.

The combination of aspects three, four and five are shown in the key diagram below, which is based on the MCR (Fig. 8.4).

### 8.2.4 The Dynamics of Structural Change

The dynamic of structural change can be seen most directly through conflict and controversy. The MCR was the archetype for study of the contradictions of capitalism (Engels 1845); it was also one of the birthplaces of the trade union and the Cooperative movements. In the 1930s, there were conflicts in the region over access to open land in private ownership (symbolised by mass trespasses on the upland area of Kinder Scout in Derbyshire Shoard 1983; Fairlie 1996). Recently the Manchester airport expansion provoked demonstrations and occupation of the site by environmental activists (Ravetz 1999). The Trafford Centre shopping mall (Kitchen 1997), road building and business parks in the Green Belt (Hyde et al. 2004), and the Greater Manchester Congestion Charge referendum were also controversial issues (Sherriff 2012). Each of these cases reveals underlying tensions in the dominant 'discourse' of neo-liberal attitudes to urban development, which is then manifested in the spatial patterns of social and economic differences

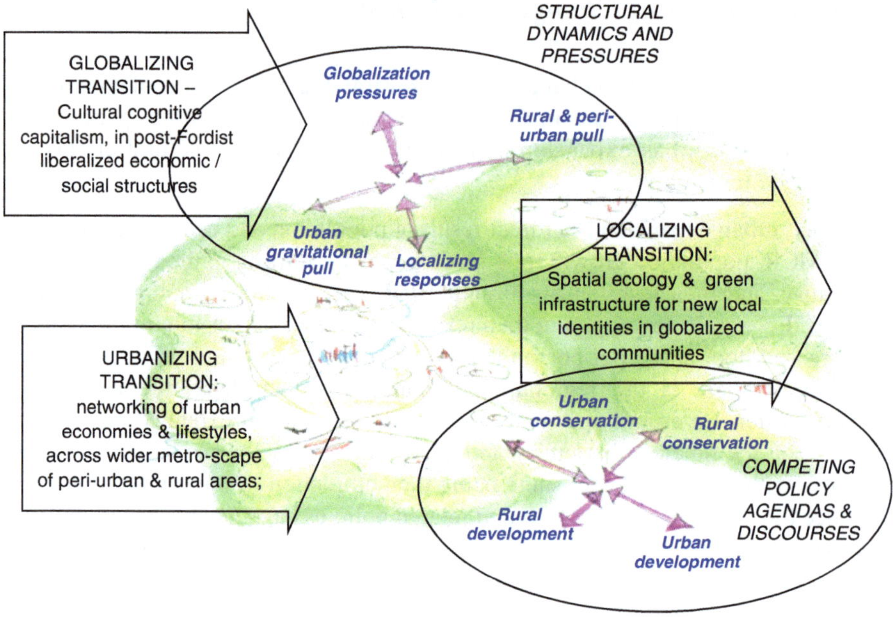


**Fig. 8.4** Structural dynamics and transitions in the Manchester city-region

(Hajer 2003; Brand and Thomas 2005). We can summarize these patterns of conflict and competition, seen through the perspective of 'urban political ecology' (Kaika 2005; Heynen et al. 2005), such as global versus local agendas, economic innovation versus social values, competition between economic or social groups and social innovation versus cultural values. Applied to the MCR case, drawing on secondary analysis of interviews and workshops, each of these are forces which shape the structure of the peri-urban city-region (Table 8.1).

### *8.2.5 Transitions and Feedback Effects*

The concept of 'transitions' is essential in understanding peri-urban change and peri-urban policy. From discussions with stakeholders it was possible to define three types of transition in the MCR: 'metropolitan', 'globalizing' and 'localizing'. Each can be studied in various ways, for instance, the 'systems thinking' approach (Rauws and de Roo 2011), the 'community resilience' approach (Berkes et al. 2003) or the 'policy innovation' approach as a transition response (Randles and Green 2006).

These transitions, and the policy responses which they generate, raise very practical questions which are debated by policy-makers and communities in the MCR, such as what or who is the peri-urban landscape for? How to find a balance

**Table 8.1** Structural dynamics and conflicts in the peri-urban

| Type of dynamic | General effects in the peri-urban | Effects in the Manchester city-region peri-urban |
|---|---|---|
| Global versus local tension and conflict | Structural 'metropolitan transition' with globalizing/localizing tensions (Soja 2000) | Policy discourse of the global competitive city-region: expanding labour and retail markets, against small local economies and SMEs, across the whole peri-urban MCR expanding transport infrastructure, centred on Manchester airport and motorways. |
| Innovation and conflict between economic groups | Peri-urban economies as creative destruction of obsolescent economies and communities | Declining town centres with local shops and services, centralized by efficiency criteria, many peripheral locations.<br>Shift of producer and consumer activity to business parks, science parks, retail parks, centred on MCR motorway network. |
| Economic accumulation and social impacts | Peri-urban land and development as capital accumulation, in the circuit of urban property investment (Harvey 1995) | Legacy of former industry and mining which continues in 'low value' locations and housing markets:<br>New outer suburbs for commuters/ retirement, with high and stable property values: property values increases from planning permission for strategic landholdings, aided by land speculation and market 'lotting' tactics. |
| Conflict between social groups | Peri-urban land use questions, on the frontier of class competition for territory, security and identity (Shoard 1983) | Historic and aristocratic country estates which dominate the lowland farmland landscapes of MCR:<br>Social housing on peripheral estates without services;<br>New high value commuter developments with security gates and fences:<br>NIMBY defence of green belt against incoming social housing. |
| Innovation on the social-ecological axis | Peri-urban community initiatives as new social movements, socio-ecological enterprise etc. (Hoggart 2005) | Urban fringe partnerships for green infrastructure, community forests: integrated catchment management on all river valleys: heritage/ cultural/eco-tourism in higher quality landscapes. |
| Innovation on the social-economy axis | Peri-urban community initiatives as business strategies with socio-cultural enterprise etc. | Town and village partnerships, events, niche branding, heritage/cultural tourism: many peri-urban locations:<br>New models for local planning/local services/social housing. |

(continued)

**Table 8.1** (continued)

| Type of dynamic | General effects in the peri-urban | Effects in the Manchester city-region peri-urban |
|---|---|---|
| Conflict between urban versus rural | Peri-urban land use as frontier of dominant urban power and wealth, over rural interests | Rural landscape dominated by road interchanges, energy, water and other infrastructure: many parts of peri-urban MCR:<br>Farmers and SMEs constrained by green belt and similar policy. |

between private profit and the public good in managing land and ecosystems? Large parts of the MCR are no longer competitive in their former original agricultural or industrial uses. New local residents may be commuters or knowledge workers, the semi-retired or retired, with little direct involvement in the local economy. Thus the identity and 'reason for being' of the place and its communities is increasingly an open question. In any one location there may be competition between many agendas, including the economic, ecological, historical, residential, or functional (Gallent 2006).

In the MCR, most direct urban expansion is constrained by the Green Belt, related landscape designations, and the general planning policy of 'presumption against development on open land'. The result is that change is directed into a wider peri-urban metamorphosis—the ***metropolitan transition*** of formerly peri-urban or rural communities. This is a transition towards urban-centred and urban-networked economic activities, social types, cultural attitudes, and spatial patterns of work and lifestyle (Soja 2000; Duany et al. 2000). The effects are seen in the peri-urban MCR, such as the gentrification of the former industrial towns and rural villages of the South Pennines. They are also seen in other areas, as a transition away from locally-based jobs and services, towards suburban development with urban-centred jobs and services, made possible by modern transport and communications infrastructure.

The metropolitan transition takes place in parallel with a wider globalizing transition, towards what some have called a state of ***cognitive capitalism.*** This is based not only on new economic functions but on a knowledge- and innovation-based global order, with changes in social behaviour, cultural patterns and public attitudes (Scott 2000, 2006). This has much in common with the creative cities idea, and also draws on the theory of 'creative classes' (Florida 2004; Ravetz 2011b). In the MCR, discussions with the Manchester Knowledge Capital agency showed how the 'creative city' theme, closely related to the 'Ideopolis' or knowledge city, could be 'spatialized' across different parts of the peri-urban landscape (Lee 2007). Stakeholders pointed to conflicts between providing for entrepreneurs or cultural consumers—through constructing science, retail or leisure parks—and providing space for other kinds of creative action. These cover both landscape-based schemes (sculpture parks and community arts projects) and town or community-based schemes (heritage trails, cultural festivals, local food and farmers' markets).

Thus, the creative agenda works not only with a globalization agenda, but also a new kind of ***localization*** which can be seen in changing preferences for places and landscapes; in the MCR as elsewhere, surveys find a growing public desire for small, safe communities with good public services and plenty of green space. This is linked to the trend of out-migration and counter-urbanization, which has led to the re-population of the former industrial peri-urban MCR, at the same time as de-populating parts of the inner urban areas. Alongside these 'creative' local schemes there is a wider concern for the enlargement and protection of the peri-urban ***green-blue infrastructure*** (GBI), of open spaces, green corridors, urban woodlands, waterways, walking, cycling and horse trails, as well as existing nature conservation sites (Ravetz 2011a; Benedict and McMahon 2001). Such an infrastructure may have social as well as environmental objectives, for example being aimed at providing a neutral and commonly accessible territory, where a diversity of social groups can enjoy nature or exercise in fresh air. The objective is more to enable social cohesion as different social groups identify with and locate themselves in a common landscape. Again there are conflicts both visible and under the surface: GBI is argued by some to be used mainly by white middle class groups and to exclude other social and ethnic groups, a topical issue in research and policy (Natural England 2009: CABE 2010). Finally, GBI is not only a local issue but strategic, as for instance the Mersey River and its valley which forms a major green corridor and ecosystem management question, right across the region (Wray and McPherson 2006).

## 8.3 Sub-Regional Case Study Areas: Creative Enterprise and Diversity

Within the size and diversity of the MCR, there is no one policy or governance unit responsible for its peri-urban areas. So, as in other case study chapters, two sub-regional areas were selected to permit closer working with stakeholders, and a more in-depth review of policy. These were the South Pennine hills in the eastern and northern parts, and the Mersey Belt and community forest area in the western part of the MCR. In each case there are active and forward looking peri-urban initiatives, led by partnership organizations, and involving networks and partnerships linking the public, private and civil sectors. Some of the wide range of places and types is shown in the sample pictures (Fig. 8.5).

### *8.3.1 South Pennines: New Forms of Urban-Rural Linkage*

In between three major conurbations in the north of England there is a 'green heart', a unique mixed town-and-country landscape set among hills rising 500–600 m

**Fig. 8.5** Geographic types in the Manchester city-region: (**a**) Manchester ship canal; (**b**) green belt north of Manchester; (**c**) green infrastructure in the Mersey Belt; (**d**) local development in Todmorden

above sea level, and with steep sided valleys containing small scale livestock farming, a rich industrial heritage, and both traditional and new 'lifestyle' communities. The South Pennines has a population of around 450,000 and encompasses the countryside fringes around Oldham, Burnley, Pendle, Rochdale and Rossendale, from the wider MCR and Lancashire (to the west of the Pennines): and Bradford, Calderdale, Kirklees and Craven in West Yorkshire (to the east of the Pennines). It is a decentralized geographical type, with no single centre or corridor, and with open moorland over the hill tops and between the east–west transport routes.

The South Pennine agenda centres on the role of landscape character, history and heritage in determining a new future. While on the surface the upland landscapes might appear to be long established and 'timeless', there have been massive changes over the last half century. Some Pennine towns lost half their population, and then since the low point in the 1960–1970s, some have regained it. The once dominant local textile industry almost completely disappeared, and is now being replaced with a more diverse mix of economic activity. Household car ownership rates doubled and then doubled again. Production from the land, from farming, forestry or mineral extraction, continues to decline while tourism is now the largest land-based industry, though still fragmented and ad-hoc compared to other areas or National Parks with larger-scale attractions, such as the Yorkshire Dales or Lake District (Pennine Prospects 2008).

Partnership working is seen as the key to harnessing the area's rich natural, cultural and built heritage: such as the world's first retail cooperative, set up in 1844 in Rochdale, and the first industrial cooperative in 1848 in Hebden Bridge. 'Pennine Prospects' (the short name for the Southern Pennine Rural Regeneration Company) is a non-profit partnership agency with a membership of local authorities, government bodies and private and voluntary sector bodies (www.pennineprospects.co.uk). Pennine Prospects works to raise the profile of the South Pennines. It aims to help regeneration by promoting natural assets and heritage, supporting the development of a sustainable local economy, protecting and enhancing the character of the area and improving the South Pennine environment and infrastructure. In 2011 it had a wide range of projects and wider partnerships, including the management of a £4 million EU-funded LEADER programme.

Some stakeholders perceive that the South Pennines is a peripheral area at the 'back of beyond', a lower priority for urban-centred local government and governance. An alternative view is that, on the contrary, it is also an essential resource, providing space and ecosystem services to the surrounding conurbations with a total population of over 6 million. This is a good example of some of the 'linkages' noted in Chap. 1, including, urban to rural, urban to peri-urban, peri-urban to rural, social to economic, economic to environmental, etc. Such linkages then provide opportunities which can be realized in creative projects led or enabled by Pennine Prospects. Examples include 'Watershed Landscapes'—which rehabilitates the moorland plateau peat bogs for multi-functional uses: and the programmes for increasing woodland cover and woodland management across the landscape.

### 8.3.2 *Mersey Belt: Community Forests in a Peri-Urban Landscape*

On the western side of the MCR, stretching from Manchester to Liverpool is an extended area of 'mixed town-and-country' landscape (Breheny and Rookwood 1993). In the land use modelling of MCR, this shows up as 'low density discontinuous urban form', but on the ground it is experienced as a diverse urban-ecological pattern. With a legacy of 250 years of industrial revolution and restructuring, this area includes former mining and landfill sites, housing of various kinds, old and new types of industry, urban infrastructure for energy and waste, major roads and freight interchanges, and business and retail parks. The area has a unique history, including the first passenger railway in the world, the largest glass factory at St Helens, and the first industrial estate at Trafford Park (Nicholas and McWilliam 1962; Lloyd 1980).

Between the different elements of the urban patchwork, there are a few remaining areas of lowland peat bog, in a mosaic with ecologically important wetlands and woodland; sometimes the woodland is secondary, often planted over mining spoil and landfill sites. After 40 years of reclamation most of the worst land contamination and dereliction has been greened and made safe and usable. This then raises more complex questions, such as how to combine urban/economic development with ecology and landscape? Green Infrastructure, or 'Green-Blue Infrastructure' (GBI), is seen as the way forward, not only for its ecological value but also to yield economic and social benefits, such as green settings for new business parks and housing, or through networks of accessible linked recreation areas.

This is the background to the community forest. One of many around the UK and part of a specific national programme, the Red Rose Forest was set up in 1994 as a freestanding agency to cover the western part of the MCR (the red rose being the symbol of the mediaeval Lancastrian royal dynasty) (www.redroseforest.co.uk). On the eastern side the Pennine Edge Forest has similar aims though a different pedigree, set up from within the local authorities. Both followed earlier work by the Groundwork Trusts and Countryside Commission on the 'Urban Fringe Experiment' (Handley and Wood 1998), and also benefit from inputs such as grant funding from Forestry Commission England. The Forests' main aim is to promote partnership schemes as a means of land reclamation and beneficial re-use, environmental improvements and community woodland. Along with its neighbour the Mersey Forest, the Red Rose Forest provides an example of how partnership working can bring the public, private and community sectors together to rehabilitate and consolidate various industrial and post-industrial landscapes which were often damaged and fragmented (Wood and Ravetz 2000).

## 8.4 City-Region Governance and Planning: Multi-Level Management of Change

### *8.4.1 Territorial Governance and Spatial Planning in England*

In the UK's efforts to manage a complex geography, there are many questions on governance. For instance, should the lead be taken at the regional, city-regional or local level? How should spatial planning be linked to economic and environmental policy? How can plans made by the public sector be implemented, when the resources to do so are in the private sector? For peri-urban areas there are extra challenges, partly because in the UK the words are hardly used, and the theme is hardly recognized yet; there is a sub-topic referred to as 'urban fringe', but otherwise policy tends to be divided into 'urban' and 'rural', with few overlaps.

The situation in the MCR case study, and its comparison with the other PLUREL regions, changed towards the end of the project with a new Coalition government in

2010. At the time of writing the whole planning and governance system in England is in a state of flux, with debate focused on the National Planning Policy Framework (NPPF) (CLG 2011a). This aims for a more pro-development, deregulated, 'new localism' approach, which is expected to enable decisions in the common interest, with fewer structures and top-down targets. When (or if) implemented, this may change many of the existing policies and structures to greater or lesser degrees:

- National level: the forthcoming NPPF is likely to replace the current set of national Planning Policy Statements and the legal framework of the Planning and Compensation Act 2004. However, neither of these have any explicit spatial dimension: in the absence of an official 'national spatial strategy', various organizations have produced their own demonstrations (Wong et al. 2000; TCPA 2006).
- Regional level: until 2010 each region (at NUTS 2 level) produced a 'Regional Spatial Strategy', and 'Integrated Regional Strategy'. This system is now history because regional administrations have been abolished, but there are still many lessons on how higher level strategies and strategic investment sites can be managed (Wray 2011):
- Sub-regional or city-region level: spatial strategic planning for conurbations such as MCR is expected to come from the proposed 'duty to cooperate' between local authorities;
- Local authority level: each local authority will continue more or less the current system of a spatial 'Local Development Framework', which is linked to its corporate 'Community Strategy' overseen by the Local Strategic Partnership of stakeholders, and includes a Local Area Agreement for services and investments.
- The new NPPF also proposes a neighbourhood planning level, which could be led by businesses, landowners or the community. This would have power to increase the rate or amount of development beyond the levels stated in the local authority plan, but not to reduce it.

The new system is likely to have major effects on peri-urban areas, but there appears to be no analysis, at the time of writing, of what these might be (some issues are explored in the Conclusion to this chapter). Although the NPPF promises to maintain Green Belts and National Parks, other local designations would be downgraded, and there is a general presumption in favour of 'sustainable development', which in the view of the government appears to be almost any development.

### 8.4.2 Governance Trends and Prospects for the Manchester City-Region

The MCR is a clearly visible geographic unit, and a ready test case for various options in governance. The Greater Manchester County of 10 Districts (i.e. municipalities) operated from 1974 to 1986 and, following its dissolution, the

10 Districts have coordinated services such as waste, airports and emergency services through the voluntary Association of Greater Manchester Authorities, and the newly launched freestanding Greater Manchester Combined Authority (Roberts 1998). In 2010 the new Coalition government decided to abolish all regional-level strategies (spatial, economic, social and infrastructure) and shift all decision-making to the local level. It is not yet clear how the Greater Manchester conurbation of 10 authorities, or the wider MCR of 16 or even 22 authorities, can work with or without either formal or informal coordination, what resources might be available, and how conflicts and disputes could be managed.

At the time of writing a new 'Local Economic Partnership' has been approved, led by the private sector but including the 10 local authorities. There is also an active policy/partnership network which includes New Economy Manchester (formerly the Chamber of Commerce and Industry), various Commissions on Environment, Transport or Housing and the production of a Greater Manchester Strategy. There are few details yet of how this will work, obtain funding, or coordinate with policies and activities in the surrounding areas.

Economic development policy has been dominated in the past by EU funding, since most of the MCR was an Objective 2 area. There is now a legacy of 'a thousand flowers', including science-based innovation, digital industries, cultural industries, a heritage-based tourism and visitor economy and local social enterprises. Environmental projects form a parallel strand: while funding was cut in 2010 from most climate change programmes, there is continuing activity on local environment and Green-Blue Infrastructure schemes.

### 8.4.3 The Policy Agenda: How to Define the Problem?

Behind formal governance and policy objectives lie a series of wider questions, such as: what are the problems which policy aims to solve? Who decides, and with what resources? In the peri-urban situation this is not a simple question: for each kind of problem there are arguments over who is responsible, for what area, and over which strategic agendas, such as urban or rural, local or regional. And at a time when much former industrial pollution has been cleaned and reclaimed, or outsourced at a global scale, surface level affluence can hide deep social and economic divides (Roberts et al. 2009). Tangible and visible problems, as identified through desk studies and stakeholder consultation, include:

- The generally diffuse structure of urban settlements, communities and local services;
- Fragmentation of GBI, decline of habitats, and poor quality/inaccessible landscapes with lack of investment;
- Traffic congestion due to commuting in peri-urban areas, contributing to pollution;
- Local housing and public services in decline and/or inaccessible to local people;

- Rapid social and economic change and fragmentation in local (rural) communities.

Such problems can be exacerbated by economic growth and development, suggesting that the challenge is more about distributions within and between social groups. Such problems are less visible, but have been identified in studies on the 'rural' (i.e. peri-urban) parts of Greater Manchester (GMCVO 2007; New Economy Manchester 2008).

Affluence and environmental improvements introduces 'gentrification' in many peri-urban locations, which, together with a lack of new build housing, causes increases in house prices. This then excludes lower income groups from the housing market, and it also tends to hide poverty and exclusion within surrounding affluence. Government socio-economic data shows that 25 % of 'rural' households are deprived either on poverty, unemployment or housing criteria: and such deprivation is exacerbated by lack of access to services such as banks, shops, health and education facilities.

### *8.4.4 Spatializing the Policy Agenda*

Spatial policy is based not only on the definition of the 'problem spaces' as above but also the 'solution spaces' (i.e. what could and should be done). These are then related to physical areas or locations on the ground. Competing spatial issues include urban versus rural and development versus conservation, combining or conflicting in the peri-urban area, each offering its own version of 'sustainability' (Ravetz 2000; CURE 2003). In the MCR such issues include:

- Urban containment: using the peri-urban as a boundary to urban expansion. In MCR the Green Belt surrounds the main conurbation and many smaller towns, although not the Pennine hills. There is often fierce debate about the effect of Green Belt policy on businesses in sectors such as leisure or tourism which can be constrained by the restrictions on development in the Green Belt.
- Urban development and expansion: this aims to supply the city with roads, airports, business and retail parks. For MCR the motorways form the primary network, with Manchester Airport as the primary hub forming a so-called 'aerotropolis' (Kasarda and Lindsay 2011).
- Rural conservation: in the densely populated MCR the countryside is seen as an asset for the urban population, where landscape and ecology attracts leisure and tourism in areas such as the Pennine hills and in the community forest area.
- Rural development and enterprise: with a focus on small and medium-sized enterprises (SMEs) and local economic development. In the MCR this can create conflict with rural landscape conservation, compete for space with large urban development, and undermine the urban containment goals.

These spatial aspects should in principle fit together as part of a far-sighted spatial strategy and integrated policy framework. In practice there are many questions and conflicts. Below, we assess three types of strategy in the MCR, looking at actors and stakeholders, resources and capacities, rules and institutions: their discourses and cultures, policies and governance.

## 8.5 Strategy and Policy Assessment

The key question for policies and strategies, as discussed with stakeholders—is: do they work? To explore this needs a view beyond the stated objectives within the stated boundary. Much of the policy process in MCR seems even to the experts and policy-makers who were interviewed, to be very complex, fuzzy, opaque,, and relying on partnerships for peri-urban issues, which are not often recognized as such. The five main assessment questions from the PLUREL Joint Analytic Framework raised general issues through interviews and workshop discussions:

- *Will the policy be robust against future changes*? In the MCR it was often difficult to answer this, in a situation of rapid change all around. However, policies such as the Green Belt are based on a longer term perspective, where part of their success is through longevity.
- *Does the policy integrate with other policies*? The MCR shows many examples where projects and partnerships are set up in order to integrate between policies which are otherwise fragmented. Both the South Pennines and Red Rose Forest areas are examples of 'integrating' organizations which bring together other policies and programmes.
- *Is there coordination between public, private and civic sectors?* Similarly, there are efforts towards coordination, as seen in both the above.
- *What are the external effects and who are the winners/losers*? In the MCR this raises some of the 'structural' questions as summarized above. For example the Green Belt is seen by some stakeholders as widening the gaps between urban versus rural, development versus conservation, rich versus poor and commuters versus local residents.
- *Does the policy encourage innovators and entrepreneurs?* A very topical question for the MCR, and the starting point for the Coalition government's reforms of the planning system, which was subsequently labelled 'the enemy of enterprise' by the Prime Minister (Cameron 2011).

The summary here shows the general themes (actors, resources, rules, discourses and strategies), mapped out for the most topical sectors discussed in the MCR case: housing, transport, tourism, agriculture, landscape and water (Table 8.2). The table also illustrates different and parallel types of knowledge, from 'hard' technical data to 'soft' strategies and discourses. The research team used this to explore three main strategies, in the sub-region case studies and in the MCR as a whole: the Green Belt, Green-Blue Infrastructure and local economic development.

**Table 8.2** Summary of key policy agendas and strategies

| | Technical data on trends and patterns | Actors and stakeholders | Resources (financial, physical) | 'Rules' and key system drivers | 'Discourses' and underlying tensions | 'Strategies'/ policies/agendas |
|---|---|---|---|---|---|---|
| *Policy agendas* | | | | | | |
| Housing | Household growth: | Private housebuilders | Land with permissions | Housing tax and finance system | "Not in my back yard" | Urban regeneration policy |
| | Housing density | Social housing providers | Private housing stock | Social security benefits system | "Not in my term of office" | Rural housing policy |
| | New housing: % of stock. | Private landlords | Social housing stock | Law on landlords etc. | "Build absolutely nothing anywhere near anyone" | Urban fringe policy |
| | Peri-urban versus urban housing | Finance companies | | | | Green belt policy |
| | | Housing estate agents | | | | |
| Transport | Private transport growth % per year | Private drivers | Public investment | Price and tax structure of transport policy | "Freedom to drive" | Regional and local transport strategy |
| | Commercial road traffic % per year | Commercial firms | Transport corridors and interchanges | Integrated transport challenges. | "Buses are for losers" | Public transport finance |
| | Air travel % per year | Public transport firms | Parking and access | | | |
| | | Public transport agencies | | | | |
| | | Walkers/cyclists | | | | |
| Tourism and Leisure | Countryside visitor trends: | Private tourism firms | Tourism/leisure sites and facilities | Personal mobility and leisure dynamics | "Whose land is this land?" | Regional tourism strategies |
| | | Tourists | Tourism/leisure landscapes | Overloading of 'honeypots' | "The tourism curse" | Countryside and access policy |

(continued)

**Table 8.2** (continued)

| | Technical data on trends and patterns | Actors and stakeholders | Resources (financial, physical) | 'Rules' and key system drivers | 'Discourses' and underlying tensions | 'Strategies'/ policies/agendas |
|---|---|---|---|---|---|---|
| | | Residents of leisure | | Competition with overseas travel | "They're not from round here" | Peri-urban partnership |
| | | Community/civil society groups | | | | |
| Agriculture | Agriculture output | Farmers | High quality farmland | Industrialization of food supply chains and quality standards | 'Eat the view' | Sustainable food and farming strategy |
| | Regional productivity | Agri-business | Soil, water, etc. | Cheap food imports | Farmers—stewards or producers? | Cap reform and stewardship schemes |
| | | Hobby farmers<br>Manufacturers/ retailers | Farming infrastructure | 'Horsiculture' and land abandonment | "Our kids are fat" | Urban food resilience |
| | | consumers | | | | |
| Ecology and landscape | Biodiversity quality index | Ecologists/nature lovers | Protected sites/ landscapes | Ecological connectivity | Suburban gardens as eco-habitats | Biodiversity strategy |
| | Site protection trends | Countryside managers | Social/cultural commitments | Inter-species co-existence and multi-functional landuse. | | Special landscape areas |
| | | Regulators/ planners | | | | |
| Water management | Water resources | Water utilities | Water catchment land | Finance and asset structure of utilities. | 'Dilute and disperse' approach to pollution control. | Integrated catchment management |
| | Flood incidence | Water consumers | Water infrastructure | Institutional gaps in drainage and flood management. | 'P-ing into the wind' | Water stewardship programs. |
| | Flood vulnerability | Residents | Flood management infrastructure | | "The poor get flooded" | Sustainable urban drainage policy |

### 8.5.1 Strategy 1: Green Belt Policy

UK planning policy for 'open land' around cities is founded on the Green Belt, which has been at the heart of national planning policy since 1949. Current guidance (at the time of writing) on Green Belts, which can be found in Planning Policy Statement 2, sets out five main objectives:

(a) To check the unrestricted sprawl of large built-up areas;
(b) To prevent neighbouring towns from merging into one another;
(c) To preserve the setting and special character of historic towns;
(d) To assist in safeguarding the countryside from encroachment; and
(e) To assist in urban regeneration, by encouraging the recycling of urban land.

Within Green Belt areas the main policy aim is to permit no new development except for specific purposes such as agriculture and forestry, essential facilities for outdoor sport and recreation, cemeteries and other uses that preserve the 'openness' of the landscape. In principle, over 600 $km^2$ or 47 % of the Greater Manchester area is protected by Green Belt, but in practice there is a long list of 'strategic deletions' for business and infrastructure. Possibly a bigger issue is that much of the landscape within the Green Belt shows neglect and degradation, with problems for farm diversification, local services and the rural economy. Other local policies such as 'areas of landscape value' have similar goals, but without the legal status of the Green Belt. A recent review of green belt policy aimed to re-think the Green Belt as more of an 'Eco-Belt'—a more responsive and multi-functional zone of low-impact, high value-added, ecological and social diversity (Elson et al. 1993; Ravetz 2000; Natural England and CPRE 2009). This aspiration has to be fitted with the realities of a fast moving complex property market, interacting with a slow moving complex policy machine (Henderson 2005).

The general consensus of policy-makers is that Green Belt policy is a success: but there are contrasting views from business stakeholders who see development and enterprise being blocked: or from residents who see failures to contain development particularly in their own 'back yard'. By applying the PLUREL analytic framework the following evaluation emerged:

- *Robustness against change*: the success of the Green Belt is in its strong legal basis, and its perceived quality of permanence. There is concern from some policy makers that a more 'responsive' green belt policy would risk opening the doors to profit-seeking speculators and developers.
- *Policy integration and external effects*: the Green Belt is a mainly static or reactive policy tool with many impacts on local communities, local enterprise and integrated landscape management. Often, the role of peri-urban partnerships, such as the Community Forests, is to bridge the gaps between the Green Belt and other policies. On a wider note, the Green Belt ideal is rooted in the English aspiration for a 'green and pleasant land', and this has a positive side (landscape protection) as well as a negative side (maintaining housing values on the edge).
- *Overall effectiveness:* there are conflicting views either of success as seen by the majority or failure as measured by the stream of 'deletions' and 'exceptions' for

larger scale developments. However, the modelling work presented in the next section shows how even a small relaxation of Green Belt policy could lead to rapid urban expansion in locations which are otherwise very attractive for development, under the different shock scenarios.

The results of interview and workshop discussions are shown in summary form in the policy analysis table (Table 8.3):

### 8.5.2 Strategy 2: Green Infrastructure and Ecosystems

The Greater Manchester Green–Blue Infrastructure (GBI) strategy points to a wider scope for spatial policy, based more in partnership working, combining the mandate of the public sector, the enterprise of the private sector and a wider scope of community involvement (TEP Consultants 2008). It also raises questions of social and cultural values which are more than financial or functional, and of timescales which are more than the short term (Ravetz 2011a).

The MCR and the northwest region have seen much activity on creating GBI. This overlaps with the Community Forest role and activity. The Red Rose Forest for instance, had 120 active green area (woodland and other habitat) creation or management projects in 2010 (although, with the abolition of the North West Regional Development Agency, it now faces shortages in core funding, even while the policy agenda is ever more keen on GBI). At the regional level the exemplary Natural Economy Northwest programme produced tools such as the Green Infrastructure Guide, and topical studies such as 'Integrating Green with Grey Infrastructure' (NWDA 2008).

Generally, GBI is seen as a positive asset at both local and regional levels, contributing to social and economic welfare, climate change and carbon capture, local food production, and helping to meet the 'sustainable communities' objectives (Roberts 2008). However in reality GBI is still on the margins: core funding is much reduced, and project finance is often difficult, as is finding land and landowners willing to make long term commitments. However, there are new directions such as:

- Local food schemes which are growing rapidly, even while the definition of 'what is local food?' is still up for debate. In some cases local food (produced by new cultivation methods, niche products, farmer's markets, health and education schemes) is seen as a generator of social and economic development (as in the Todmorden example below).
- In the longer term, the possible policy implications of climate change on peri-urban areas may be many (impacts and vulnerability, mitigation policy, adaptation and resilience programmes etc.) (Gill et al. 2007). This suggests an increasing need for policy integration, where the linkages between social, economic, environmental and infrastructure activities can be combined (Fig. 8.6).

**Table 8.3** Green Belt policy analysis and assessment

| | Goals/objectives | Resources/inputs | Effects/outputs | Impacts/outcomes | Metrics |
|---|---|---|---|---|---|
| Ideology/ critique/ discourse | 'England's green and pleasant land': 'dig for victory' | | | 'Rich man in his castle'? | |
| Direct policy/ program/ project | Contain urban areas and encourage regeneration | Strong legal powers to control development | (+) most development controlled:<br>(−) many larger land uses go ahead | (+) most housing areas contained:<br>(−) urbanization of rural areas continues | % Change in GB area: % development on recycled urban land |
| Indirect effects/ external factors | (sustainable land management not an objective) | Rural village inset/ washing over policies:<br>Rural affordable housing policies | Boundary effects: displacement: market distortion:<br>Land speculation: land under-use: | Farm/business diversification is more difficult:<br>Urban/rural areas are polarized: | Social mix in GB areas |
| Process factors | Relies on permanence to avoid land speculation | Review of GB boundaries is long and expensive | Successful challenges to the policy can claim legal costs from local authorities | | % applications successful/on time |
| Context factors | Fragmentation between urban and rural governance:<br>Lifestyle and location choices: | Rural/fringe development policies:<br>Transport road/rail/ interchanges and infrastructure | Transport congestion: commuting growth: | Property regime:<br>Urbanization of rural areas | |

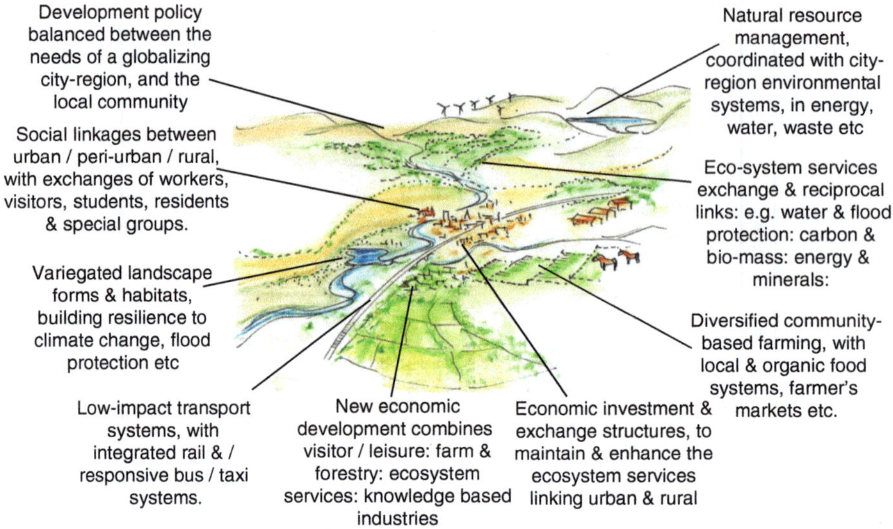


**Fig. 8.6** Green infrastructure policy integration

## *8.5.3 Strategy 3: Local Development*

There is an established (although not always well-funded) policy regime for 'rural development', with a patchwork of schemes and programmes aimed at peri-urban areas, often with larger populations than the more remote rural areas. In their own terms these programmes can be successful. However, it can be difficult to coordinate them, in the face of the trends for the urbanization/metropolitization of peri-urban areas by incoming commuters, high value housing, globalized businesses, and expanding activities such as tourism, conferencing, leisure and education.

The main local development framework comes from the Rural Development Plan for England (RDPE), funded by the European Agricultural Fund for Rural Development and UK Government. This has a budget of £3.9 billion (for the period 2007–2013), double that of the previous programme, over 80% being allocated directly to agri-environment and other land management schemes under 'Axis 2'. Over £600 million is being made available to make agriculture and forestry more competitive and sustainable, and to enhance social and economic opportunity in rural areas. The RDPE is the main platform for national level funding streams in rural areas, but the investments are spread thinly: the largest funding is Pillar 2 'High Level Stewardship' for special landscapes. However, this does not cover most peri-urban areas, which have no special recognition or framework, compared with more remote rural areas, and this is a major gap.

The local development strategy highlights the question of local governance, a very topical issue in peri-urban areas, where there may be major divisions between long-term local and newly incoming residents. One good example is the pilot programmes in the outer areas of Bradford (in the South Pennines on the eastern edge of the MCR) which have set up active Parish Councils at the village level, with real decision making powers and financial resources. There are also area-based Forums for businesses, landowners, tourist developers, lobby groups, residents, and transport operators: these organize village and town schemes which help to build social cohesion through joint cultural, education and leisure activities.

The EU-funded LEADER programme which is now in operation in the South Pennines is also based on locally-led actions covering three strands: provision of basic services for the economy & rural population: village renewal and development: renovation and interpretation of local culture and heritage (Pennine Prospects 2008). Generally, this was seen by stakeholders to be successful, as being more locally directed and responsive to local concerns.

Although spatial planning, GBI and local development strategies appear in separate strands, greater added value can be obtained from integrated schemes. One excellent example is the award-winning "Incredible Edible Todmorden" local food scheme (www.incredibleedibletodmorden.org.uk). This has made links on a local economic development and small business/tourism agenda in a very creative way. It enhances social capital and enterprise, it produces local food, and it contributes to public health and education, while also enhancing GBI, landscape protection and climate adaptation. In this way it shows how divisions and possible conflicts between economic, social, environmental and spatial planning can be bridged by creative partnerships for multi-functional agendas.

## 8.6 Towards the Future

Scenarios help to explore alternative 'stories of the future', as explained in Chap. 1, which help to anticipate potential risks and opportunities and to test possible policy responses. In the MCR the PLUREL scenario framework was applied to a process of 'policy-scenario testing', including results from the Metronamica land use model. This process had three main stages: application of the 'locally adapted' scenarios to the MCR, modelling the results of different policy mixes, as far as possible and future-proofing the policies for robustness under each of the scenarios.

The adapted scenarios for the MCR include some topical points:

- *A1: 'Hyper-tech' scenario: globalized and privatized:* In the MCR, the city centres are networked outwards, to the airports and motorway junctions around the region. Unrestricted peri-urban development turns these nodes into booming business and residential clusters. A new range of private health and education facilities accelerates the transition to a 'gated society'.
- *A2: 'Extreme water' scenario: localized and privatized:* In the MCR this sees a potential resurgence of local identities—not all of them benign or tolerant to outsiders. In the South Pennines this builds on the historic tradition of local cooperation and self-reliance. On the Mersey plains the GBI programme runs

into challenges from communities and landowners preferring local short term benefits.

- *B1: 'Peak oil' scenario: globalized communities*: In the MCR, this reinforces the strategic planning of the city-region, bringing housing, jobs and services closer together, and connected by GBI for walking, cycling and horse-riding. Large peri-urban areas of under-used land are claimed for production of food or energy crops.
- *B2: 'Fragmentation' scenario: localized communities:* In the MCR power and resources shift towards the local level: the result is that more affluent areas erect barriers and withdraw resources from poorer areas. State provision of housing, health and education declines, while social enterprise takes over public services.

Overall, these scenarios show medium term possibilities (2020–2050): they show that prospects for peri-urban areas, raise many questions and uncertainties such as location choices, employment patterns, lifestyles and cultural shifts, globalization and technology networks, climate change impacts and adaptations, and so on. There are also dilemmas, contradictions, and resistance to the 'official' version: e.g. could speculators profit from flooding of low-lying areas? Or, could underground movements resist the corporate land-grabs? Such questions are not easy to replicate in technical models, but there is resonance with the wider policy debate. For instance, a PLUREL stakeholder workshop discussed the future of GBI, on a 'what-if' basis: (a) if there is no public money, or (b) if all land is community owned, or (c) if new schemes need security fences to keep out 'undesirables'. There are questions of resistance: how far the MCR fights against the tide of national/global change: e.g. the response of some to a 'global business' future is to strengthen local opposition and resilience. There are also questions of 'countervalence', in how far the MCR is one unit, or shows internal splits, conflicts, opposition movements etc. Generally the most challenging question is that of 'transition'—how far the future of the MCR can be forecast with 'trend projection', or anticipated creatively through a wider range of structural changes.

### 8.6.1 Urban Development and Land Use Modelling

Land use modelling in an older industrial area such as the MCR, is rarely simple: many land use changes are about multiple functions, or the quality of mixed land use, or 'indeterminate' land uses, or the detailed pattern which is beyond the resolution of the model. There are external forces to be translated into model terms: for instance the CAP regime may decide whether marginal land is kept in pasture: converted to habitats: opened for economic development: or left as unmanaged. There are internal policy agendas to be translated: for instance whether the MCR should aim to feed itself, a question with potentially large effects on the peri-urban. In this way land use modelling is not so much a forecast, more a way of exploring (a) different external conditions, and (b) alternative internal responses.

The key issues are shown in the scenario summary table, where population and economic trends are based on the NEMESIS and IIASA modelling for PLUREL (Table 8.4) (Box 8.1).

**Box 8.1 Modelling Built-Up Areas in Manchester City Region: RUG Results**

The four scenarios differ in the intensity and patterns of change in artificial surfaces (Fig. 8.7).

In the *Hyper-tech* scenario, rapid technological change, new transport technologies and few planning constraints lead to peri-urbanisation, particularly in the areas to the south and west of the city. There is less growth in rural areas compared to the other case studies, as the most of these are part of the Peak District national park (east of the city) and therefore less suitable for development.

The *Extreme water* scenario shows a similar picture to the Hyper-tech scenario, but with slightly lower and more compact growth. Extreme events such as floods and drought may affect most of the area.

In the *Peak oil* scenario, high fuel costs and strict planning policies concentrate growth in urban centres, though not in Manchester itself, which remains a shrinking city. Towns such as Warrington, on the railway line between Manchester and Liverpool, see the most growth (subject to Green Belt zoning policies).

The *Fragmentation* scenario shows a similar pattern to that of Peak oil, but the areas of high urban growth are more scattered. This reflects the clustering of communities by different age group, ethnicity, etc.

### 8.6.2 MOLAND Modelling Results

The MOLAND land use modelling enabled a finer grained analysis, with locally adapted scenarios from stakeholder inputs, higher resolution and more detailed land use classification. There was an issue in the calibration of the model: this requires a time series of compatible base-maps: but in the UK only the 2001 CORINE land use classification was available. In the event the calibration settings were adapted from modelling in the parallel project 'Eco-cities', and the three selected policies were analysed as follows:

- Green Belt and similar zoning policy: the results show the sensitivity of land use outcomes to policies which control development in very desirable locations. This is topical for the current UK debate, not just in the Green Belt but beyond, in areas with less formal zoning policy. This is also relevant to 'brownfield' and other urban regeneration policy: although there are many sites (as listed in the UK National Land Use Database) which can be prioritized by policy, the total areas are small in relation to other population and household density trends.

**Table 8.4** Scenario modelling parameters for the Manchester city-region

| General demographic parameters | 'Hyper-tech'<br>A1<br>Globalizing/privatizing | 'Extreme water'<br>A2<br>Localizing/privatizing | 'Peak oil'<br>B1<br>Globalizing/public | 'Fragmentation'<br>B2<br>Localizing/public |
|---|---|---|---|---|
| General PLUREL scenario | Rapid technology innovation and counter-urbanisation | Rapid climate change and defence of the cities | Energy price shock and retreat from the peri-urban | Communities in retreat with polarisation of cities |
| Population growth per year | 0.41 % | 0.30 % | 0.19 % | 0.30 % |
| Urban/peri-urban pop. growth | Low/high | High/medium | Medium/low | Medium/medium |
| GDP growth *(figures from 2007)* | 3.02 % | 2.62 % | 2.12 % | 2.00 % |
| "Shock" storyline | Rapid technology advance | Extreme water and weather events | Peak oil and effects on road transport | Fragmentation, social exclusion |
| General migration pattern | Out-migration to peri and rural areas (+ metro-lifestyles) | Stagnation and clustering towards growth areas | In-migration to larger urban centres | Fragmentation into smaller communities |
| Spatial patterns in Manchester city-region | Manchester CR grows as a global hub, with waves of global capital spinning out from the centre and the peri-urban hubs | Manchester CR urban economy stagnates, along with peri-urban retail and business parks | Manchester MCR becomes a national/EU centre of governance, public services and social innovation | Manchester CR reverts to 500 neighbourhoods, gated enclaves spring up in the peri-urban zone |
| Implications: for peri-urban | Peri-urban becomes segmented by carefully graded differences in value and status, coupled with risk and opportunity | Chaotic zone of hazardous areas, private enclaves, and informal/illegal landuses: floods and storms, while planning and investment reduces | Carefully planned at national and regional level, with green infrastructure and multi-functional land for food, biodiversity and climate adaptation | Ideal space for self-contained communities to grow, with many functions of food, energy, water etc., in an 'archipelago of enclaves' |

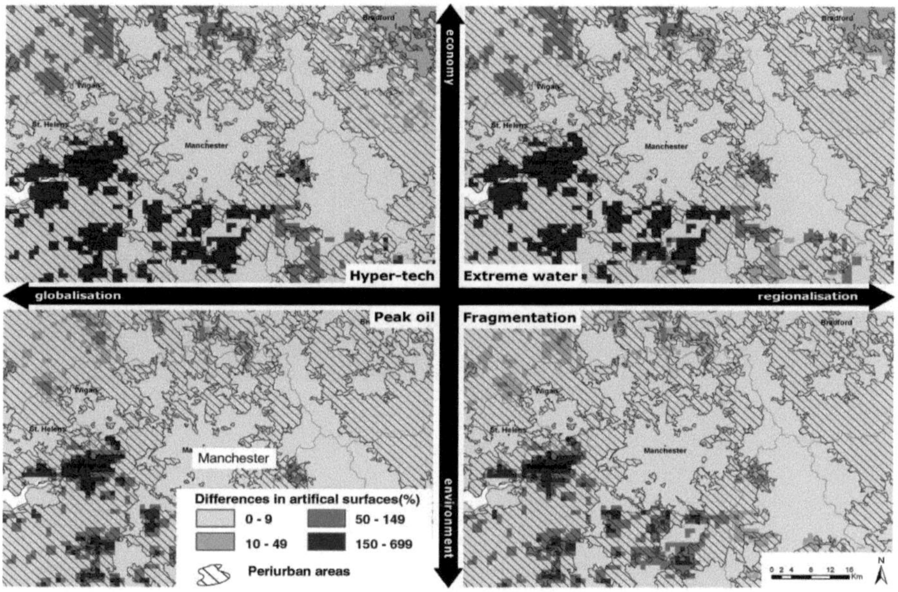


**Fig. 8.7** Shows the development of artificial surfaces in the Manchester region according to the RUG (Regional Urban Growth Model) model for the year 2025 (Source: University of Edinburgh)

The deregulated approach of scenarios A1 and A2 shows as a chaotic diffusion of development across the landscape.

- GBI is modelled approximately by 'urban green' and 'forest/community woodland' categories, although in reality much GBI is at a much finer scale than 100m cells. The modelling shows some contradiction between urban density and GBI: i.e. increases in GBI areas are most visible in the Hyper-tech scenario of counter-urbanization. Designated ecological sites are relatively small in the MCR, and the total makes little difference to the stock of land for development. However, National Park policy is crucial: any policy relaxation opens the door to rapid development, as seen in the A2 'extreme water' scenario.
- Local development policy can be modelled indirectly by enabling smaller scale growth in more remote towns and villages: the results depend on spatial planning, as to whether there are very small clusters on open land (which may or may not be eco-villages or commuter villages), seen in scenarios A1 and A2: or instead, expansion and densification of existing settlements, as in scenarios B1 and B2.

Overall there is a topical question on which is the 'baseline' scenario, closest to existing trends: in the UK, as of 2011, this looks like the A2 *'Extreme Water'* scenario, of low growth, deregulated policy, private enterprise and climate-related problems. This is shown in more detail (Fig. 8.8). Higher density urban form shows a few increases around city centres: lower density increases in suburban areas, and

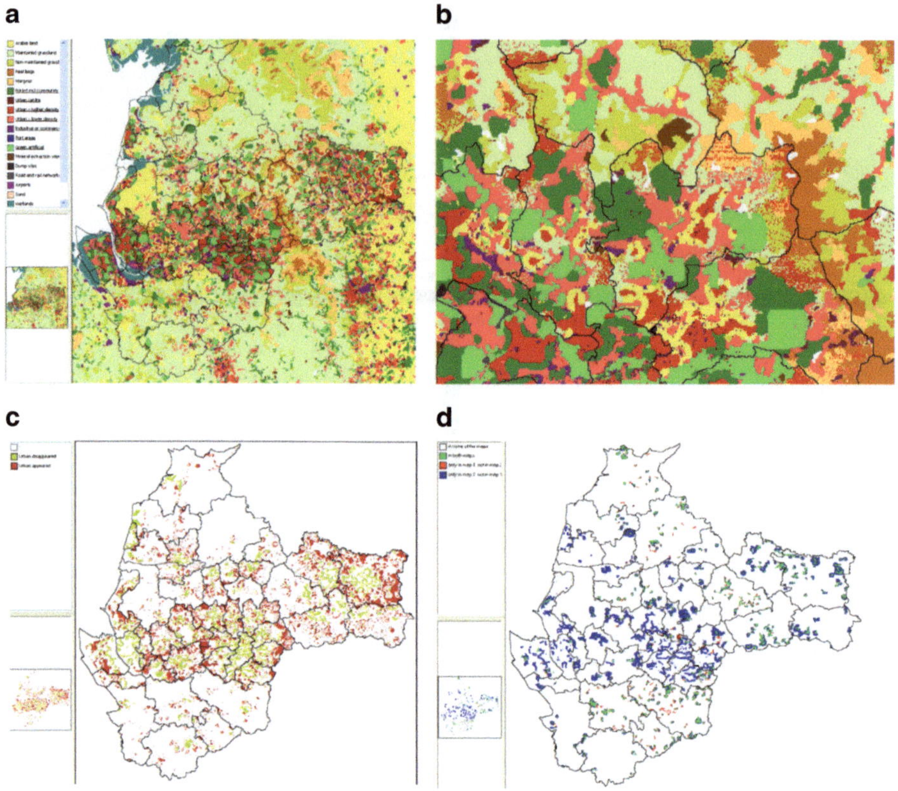

**Fig. 8.8** MOLAND model results: (**a**) landuse map 2050; (**b**) landuse map 2050—detail; (**c**) urban expansion 2050; (**d**) forest land—comparison 2000–2050

locations not covered by Green Belt or ecological protection. In this scenario the Green Belt was modelled as 'negotiable', so where conditions of proximity or accessibility are favourable, then the pressure for low density development breaks through. Development is relatively unrestricted in floodplains and national parks: while industrial brownfield land tends to remain derelict and vacant.

### *8.6.3 Quality of Life Impacts of Lands Use Change*

The attitudes and preferences of the population are the other half of the policy agenda. These were tested as for some of the other case studies using the conjoint method and the Quality of Life simulator (see Chap. 3). Box 8.2 summarises the results of the survey for Manchester.

**Box 8.2 Results of the Quality of Life Indicator Analysis for MCR**

As part of the quality of life indicator work (see Chap. 3), a sample of people in MCR were asked about their perceptions of different factors of the environment in relation to their perceived quality of life. The results are statistically valid owing to the good sample size and they offer some insights into the factors which people view as important. The most important issue as an indicator of environmental quality of life is a feeling of safety and security which might change if areas change their character, perhaps more densely populated and less well managed. The second factor in importance is suitability of housing, this being an important factor in a country with high levels of owner-occupancy where a house is a major investment. Noise is the next most important factor—traffic noise, aircraft and noisy neighbours may all have an impact. Convenience of transport is the next priority—if land use change results in slower and less convenient transport or commuting times then people are less happy with their living environment. Air quality comes next—pollution from cars and industry may be important in inner city areas and next to major highways. The availability of shops comes next on the list—this may be connected to the changes in distribution of shopping opportunities. Thus, if land use change as a result of urban densification or further shrinkage starts to affect the higher level indicators negatively people may try to move to somewhere better. Waste collection service comes second to last with—surprisingly—access to green space being the bottom of the list, despite its perceived importance in other surveys. A more sophisticated analysis of the variability amongst the population and their current environment—such as whether they lived in a leafy suburb or inner city is contained within the Quality of Life Simulator which enables some of these factors to be tested along with predicted land use changes (Fig. 8.9).

### *8.6.4 Environmental-Economic Valuation*

A topical question for peri-urban MCR, as elsewhere, is the value of landscapes or ecosystems, or the costs/benefits of land use changes and policies. Such values should, in principle, be used to justify community forestry or GBI programmes, which might appear to generate large added value for modest investment. A new approach, 'The Economics of Ecosystems and Biodiversity', is based on a functional analysis of 'ecosystems services' (TEEB 2010). However, there are many problems with putting money values on functions which are not in the market place, as outlined in Chap. 3. Such questions point towards new concepts in environmental economics, going beyond a reactive cost-benefit analysis of fixed ecosystem relationships, towards a creative 'institutional design' approach which focuses on collaboration and added value (Everard and Ravetz 2009). This was the starting point for a pilot study in the MCR (Chou and Taylor 2010).

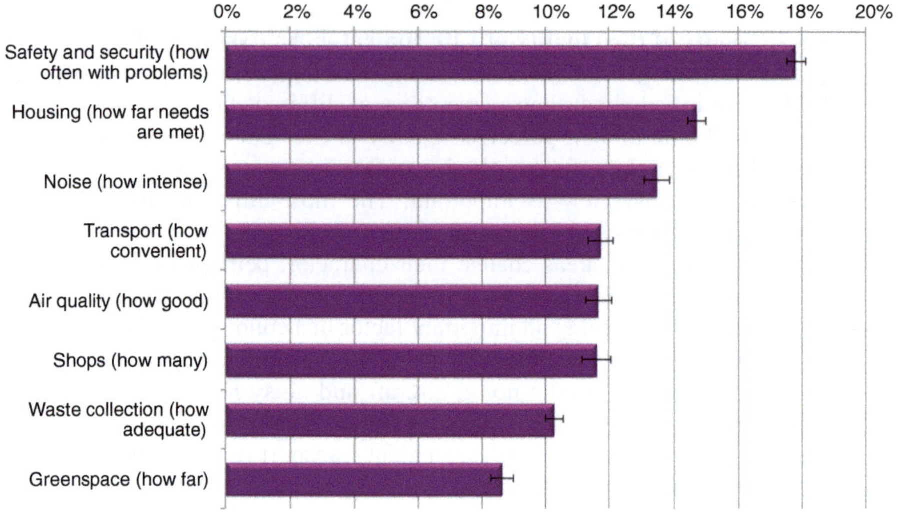


**Fig. 8.9** Social group and environmental economic analysis

The study looked at alternative policy options for the Mosslands area in Salford, one of the largest Red Rose Forest sites. This is a unique ecological habitat, now under pressure and needing a strong planning and management strategy. The land use mix includes zones for biodiversity, hydrological management, agricultural, and public access. Three policy options (spin-offs from the main PLUREL scenarios) included 'no intervention', 'maximum wetland' and 'integrated vision'.

In each case the economic valuation looked at the direct costs of land tenure, reclamation and restoration, planting, and maintenance and direct benefits from ecosystems services, e.g. production, flood management and soil conservation. The outcome suggested that indirect factors are both larger and more uncertain than direct factors, and with a 30-fold difference between the highest and lowest estimates. Assuming a median point between the highest and lowest values, the initial results are very topical: by far the greatest benefit (between 14 and 21 million GBP per year) is found in the 'maximum wetland' policy option, under the conditions of the 'Hyper-tech' scenario.

Further economic analysis could aim to explore more extended value chains, as often found in local development with multi-functional land use. For instance, the Incredible Edible Todmorden scheme described above involves the public, private and social sectors, generating values which are financial, social and environmental. This is achieved by using marginal land at minimum public cost with spin-off benefits such as improved health and well-being. This suggests a 'relational economics' approach to added value which is realized through collaboration between stakeholders (Bathelt and Gluckler 2011). This approach potentially generates much greater values than expressed in a conventional 'bottom line' figure, and could be the start of a wider socio-economic-political debate.

## 8.7 Conclusions and New Directions

### *8.7.1 Sustainability and Peri-Urban Linkages*

Settlements and landscapes continuously change and evolve into new kinds of structures and patterns, providing new types of functions. In a peri-urban area which is simultaneously globalizing and localizing, such changes are multi-level and inter-connected. So how do policy makers decide what is 'sustainable'? This is not only an academic question, but at the heart of the policy discourse. In the MCR of 2011, 'sustainability' is often stated to be the ultimate policy goal, but is rapidly becoming devalued as a term, and is in need of more clear and practical criteria.

In the case of MCR, a previous study of 'integrated planning for sustainable urban development' explored the inter-connections between sectors and locations across the city-region (Ravetz 2000). Applying this thinking to the peri-urban MCR, there are added dimensions to be considered, such as competition or conflict between urban and rural, between local and regional, and between communities and landscapes. The peri-urban is at the interface between each of these, and is both a territorial and a functional system. So we can propose here a working definition for 'peri-urban sustainability', based on the MCR experience (Ravetz 2011b):

> ... a spatial pattern which enables positive interactions between economic, social and environmental systems in both settlements and landscapes, which promotes self-organization, multi-functionality, diversity and resilience, with low impact and high value added to all stakeholders, internal and external.

To implement such a concept through policy needs some kind of territorial definition: and this was one of main themes of the PLUREL research: that the 'Rural–urban-Region' is the optimum territory for the integration of policy between urban, peri-urban and rural areas. However, in the MCR as elsewhere, the actual boundaries of ecosystems, watersheds, economic markets or social communities, are rarely the same as governance units. So a governance system based on the Rural–urban-Region does not often work well in fixed boundaries, but is more likely to be effective in a flexible multi-level framework of functions and linkages (as discussed in Chap. 1). The MCR experience points to such linkages, and the opportunities for policies to align them towards the 'sustainable peri-urban':

- Direct rural–urban linkages: efforts in the MCR on Green Blue Infrastructure, flood management and climate change adaptation are pointing in this direction. However, it is often difficult if not impossible to raise the investment and secure the returns. A strategic approach to Rural–urban-Region governance would ensure that investment would be available wherever it adds value.
- Indirect rural–urban linkages: if the urban resources of the MCR are mobilized for re-investment in the peri-urban surroundings, then there would be added-value on both sides. This might be realized through ecosystems services valuation and payment/exchange systems.

- Social-cultural linkages: in the MCR there are visible problems in peri-urban settlements, such as rural housing shortages, deprivation in peripheral public housing estates and fragmentation of growth and decline. For each of these there are also opportunities which can be mobilized by looking at new kinds of linkages: ecological enterprise, economic diversification, social interaction and cultural exchange.

### 8.7.2 Update: Spatial Development Post-Planning in the UK

The UK planning and development community is (at the time of writing) in flux: the Coalition government of 2010 brings not only radical changes in structures, but a different mindset to the underlying concepts of spatial planning. The shift from the previous 'managerial' approach, to a more bottom-up 'localism' approach, comes alongside major cuts to local government and regeneration programmes, particularly in deprived areas. Many areas of policy, legislation and funding are still uncertain and lack details. So, the results of the MCR research are now pointing towards a rather different situation from the one which was studied.

The Coalition agenda is based on the 'open source planning' concept, of small state and 'big society' (Conservative Party 2009). The draft National Planning Policy Framework, having abolished regional planning, rests for coordination on a so-called 'Duty to Cooperate' between 432 local authorities. The 'Local Enterprise Partnerships' aim to coordinate city-regions, but with ad-hoc and uncertain powers and resources, and split between economic growth and the 'new localism' (CLG 2011b).

The question here is: what are the implications of the new regime for peri-urban areas, and how to move forwards? The new system may in some cases bring opportunities for local resources: for example there are proposed powers for communities to take over local services such as pubs or shops, and this reflects some of the MCR's best practice examples. However the withdrawal of public funding from nearly all forms of local voluntary sectors and social enterprises, raises the threshold over which such local actions must cross.

There is also an agenda for 'marketization' which is showing up in various ways: a 'New Homes Bonus' to provide incentives to allocate housing sites: land auctions for realizing the appreciation of value on development: and the possibility of direct payments to local planning authorities—which some argue, opens the door to the 'sale of planning'. For the peri-urban, this might appear a realistic and entrepreneurial approach to the realities of development, i.e. if finance is the main driver of action, then success depends on its mobilization. Others foresee chaos and confusion, where neighbourhood plans in the peri-urban are dominated by wealthy landowners, while in less desirable locations, larger developers effectively 'buy the plan'.

At the strategic level, there is a belated recognition of the need for infrastructure planning across city-regions, and the MCR in many ways appears to be a demonstration case. At national policy level, and also in the 'City Deal for Greater Manchester' (Greater Manchester Combined Authority 2012), major schemes for road and rail, water and energy are all proceeding rapidly. Generally it is likely that the growth lobby for infrastructure, business and housing, coupled with the increasing privatization of health and education, and the regulation of national targets for brownfield recycling and retail hierarchies, will have major impacts on development in the peri-urban. It also proposes to "reduce unnecessary cost and delay to developers by setting up a Major Infrastructure and Environment Unit; streamlining guidance; setting clearer standards for evidence; and changing the culture of statutory bodies" (HM Treasury 2012, pp 44, 45). The effect on the peri-urban could then be shaped through socio-economic patterns: wealthy educated communities are likely to resist development through active neighbourhood planning: deprived or declining communities are likely to be over-ruled or bought off by powerful interests.

Both opportunities and threats can be seen in the largest private sector property portfolio in the UK, based on the corridor of the Manchester ship canal. The 'Atlantic Gateway' promoted by Peel Holdings, provides much of an integrated sub-regional plan for economic development and GBI, and unlike most public bodies, has the resources to carry it out (Peel Holdings 2010).

Finally, there is a topical agenda on 'ecosystems services' (ESS)—proposed by the Millennium Ecosystem Assessment (UNEP 2005, 2010), as the functions provided to human society by natural ecosystems. 'The Economics of Ecosystems and Biodiversity' ('TEEB') now takes these functions into the economic realm with a valuation and exchange approach. Again, for the UK peri-urban agenda, there is great uncertainty. If we assume a re-emergence of strategic planning, then the ESS concept could help to shape the relationships and linkages between urban, peri-urban and rural. However if we assume a" localist" and entrepreneurial approach, then ESS become more like financial commodities, subject to speculation, profiteering and moral hazards of all kinds. There may be opportunities where communities, towns or cities can self-organize the management and purchase of ESS resources with their hinterlands: alternatively there may be stalemate or open conflict between opposing interests.

### 8.7.3 Recommendations and Next Steps

Overall, the lessons from the MCR have wider significance, particularly for more mature and knowledge-based city-regions. Although the spatial expansion of the physical urban structure has slowed, the restructuring process continues, with many forms of transitions—economic, political, technological, social and cultural—and with new constraints and opportunities. Three main policy recommendations for the peri-urban MCR can be summarized as follows:

- To promote strategic planning and investment (whether managerial or entrepreneurial), which works across the wider city-region (or, 'Rural–urban-Region').
- To enhance the linkages (urban-rural, social-economic-environmental, and local-strategic), and the opportunities for enterprise which combines social, economic and environmental values.
- To enable self-organization, diversity and resilience (social-economic-environmental), at the community and landscape levels.

Overall it can be argued that policy for peri-urban areas is as important as for urban core areas, which tend to demand most policy attention and receive most of the available funding. What stands out is the need not only for *better* governance (e.g. strategic planning authorities, or IT-enabled town halls) but *new forms* of governance. Such new forms can be framed as 'shared intelligence and learning capacity, to respond creatively to complex multi-level and multi-lateral problems and opportunities' (Ravetz 2011b). Through this a more advanced and pro-active concept of a 'sustainable peri-urban region' begins to emerge. This aims to work with the challenges raised by the UK coalition's 'brave experiment' on the front line between localism and economic development. In this way the MCR experience highlights an upcoming agenda, for others in Europe and beyond to build on.

**Acknowledgements** This is to say thanks for excellent contributions from Graham Joyce (Pennine Prospects), Tony Hothersall (Red Rose Forest): the Manchester research team, including at various times, John Handley, Sally Randles, Jeremy Carter, Nick Green; modelling work by Yi Gong and Chris Kenny; modelling advice from Hedwig van Delden and Jasper van Vliet (RIKS), with data from Olga Little (HCA); advice from Ian Wray (Liverpool University, formerly North West Development Agency) and Vince Goodstadt (RTPI Past President); the participants in the PLUREL workshops and interviews; the PLUREL Module 3 researchers, and the review by Steve Littlewood of CUDEM.

## References

Bathelt H, Gluckler J (2011) The relational economy: geographies of knowing and learning. Oxford University Press, Oxford

Benedict MA, McMahon ET (2001) Green Infrastructure: smart conservation for the 21st century. The Conservation fund. Sprawl Watch Clearinghouse, Washington, DC

Berkes F, Colding J, Folke C (2003) Navigating social–ecological systems: building resilience for complexity and change. Cambridge University Press, Cambridge

Bibby P, Shepherd J (2004) Developing a new classification of urban and rural areas for policy purposes: the methodology. DEFRA, London

Blair AM (1987) Future landscapes of the rural-urban fringe. In: Lockhart DG, Ilbery B (eds) The future of the British rural landscape. Geo Abstracts, Norwich

Brand P, Thomas MJ (2005) Urban environmentalism: global change and the mediation of local conflict. Routledge, New York/Oxford

Breheny M, Rookwood R (1993) Planning the sustainable city region. In: Blowers A (ed) Planning for a sustainable environment. Earthscan, London

CABE (Commission on Architecture and the Built Environment) (2010) Community green: using local spaces to tackle inequality and improve health. CABE, London

Cameron D (2011) Speech to the conservative party conference. Reported as: Cameron labels planning officials 'enemies of enterprise': Planning, 7 March 2011
Chou WJ, Taylor T (2010) Applying environmental valuation methods to assess benefits of Peri-Urban Land Use Policy—the Mosslands project. PLUREL working paper. www.manchester.ac.uk/plurel. Accessed 12 Jan 2012
CLG (Communities and Local Government) (2011a) Draft national planning policy framework. CLG, London
CLG (Communities and Local Government), (2011b) A plain English guide to the localism bill. CLG, London
Conservative Party (2009) Open source planning. Policy green paper 14. Conservative Party, London
CPRE (Campaign for the Protection of Rural England) (2007) 2026: a vision for our countryside. CPRE, London
CURE (Centre for Urban & Regional Ecology) (2003) Sustainable development of the countryside around towns: main report to the countryside agency. Natural England, Wetherby, UK
Douglas I, Ravetz J (2011) Urban ecology: the bigger picture. In: Niemelä J, Breuste JH, Guntenspergen G, McIntyre NE, Elmqvist T, James P (eds) Urban ecology: patterns, processes, and applications. Oxford University Press, Oxford
Duany A, Plater-Zyberk E, Speck J (2000) Suburban nation: the rise of sprawl and the decline of the American dream. North Point Press, New York
Elson M, Walker S, MacDonald R, Edge J (1993) The effectiveness of green belts. DOE, HMSO, London
Engels F (1845) The condition of the working class in England, 1958th edn. Blackwell, Oxford
Everard M, Ravetz J (2009) Ecosystem services—joined up thinking in an interdependent world. Environ Sci Spec Issue Environ Futures 18(2):15–20
Fairlie S (1996) Low impact development: planning and people in a sustainable countryside. Earthscan, London
Florida R (2004) The creative talent: cities and the creative class. Routledge, New York
Gallent N (2006) Planning on the edge. Routledge, Abingdon
Gill S, Handley J, Ennos R, Pauleit S (2007) Adapting cities for climate change: the role of the green infrastructure. Built Environ 30(1):97–115
GMCVO (Greater Manchester Council of Voluntary Organizations) (2007) On the edge? Perceptions of Greater Manchesters rural and fringe communities. GMCVO, Manchester
Hajer MA (2003) The politics of environmental discourse: ecological modernization and the policy process. Oxford Scholarship Online, Oxford
Handley J, Wood R (1998) Defining coherence for landscape planning and management: a regional landscape strategy for North West England. Landscape Res 23(2):133–158
Harvey D (1995) The condition of post-modernity: an enquiry into the conditions of cultural change. Blackwell, Oxford
Henderson K (2005) Tensions, strains and patterns of concentration in Englands city-regions. In: Hoggart K (ed) The Citys Hinterland: dynamism and divergence in Europes peri-urban territories. Ashgate, Aldershot
Heynen N, Kaika M, Swyngedouw E (eds.) (2005) In the nature of cities: urban political ecology and the politics of urban metabolism. Routledge, London and New York
Hoggart K (2005) City hinterlands in European space. In: Hoggart K (ed) The City's Hinterland: dynamism and divergence in Europe's peri-urban territories. Ashgate, Aldershot
Hyde M, O'Rourke A, Portland P (2004) Around the M60: Manchester's orbital motorway. AMCD Publishers Ltd, Manchester
Kaika M (2005) City of flows: modernity, nature, and the city. Routledge, New York
Kasarda JD, Lindsay G (2011) Aerotropolis: the way well live next. Farrar, Straus and Giroux, New York
Kitchen T (1997) People, politics, policies & plans: the city planning process in contemporary Britain. Paul Chapman Publishing, London

Lee N (2007) Ideopolis: the knowledge city-region—distinctiveness and knowledge cities. The Work Foundation, London

Lloyd PE (1980) Manchester: a study in industrial decline & economic restructuring. In: White P (ed) The continuing conurbation: change & development in Greater Manchester. Aldershot, Gower

GMCA (Greater Manchester Combined Authority) (2012) Greater Manchester city deal. AGMA, Wigan

Natural England (2009) Green infrastructure guidance: (NE176) Natural England, Wetherby, UK

Natural England and CPRE (2009) Green belts: a greener future. Natural England and CPRE, Cheltenham, UK

New Economy Manchester (2008) Rural economic baseline study. Manchester, New Economy Manchester

Nicholas R, McWilliam GA (1962) Planning the city of the future. In: Carter CF (ed) Manchester and its region: a survey prepared for the British association. Manchester University Press, Manchester

Nicholson-Lord D (1987) The greening of the cities. Routledge, London

NWDA (North West Development Agency) (2008) Developing an outline strategy for linking green and grey infrastructure: Warrington, NWDA. Available on http://www.naturaleconomy-northwest.co.uk/resources+reports. Last Accessed 12 Jan 2012

Peel Holdings (2010) Accelerating growth across the Manchester and Liverpool City regions: framework for a global growth opportunity. Peel Holdings, Liverpool

Pennine Prospects (2008) South Pennines LEADER: local development strategy: core strategy: 2008–2013. Bradford, Pennine Prospects. Available on http://www.pennineprospects.co.uk/leader. Last Accessed 15 Mar 2012

Randles S, Green K (2006) At the interface of innovation studies and industrial ecology. In: Randles S, Green K (eds) Industrial ecology & spaces of innovation. Edward Elgar, Cheltenham

Rauws WS, de Roo G (2011) Exploring transitions in the peri-urban area. Plann Theor Pract 12(2):269–284

Ravetz J (1999) Citizen participation for integrated assessment: new pathways in complex systems. Int J Environ Poll 11(3):331–350

Ravetz J (2000) City-region 2020: integrated planning for a sustainable environment. Earthscan with the TCPA, London

Ravetz J (2011a) Peri-urban ecology: green infrastructure in the twenty-first century metro-scape. In: Douglas I, Goode D, Houck MC, Wang R (eds) The Routledge handbook of urban ecology. Routledge, Abingdon

Ravetz J (2011b) Exploring creative cities for sustainability with deliberative visualization. In: Girard LF, Nijkamp P (eds) Creativity and sustainable cities. Heinemann, Oxford

Roberts P (1998) Retrospect and prospect. In: Roberts P, Thomas K, Williams G (eds) Metropolitan planning in Britain: a comparative study. Jessica Kingsley, London

Roberts P (2008) Sustainable communities policy, practice and professional development: a model for Europe. In: Cooper I, Symes M (eds) Sustainable urban development: changing professional practice. Routledge, London

Roberts P, Ravetz J, George C (2009) Environment and city: critical perspectives on the urban environment around the world. Routledge, Oxford

RTPI (Royal Town Planning Institute) (2006) Uniting Britain: the evidence base—spatial structure and key drivers. RTPI, London

Scott AJ (2000) The cultural economy of cities: essays on the geography of image-producing industries. Sage, New York

Scott AJ (2006) Creative cities: conceptual issues and policy questions. J Urban Aff 28(1):1–17

Sherriff G (2012) From burden to asset: the political ecology of sustainable transport: town and country planning, (forthcoming)

Shoard M (1983) This land is our land. Paladin, London

Soja E (2000) Postmetropolis: critical studies of cities and regions. Blackwell, Oxford
TCPA (Town and Country Planning Association) (2006) Connecting England. TCPA, London
TEEB (2010) The economics of ecosystems and biodiversity in local and regional policy makers. UNEP, New York
TEP (Consultants with Natural England) (2008) Towards a green infrastructure framework for Greater Manchester. AGMA, Wigan
HM Treasury (2012) Budget 2012, HC 1853. HM Treasury, London
UNEP (2005) Living beyond our means—natural assets & human wellbeing: a statement from the Board. UNEP, New York. Available on www.milleniumassessment.org
UNEP (2010) The economics of ecosystems and biodiversity: mainstreaming the economics of nature: a synthesis of the approach, conclusions and recommendations of TEEB. UN Environment Programme, New York
Ward K, Fagan C, McDowell L, Perrons D, Ray K (2010) Class transformation and work-life balance in urban Britain: the case of Manchester. Urban Stud 47:2259–2278
Wong C, Ravetz J, Turner J (2000) The UK spatial planning framework. Royal Town Planning Institute, London
Wood R, Ravetz J (2000) Recasting the urban fringe. Landscape Des 294:13–16
Wray I (2011) In search of strategic sites: North West England, 1990–2010. Town Plann Rev 82(3):1–10
Wray I, McPherson C (2006) Mersey: the river that changed the world. Bluecoat Press, Liverpool

# Chapter 9
# Leipzig-Halle: Ecosystem Services in a Stagnating Urban Region in Eastern Germany

**Annette Bauer, Dietmar Röhl, Dagmar Haase, and Nina Schwarz**

## 9.1 Introduction

Land use change makes an impact on ecosystem services and quality of life in urban regions. The dynamics of land use change as well as their effects are more clearly understood by taking local conditions into account. In the case of Leipzig-Halle, a period of heavy commercial and residential suburbanisation has recently come to an end. With uneven economic and population development, the region is characterised by stagnation and internal disparities. A clear-cut future development path is not self-evident. This raises questions for regional decision makers, planners and scientists alike: What are the potential land use developments of the future? Which of them are

A. Bauer (✉)
Department Comptational Landscape Ecology, Helmholtz Centre for Environmental Research GmbH - UFZ, Permoserstr. 15, 04318 Leipzig, Germany

Sachgebiet Raumordnung, Landes- und Regionalplanung, Regierung von Mittelfranken, Promenade 27, 91522 Ansbach, Germany
e-mail: annette.bauer@reg-mfr.bayern.de

D. Röhl
Landesdirektion Sachsen, Referat 34 / Raumordnung, Stadtentwicklung, Braustr. 2, 04107 Leipzig, Germany
e-mail: dietmar.roehl@lds.sachsen.de

D. Haase
Department Comptational Landscape Ecology, Helmholtz Centre for Environmental Research GmbH - UFZ, Permoserstr. 15, 04318 Leipzig, Germany

Humboldt University Berlin/Institute of Geography, Unter den Linden 6, 10099 Berlin, Germany
e-mail: dagmar.haase@geo.hu-berlin.de; dagmar.haase@ufz.de

N. Schwarz
Department Comptational Landscape Ecology, Helmholtz Centre for Environmental Research GmbH - UFZ, Permoserstr. 15, 04318 Leipzig, Germany
e-mail: nina.schwarz@ufz.de

K. Nilsson et al. (eds.), *Peri-urban futures: Scenarios and models for land use change in Europe*, DOI 10.1007/978-3-642-30529-0_9,

desirable and which are not? What, therefore, are appropriate and effective steering options? These questions are examined in this chapter.

## 9.2 The Leipzig-Halle Region

### *9.2.1 An Urbanised, Polycentric Region with a Rich Heritage*

With a population of just over one million in 2008, the rural–urban region of Leipzig-Halle is an important agglomeration in eastern Germany (Saxon Office for Statistics 2010a). It is located partly in the federal state of Saxony and partly in the state of Saxony-Anhalt and covers a total of 4.390 $km^2$. Its main urban cores, Leipzig and Halle, are encircled by small towns and rural areas (Loibl and Köstl 2008, see Chap. 2).

The region has a rich cultural and industrial heritage. Located at a crossing of trade routes dating back to mediaeval times, the cities of Leipzig and Halle prospered early on. The Leipzig fair developed as a magnet for mediaeval merchants. Because of valuable soils in the cities' hinterland, the local agriculture-based economy prospered. The foundation of the University of Leipzig, followed by the University of Halle, prepared the ground for a flourishing printing and publishing industry.

In the more recent past, Leipzig and Halle were the setting of prominent events such as the defeat of Napoleon's army south of Leipzig and the construction of the first long distance railway in Germany between Leipzig and Dresden, soon followed by connections to Halle and Magdeburg. This development stimulated manufacturing, the rise of enterprises in mechanical engineering and led to an early industrialisation of the whole region. At the turn of the twentieth century, open cast lignite mining also developed nearby. This became more and more large-scale due to shortages of imports during the First World War, leading to considerable changes to the Leipzig-Halle peri-urban area.

The severe disruptions of the two world wars were followed by the foundation of the communist German Democratic Republic (GDR) in eastern Germany. During GDR times, following the central planning processes applied in that regime, the Leipzig-Halle region was developed as a centre for the chemical industry. Major urban reconstruction after the extensive war damage by bombing and fighting was followed by urban expansion as new housing areas were created consisting of large blocks of flats constructed to standardised designs. From having been part of the GDR and its relatively closed economic system with uneconomic and inefficient state-owned industries, major de-industrialisation occurred soon after the German reunification of 1989, when the region's economy suddenly faced national and international competition. This led to massive unemployment and a flight of people to western Germany in search of work and a better life, leading to a shrinking population (see section 9.2.3). Today, tentative new developments in the service sector are taking place, such as in the fields of education, transport, local recreation and tourism.

Thus, historically, the cities of Leipzig and Halle were characterised by many interrelationships, many of which continue to the present-day. For the research described in this chapter, the strong commuter relations between Leipzig, Halle and the surrounding areas were important for the delineation of the case study area (Pütz 2010). The case study encompasses the Halle planning region in Saxony-Anhalt and the urbanised parts of the planning region of Western Saxony. The area is divided in the three districts: Saale, Leipzig and Northern Saxony. There is thus no overall regional-level administration in charge of this rural–urban region which is defined more by its functionality than by administrative borders.

### *9.2.2 Urban, Peri-Urban and Rural Parts of the Region*

The two cities comprise a system of urban neighbourhoods grouped around their historic centres. In the case of Leipzig, the historic centre is a relatively small area surrounded by a belt of "Wilhelminian" style housing areas and further out from this there are housing areas dating from the inter-war period. The main developments of GDR times comprised extensive areas of standardised prefabricated apartment buildings, the so called *Plattenbausiedlungen.* Since building individual houses for owner-occupation was prohibited in the GDR era no significant suburbanisation developed. Much of the construction after reunification in 1990 took place at the outer edges of the city and was divided into either single family housing or apartment blocks (Nuissl and Rink 2005).

For analytical purposes and in particular for evaluating the impacts of land use change, the rural–urban region of Leipzig-Halle was divided into four different sections, based on its administrative structure, population and building density:

- The urban sections comprising the most densely built-up areas of the two core cities;
- The inner peri-urban areas, being the outer parts of the two core cities' territory, which are characterised by lower population densities, discontinuous urban fabric and primarily commercial activities;
- The outer peri-urban areas, represented by a ring of the surrounding and directly neighbouring municipalities around the two core cities;
- The rest of the region, composed of rural areas with the lowest population and settlement densities (Herfert 2007) dominated by arable production and small towns serving as local centres (Fig. 9.1).

### *9.2.3 Population and Economic Development*

The Leipzig-Halle region has experienced population loss since the 1970s which is still continuing (Nuissl and Rink 2005). A spatially differentiated pattern of

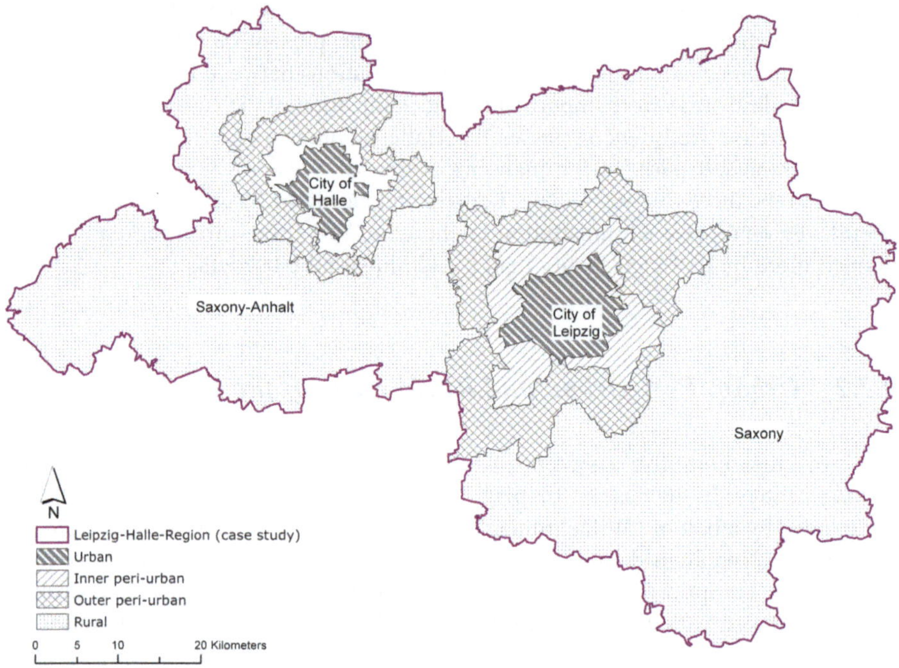


**Fig. 9.1** Delineation of the urban, inner and outer peri-urban and rural parts of the case study region Leipzig-Halle. The urban plus the inner peri-urban parts together constitute the administrative areas of the two cities Halle and Leipzig. Thus, the two cities have the immediate planning authority over these areas (Saxon Office for Statistics 2010b; Saxony-Anhalt Office for Statistics 2010b)

population change can be seen since 1990 with different developments in Leipzig and Halle. Due to suburbanisation and out-migration Halle lost 20% of its inhabitants between 1990 and 2000 (−62,956 inhabitants) and has a continuing declining tendency (City of Halle 2010). Leipzig's population decline of 12 % (−67,179) between 1990 and 2000 changed into a gentle population increase rate of 2.7 % between 2000 and 2006 (+13,370) (Saxon Office for Statistics 2010a). The opposing tendencies of peri-urban and urban population development are also striking between the two cities. In the Saale district, which surrounds Halle, the population rose slightly from 208,604 to 217,487 inhabitants (+4.2 %) between 1990 and 2000 and only thereafter beginning to shrink constantly to 206,146 inhabitants by 2006 (Saxony-Anhalt Office for Statistics 2010a). The population of the neighbouring district to Leipzig increased slightly by 0.3 % (+741) between 1990 and 2000—as opposed to a simultaneous population decline in the city of Leipzig itself. From then onwards, a downward trend is visible. Up to 2006 this district lost 13,491 inhabitants (−4.8 %) which equals a loss of 4.5 % in total (−12,767) between 1990 and 2006 (Saxon Office for Statistics 2010a).

In short, the focal points of population growth and shrinkage shifted during the last couple of decades. Today, with the exception of Leipzig city centre, a general shrinking tendency can be seen. Thus, growth and shrinkage occur next to each other, conveying the overall impression of a stagnating region.

In economic terms, the region shows a decline in investment rates and public finances with high unemployment rates. In Leipzig, for example, 90,000 of the 100,000 industry jobs in 1990 disappeared shortly afterwards; in and around Halle the large Leuna and Buna petro-chemical plants almost completely shut down (IfL 2001). The unemployment rate in the entire area remained in 2010 at about 20 % (Saxon Office for Statistics 2010c).

### *9.2.4 Land Use Change*

The demographic growth-shrinkage patterns described above are accompanied by changes in residential, commercial and infrastructure patterns. While the post-communist transformation period with heavy urban sprawl has passed, moderate development in the peri-urban zone continues (Haase and Nuissl 2010). At the same time, considerable parts of the inner city faced a population outflow leading to residential vacancy, large urban brownfields following demolition of vacant and redundant buildings and massive under-utilisation of urban infrastructure such as the water supply system (Haase and Nuissl 2007).

In the urban cores, land use changes in the period from the early 1990s until today predominantly meant in-fill development—mostly residential—next to vacant sites and urban brownfields (Haase and Nuissl 2007). In the inner peri-urban and outer peri-urban zones, different forms of residential and commercial sprawl can be found. Conflicts have arisen between the inner peri-urban areas which belong to the city and the outer ones belonging to the municipalities surrounding the core of each city in terms of land preparation for development and attracting people to live there in order to counteract population shrinkage (Haase et al. 2007; Sinn et al. 2008). Commercial development has been concentrated along the axis between Leipzig and Halle close to the Leipzig-Halle airport and the A14 autobahn, which lie in the outer peri-urban area (Fig. 9.1). Since the late 1990s the rate of both residential and commercial sprawl around Leipzig-Halle has abated considerably. Finally, rural land use change mainly took the form of a reduction in or intensification of arable production and, but only occasionally, a loss of arable land due to residential or commercial sprawl. The biggest changes took place in the former lignite opencast mining areas in the south of Leipzig where large areas have been converted into lakes arable farmland or forest (Schwarz et al. 2010). Figure 9.2 and Table 9.1 show these patterns using Corine Land Cover data.

## 9.3 The Formal Planning System

### *9.3.1 German Statutory Framework*

The key features of the German planning system are a central legal framework and a decentralised decision-making structure. The full responsibility for

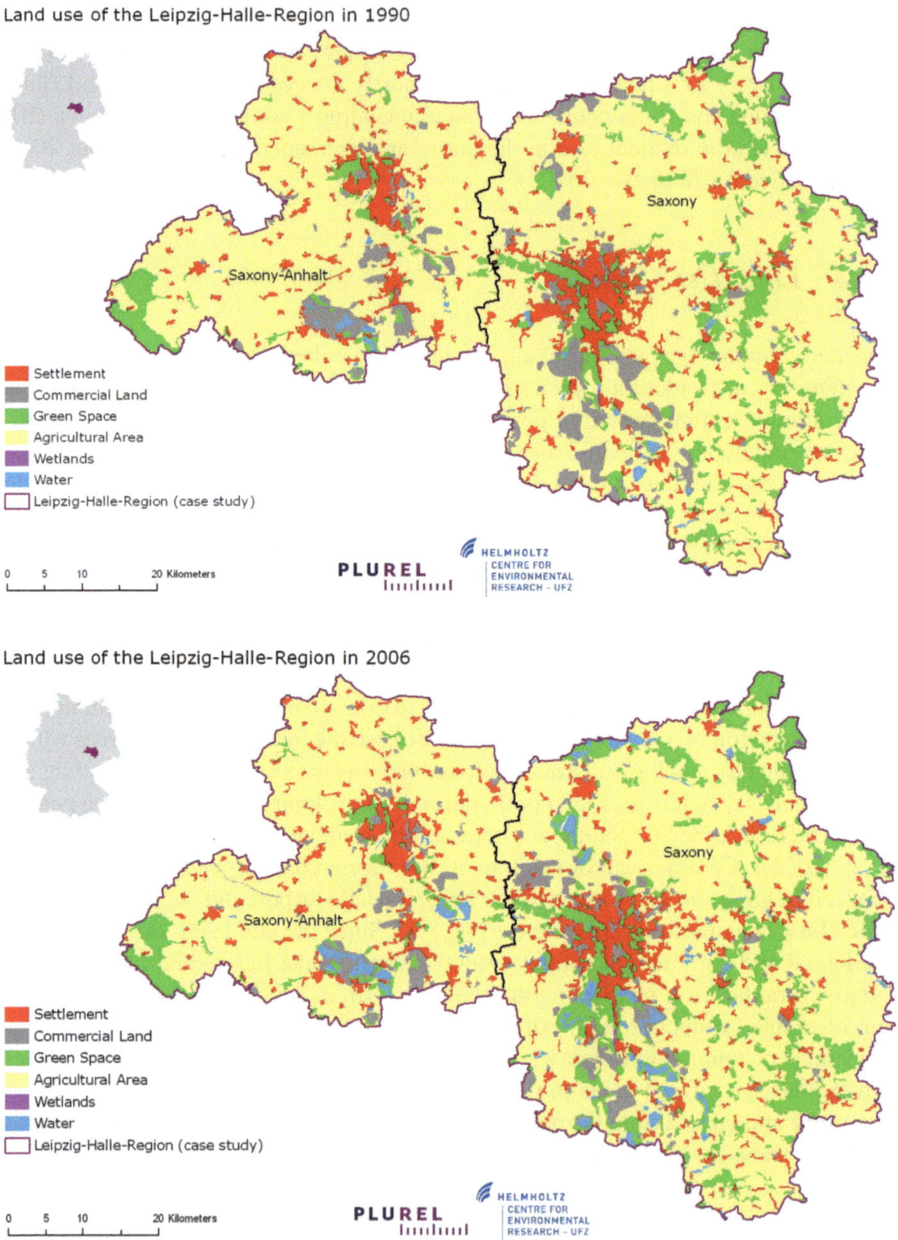


**Fig. 9.2** Land use patterns in the region of Leipzig-Halle in the years 1990 and 2006 as detected by Corine Land Cover (EEA 2006b)

development and land use control is at the local level, derived from the established self-government of communities (Eggers and Schüttlohr 2005). The supra-local level, the districts, usually have limited, more executive functions in

**Table 9.1** Table of land use changes detected by Corine land cover in 1990, 2000 and 2006 in the region Leipzig-Halle (in km$^2$)[a]

| Year | Settlement | Commercial land | Green space | Agricultural area | Mineral extraction | Wetlands | Water |
|---|---|---|---|---|---|---|---|
| | km$^2$ | | | | | | |
| 1990 | 37,772 | 9,954 | 69,953 | 882,151 | 18,759 | 163 | 2,849 |
| 2000 | 39,439 | 13,036 | 76,748 | 873,875 | 12,717 | 218 | 5,570 |
| 2006 | 40,774 | 13,960 | 80,514 | 954,758 | 8,301 | 192 | 9,632 |
| *Percentage change* | | | | | | | |
| 1990–2000 | 4.4 | 31.0 | 9.7 | −0.9 | −32.2 | 33.7 | 95.5 |
| 2000–2006 | 3.4 | 7.1 | 4.9 | 9.3 | −34.7 | −11.9 | 72.9 |
| 1990–2006 | 7.9 | 40.2 | 15.1 | 8.2 | −55.7 | 17.8 | 238.1 |

[a]The land use simulation contains more land use classes than those listed but all land use classes of the "Rest" category do not change as they were defined as inactive or inert in the MOLAND model (layout: D. Haase, based on EEA (2006b))

the field of infrastructure planning and public service provision. In the larger German federal states, such as Saxony and Saxony-Anhalt, an intermediate level of regional planning administration also exists. These administrations set out spatially explicit planning aims for sub-regions within regional plans.

The main instruments at the federal state (Land) level are state-wide spatial structure plans and programmes, in German, the *Landesentwickungspläne* or *Landesentwicklungsprogramme* (Turowski 2002). They contain broad statements on spatial development intentions with the exception of more specific targets for large retail services and the protection of river banks. Furthermore, settlement hierarchies and priority areas are defined based on population projections. Plans have to be sufficiently detailed to serve as a baseline for the next levels in the hierarchy, regional and municipal planning (Turowski 2002). Local and regional planning is controlled by the federal states (*Länder*). The implementation of spatial structure plans and regional plans is supported by a number of legal instruments. One example is the state's option not to approve development deemed detrimental to regional and state plans (§ 12, Raumordnungsgesetz (ROG 2005)).

Spatial planning at the different levels is interlinked vertically. State-wide development plans and regional plans set the frame for municipal planning as noted above. However, the diversity of local contexts should also be considered in planning at the state and regional level (the so called "feedback principle", Saxon Ministry of the Interior SMI 2004; Turowski 2002). Consequently, land use planning is organised along the hierarchy but the levels of government are interrelated.

A comparison of European planning systems shows that spatial planning in Germany allows for a medium level of public control over land use change (Tosics et al. 2010). This is due firstly to a limited degree of spatial planning control attributed to the supra-local level, the districts and regions, which are considered most appropriate for effectively steering land use and secondly to limited administrative fragmentation of urban regions (Tosics et al. 2010).

### *9.3.2 The Application of Planning in the Leipzig-Halle Region*

Spatial planning in the Leipzig-Halle region has some specific qualities: As mentioned above, an intermediate level of regional planning administration exists in both the states of Saxony and Saxony-Anhalt. The planning systems of the two became more similar over time. Initially, regional planning was a federal state responsibility in Saxony-Anhalt. More recently, separate regional planning bodies were created at the district-level. In Saxony, regional planning associations, composed of municipal representatives, are in charge of slightly larger regions. The Saxon state shows a strong presence in regional planning through intermediate level authorities responsible for planning control in the Saxon sub-regions (the *Landesdirektionen* or regional councils). These federal-level authorities implement the spatial and regional planning aims. In Saxony-Anhalt, where the city of Halle is located, an institutional counterpart to the "regional councils" does not exist. A state-wide institution with comparable responsibilities but a more indirect role in planning control is based in the city of Halle. Furthermore, Saxony started a small-scale funding programme for regional development with a focus on inter-municipal cooperation in 1997 (funding guideline "Region"), which has no counterpart in Saxony-Anhalt. Consequently regional planning and development in Saxony are somewhat better equipped than in Saxony-Anhalt and regional development cooperation across the federal state border is rare.

### *9.3.3 Boundaries and Unitary Reform*

In the case of Leipzig-Halle, a formal planning arena with responsibilities for the urban region as a whole does not exist because the urban region is divided by a state boundary. This leads to differences in building legislation, funding programmes and spatial planning in the two sub-regions. Therefore, neither a common planning policy nor joint governance structures are very developed or available for guiding land use changes towards greater sustainability (Sinn et al. 2008).

At the municipal level, the region is divided into core cities and their surrounding municipalities. In 1999, Leipzig was able to annex much of its suburban surroundings. It was argued that this change of municipal boundaries reconfigured the city closer to a functioning unit (City of Leipzig 2007). In the city of Halle, municipal boundaries were extended once in 1990 to include a major town extension constructed in the 1960s. Nevertheless, according to local and regional planners, administrative fragmentation continues and the problem of synchronising land use planning between core and suburb simply shifted in space (Sinn et al. 2008). The coordination of land use planning between the city cores and surrounding municipalities, it is claimed, leaves much to be desired (Sinn et al. 2008).

## 9.4 Strategies for Nature Protection

In the next section, three strategies for nature protection in Leipzig-Halle are outlined and their performance is discussed (see Chap. 3 for details on the methods used). The evaluation results reflect the subjective viewpoints of interviewees and focus group participants (Sinn et al. 2008).

The selection of spatial planning strategies was informed by:

Firstly, an indicator-based evaluation of potential land use configurations, which identified that the inner peri-urban areas of the Leipzig-Halle region in particular need more attention in the future. These areas are especially prone to fragmentation and loss of valuable land for ecosystem services. Therefore, planning strategies should focus spatially on these areas and also aim at protecting or even enhancing green open space;

Secondly, an analysis of spatial planning in the Leipzig-Halle region concluded that besides their formal responsibilities, spatial planning authorities in Saxony and Saxony-Anhalt also fulfil diverse facilitating tasks, ranging from local tourism development to regional landscape management and strategic planning. For such facilitating, optional tasks, spatial planning authorities cooperate with the private sector and civil society organisations or other authorities. The relevant agents are organised more or less loosely, with different degrees of statutory support, in networks, inter-municipal cooperation boards or formal planning bodies (see Fig. 9.3).

To give an impression of these differences, spatial planning strategies with varying statutory support were selected:

- The regional plan's Green Corridors,
- The inter-municipal Parthe Floodplain Cooperation,
- The Green Ring regional development concept.

Wherever it makes sense, the synchronisation of these Saxon-based strategies with spatial planning strategies in the Saxony-Anhalt section of the case study region is considered.

### 9.4.1 *Green Corridors*

The protection of green corridors of unsealed land in the vicinity of settlements is an important regional planning aim in western Saxony. Green corridors link urban and peri-urban open spaces. The Saxon Green Corridors perform a range of ecological and recreational functions; they cover natural landscapes, in particular rivers, floodplains and forests, as well as "corridors" of unsealed land with various land cover types in-between. The aim of the strategy is that these areas should be kept free of development and disruptive land uses (RPA Western Saxony 2008).

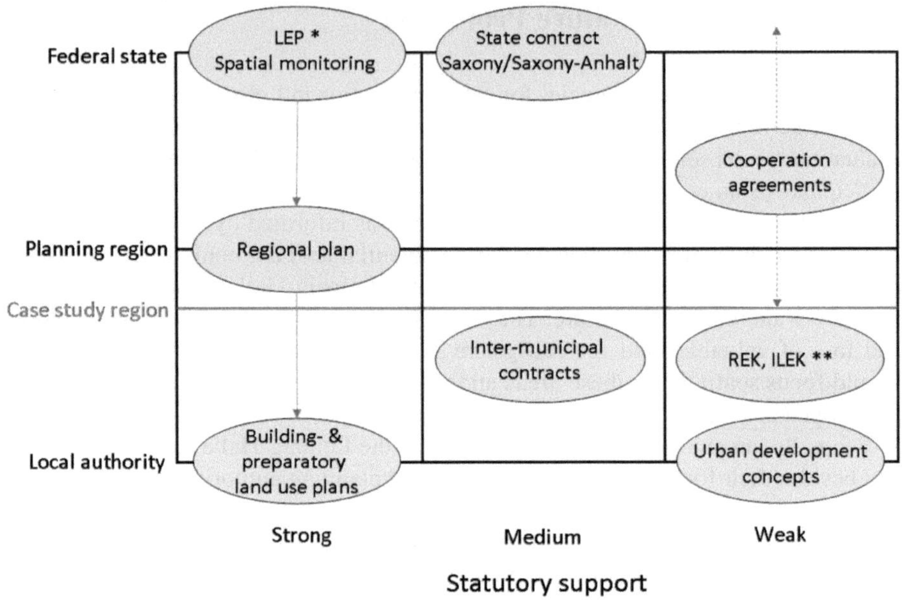


**Fig. 9.3** Levels of spatial planning and statutory support of strategies, Leipzig-Halle case study. Based on this matrix, types of spatial planning strategies with strong, medium and weak statutory support for their adoption and implementation can be differentiated (layout: A. Bauer)

Green Corridors are connected to a green space network in the nearby Halle region as defined in the Halle regional plan (see Fig. 9.4). While the Green Corridors serve ecological and recreational functions, the green space network also explicitly connects otherwise isolated ecosystems or biotopes in both natural and cultural landscapes. According to the strategy, land use conflicts within these areas should be resolved with particular attention to the requirements of environmental protection, landscape management and nature forestry (RPA Halle 2009).

Regional planning aims, such as the Saxon Green Corridors and the Saxony-Anhalt network of green spaces, are set out through a multi-stage procedure including stakeholder participation and public comment. The participants of the stakeholder group belonged to local authorities, inter-municipal cooperation organisations affected by the plan and nature conservation associations, among others (§1, No. 2, Raumordnungsgesetz (ROG 2005)). The final version is prepared by regional planning associations and authorised by its general assembly and the Saxon Ministry of the Interior (SMI 2004). The regional plans of Western Saxony and the Halle region have been recently updated (RPA Western Saxony 2008) or are in the process of revision (Halle region).

Regional planning aims, such as the Green Corridors, have exerted an indirect influence on land use. Because regional plans are legally binding, municipalities implement the Green Corridors by considering them in preparatory land use and development planning. This is aggravated by the fact that many municipal land use

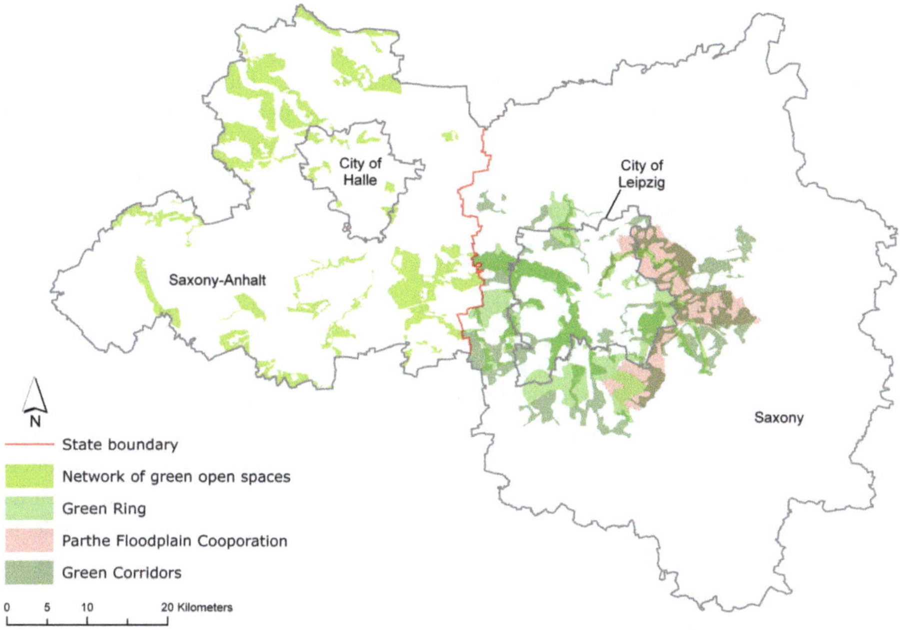


**Fig. 9.4** Delineation of the Green Ring, the Parthe Floodplain Cooperation and the Green Corridors with the adjacent network of green open spaces, Leipzig-Halle region (layout: J. Weinert based on RPA Halle 2009; RPA Western Saxony, 2009 and Parthe Floodplain Cooperation 2009; Saxon Office for Statistics 2010b; Saxony-Anhalt Office for Statistics 2010b)

plans often predate the regional plan (Bauer 2010). Furthermore, developers seem to be capable of applying pressure against the maintenance of Green Corridors. A frequently cited example is the case of an industrial investor who was granted a building permit in a Northern Leipzig Green Corridor by a municipality (Bauer 2010).

## *9.4.2 Green Ring*

The regional development concept "Green Ring of Leipzig" has been implemented in the area of the Leipzig urban fringe. Initiated in 1996 by Leipzig planning officials, the Green Ring predominantly focused on the restoration of disused open-cast lignite mines and derelict industrial estates. Further areas of responsibility were the management of the remaining cultural landscapes and development of educational activities. These tasks proved a strong incentive for inter-municipal cooperation (Sinning 2002). To date, 26 key projects have been implemented. Regarding land use change, project number 8 "Inter-municipal pool of compensation areas" is particularly interesting for this study. The participants in this project manage the legal requirement for compensation in the process of land development

(§ 8a, Bundesnaturschutzgesetz (BnatschG 2009) in conjunction with §§ 1a, 9, 135a, 200a, Baugesetzbuch (BauGB 2005)). This is realised by directing compensation measures towards land which belongs to the "Green Ring". Furthermore, a series of projects aim to encourage residents to get to know their nearby natural and cultural landscapes. To this end, infrastructure for local recreation and tourism is constructed and leisure activities are organised, e.g. guided walks (Green Ring 2010).

The Leipzig-based constituent members of the Green Ring are 14 municipalities, including the city of Leipzig, two rural districts and a number of civil society organisations, private firms and individual citizens. It is considered to be an important cooperation arena for urban and peri-urban as well as northern and southern municipalities of Leipzig (Bauer 2010). The decision-making body is a regional conference composed of delegates of member organisations.

The assessment of the Green Ring highlights the strategy's strengths and weaknesses. Interviewees emphasised the fact that the Green Ring performs well because it creates many winners among municipalities in terms of better prospects for regional tourism and recreation. The strategy provides regional agents with an opportunity to get to know each other. Furthermore, the Green Ring was able to gain public support through concrete outcomes such as the establishment of footpaths. The success of the strategy is partly attributed to its full-time, professional management (Bauer 2010).

The strategy's main weakness is that tough decisions are deferred or not taken at all; there is no preference or prioritisation of projects, for example. This, it is argued, is a future risk because of the Green Ring's potentially conflicting aims regarding nature protection and tourism development. Furthermore, not all municipalities are equally committed and free-riding behaviour by some is seen as a problem (Bauer 2010).

### *9.4.3 Parthe Floodplain Cooperation*

The aim to protect the floodplain of the River Parthe and its riparian forests led to cooperation between the municipalities of Leipzig, Taucha and Borsdorf in 1992. Taucha and Borsdorf, two medium-sized towns, are located in the peri-urban area north and north-east of the centre of Leipzig. The Parthe Floodplain Cooperation or *Zweckverband* is legally constituted; its members form a contractual agreement. It receives its main financial support from the member municipalities, although other sources, such as state-wide funding programmes, are also used. The three municipalities, aided by a general assembly and a chairperson, synchronise preparatory landscape planning for the River Parthe which crosses the area. Since 1992, the cooperation increased its influence by degrees, taking over new responsibilities such as the management of a nature station and a local pool of compensation areas. Their personnel increased by hiring former volunteers.

Apart from the three core members, the neighbouring municipalities of Brandis and Großpösna also participate in the protection and management of floodplains and water bodies. The Parthe Cooperation also collaborates with other regional development strategy organisations, such as the Green Ring, for specific projects (Parthe Floodplain Cooperation 2009). Furthermore, farmers, tourism associations and the German Association for Environmental Protection are involved in the project. Short-term cooperation may also take place with civil society organisations, e.g. village councils and sports clubs, and the local economy, e.g. inns, for specific projects.

The Parthe Floodplain Cooperation is considered by interviewees to be a resilient approach because

1. It provides incentives to the members to stay involved, such as sharing professional personnel in landscape planning, pooling resources to finance projects and to submit funding bids and
2. It has so far managed to set itself achievable aims and has been, within these limits, successful.

This, it is argued, has created trust among the membership and has increased confidence in the cooperation's ability to cope with future challenges (Bauer 2010). The strategy's weaknesses are the need to balance interests between municipalities of different size and degrees of influence. Because there are no means of enforcement and because decisions are based on consensus, political support has to be maintained. An example is the cooperation's frequent repetition of economic arguments in order to stress the floodplain's qualities vis-à-vis municipalities (Bauer 2010).

### *9.4.4 Strengths and Weaknesses*

Similarities and differences between the three strategies are highlighted below (see Table 9.2). The comparison focuses on land use, implementation principles, strengths, weaknesses, opportunities and threats.

Though operating at different scales, the Green Corridors, Green Ring and the Parthe Floodplain Cooperation have common aims regarding landscape protection and management. They put their main focus on similar landscapes, though at different scales. However, there are differences with regard to their main implementation principles: The Green Corridors rely on statutory support, the Green Ring on financial support and the Parthe Floodplain Cooperation is implemented with cooperation/consensus. Because there is spatial overlap between them, the strategies are mutually reinforcing. The statutory support for the Green Corridors is

**Table 9.2** Characteristics of the selected strategies and their evaluation in terms of strengths, weaknesses, opportunities and risks

| Strategy criteria | Green corridors | Green ring | Parthe floodplain cooperation |
|---|---|---|---|
| Aim | Protection of unsealed land | Conservation, management and promotion of the cultural landscape | Floodplain conservation and management |
| Land use | Floodplains, farmland, brown fields... | Floodplains, cultural landscapes and water bodies | Floodplains |
| Main implementation principle | Statutory support | Financial support | Cooperation and consensus |
| Strengths | Spatially explicit aims,<br>Means of enforcement,<br>Communicability. | Fairly clear aims and related projects,<br>Focus on few, selected projects,<br>Visible outcomes,<br>Incentives for cooperation,<br>Full time management,<br>Bottom-up approach. | Well-defined aims which are easily controlled,<br>Incentives for cooperation,<br>Voluntary commitment,<br>Efforts to balance interests,<br>Baseline funding,<br>Statutory support for landscape planning. |
| Weaknesses | Lack of financial support,<br>Little scope for negotiation,[a]<br>Strategy doesn't create revenues/ jobs. | Not all municipalities equally committed,<br>No means of enforcement,<br>Unpopular decisions are deferred. | Constant need to balance interests,<br>No means of enforcement. |
| Opportunities | Future relevance of strategy regarding climate regulation. | Regional agents get to know each other,<br>Implementation of projects that are beneficial for all. | Complementary funding schemes,<br>Generation of trust,<br>Mutual learning. |
| Risks | Costs and benefits aren't balanced. | Sometimes conflicting aims: environment vs. tourism. | Changing political majorities, predominance of economic arguments. |

[a]With the notable exception of the "Zielabweichungsverfahren", the allowance of deviations from spatial planning aims (see 3.2.2)
Bauer (2010)

combined with the financing of the Green Ring. In this context, local, inter-municipal cooperation in green space planning, such as the Parthe Floodplain Cooperation, can thrive (Bauer 2010).

In spite of such synergies, the strategies do not prevent commercial, industrial or residential development in peri-urban areas. They do not offer a solution to ongoing land consumption. However, they guide such developments so that natural and cultural landscapes are compromised as little as possible. If this level of area protection is considered unsatisfactory, then more emphasis should be laid on the enforcement of regional planning aims.

**Box 9.1 Quality of Life in The Leipzig-Halle Region: Indicators of Land Use Change and Its Effect**

As part of the quality of life indicator work (see Chap. 3), a limited sample of people in Leipzig were asked about their perceptions of different factors of the environment in relation to their perceived quality of life. Although the results are not statistically valid owing to the restricted sample size, they offer some insights into the factors which people view as important. The most important issue as an indicator of environmental quality of life is a feeling of safety and security which might change if areas change their character, perhaps less populated and less well managed. The second factor in importance is air quality, possibly because people are concerned about increased pollution and dust from cars, for example. Noise is the next most important factor—traffic noise, aircraft and noisy neighbours may all have an impact. Housing suitability comes next followed by the accessibility to green space which is not as high as might be expected but is higher in priority than in other regions. Convenience of transport is the next priority—if land use change results in slower and less convenient transport or commuting times then people are less happy with their living environment. Adequacy of waste collection is just above the availability of shops which is at the bottom of the scale of these indicators. Thus, if land use change as a result of urban densification or further shrinkage starts to affect the higher level indicators negatively people may try to move to somewhere better. More samples would have enabled a more sophisticated analysis of the variability amongst the population and their current environment—such as whether they lived in a leafy suburb or inner city. The Quality of Life Simulator enables some of these factors to be tested along with predicted land use changes (Fig. 9.5).

## 9.5 Future Land Use

Modelling of future potential land use changes was carried out for Leipzig-Halle using two different systems, the Regional Urban Growth (RUG) model and MOLAND (see Chap. 3 for details on the methods used in each). The modelling was carried out using the four scenarios used in PLUREL and described in Chap. 1. For the RUG modelling see Box 9.2. The MOLAND modelling used locally adapted scenarios following stakeholder discussions (see below).

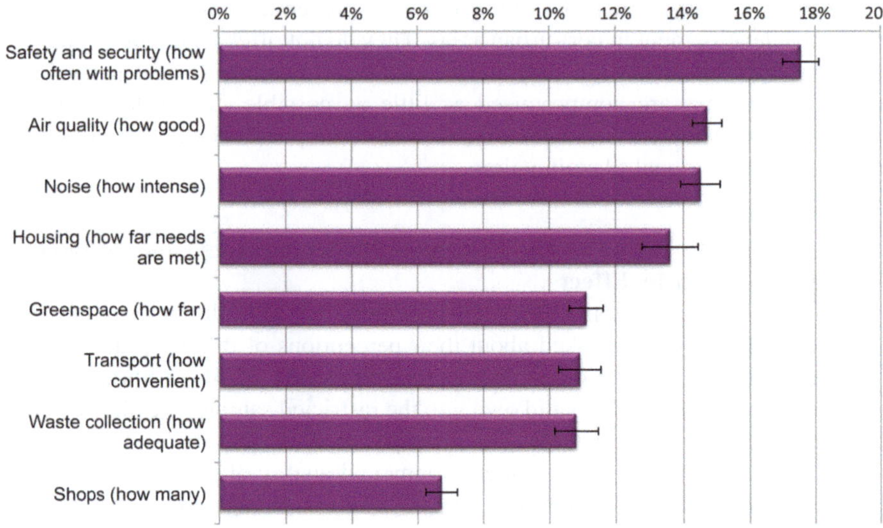


**Fig. 9.5** A graph showing the ranking and percentages of importance of different indicator values for Leipzig-Halle (Source: OPENspace research centre, Edinburgh College of Art)

**Box 9.2 Modelling Built-Up Areas in Leipzig-Halle: RUG Results**

Compared to the other case study areas, urban growth is very low in Leipzig-Halle (N.B. RUG is a "growth-only" model, i.e. shrinking cities will appear as having no growth).

In the Hypertech scenario, rapid technological change, new transport technologies and few planning constraints lead to counter-urbanisation. Although the population decreases in this region, economic growth leads to some expansion of built-up areas as dwellings become larger. In the Extreme Water scenario, the larger decrease in population and slower economic growth lead to less urban expansion than in the Hyper-tech scenario. Stricter planning policies and a higher risk of drought encourage a more compact urban form. In the Peak Oil scenario, high fuel costs and strict planning policies concentrate growth in the urban cores of cities. There is some increase in the peri-urban zone, mostly along the main transport axis between Leipzig and Halle, and very little in the rural areas. In the Fragmentation scenario, the shrinking population and slower economic growth lead to a very moderate increase in artificial surfaces. Urban growth occurs mostly in urban areas and very little in peri-urban areas (Fig. 9.6).

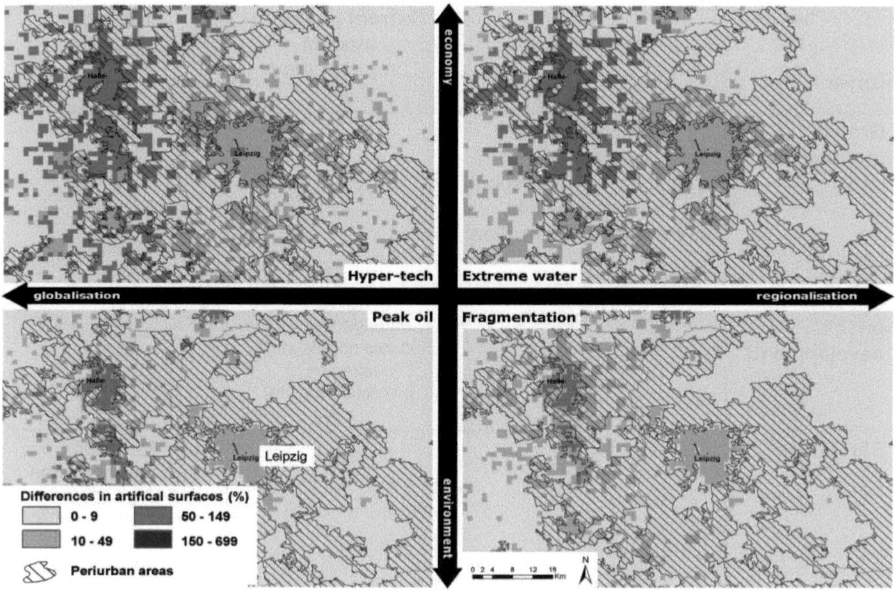


**Fig. 9.6** Shows the development of artificial surfaces in the Leipzig-Halle region according to the RUG (Regional Urban Growth Model) model for the year 2025. Compared to the other case studies, urban growth is very low in the Leipzig-Halle case study (Source: University of Edinburgh)

## *9.5.1 Modelling Land Use Change*

Compared to the RUG model, the MOLAND light model conveys a more detailed insight into the future development of land uses in the Leipzig-Halle region (see Chap.3). Based on MOLAND, a scenario framework was developed including a Business-as-Usual, a Hypertech and an Eco-Environmental scenario. These scenarios are consistent stories of how futures could look. In a stakeholder workshop in the Leipzig-Halle region the basic PLUREL framework scenarios were locally adapted and specified to capture the possible trends in regional development better. In particular, population decline was added to population growth and a shrinkage scenario was explored (Haase et al. 2009; Petrov et al. 2009) since this is a key characteristic of the region. To make these locally adapted scenarios comprehensible and not over-complex, different assumptions on general driving factors of land use change were agreed in discussion between researchers and stakeholders, these being population, economics and spatial planning. Using these factors, a matrix was developed combining assumptions on demographic and economic development (growth and shrinkage) with the strength of spatial planning (strong and laissez-faire). Possible combinations of the matrix are shown in Fig. 9.7.

The simulation results (Figure 9.8) show significant growth of settlement and commercial land—regardless of population decline or growth. This development

| Planning / Drivers | No restrictions | Current planning system implemented | Strong planning |
|---|---|---|---|
| Growth (population, GDP) | Hypertech (unrestrained growth, in-migration, new transport axes, commercial investment, gentrification) | | |
| Baseline (stable development) | | BAU (managed growth, stable population, weakening of protection zoning) | |
| Shrinkage | | | Peak Oil (managed shrinkage, declining population, investment in public transport, urban renewal) |

**Fig. 9.7** Scenario matrix including the locally adapted scenario storylines for the Leipzig-Halle region showing the selected land use scenarios "Hypertech", "Business as Usual" (BAU) and "Eco-Environmental" (layout: D. Haase)

would be at the expense of open and arable land (Haase et al. 2009). While in the "Business as Usual" (BAU) scenario the amount of discontinuous, low density residential land increases most, in the "Hypertech" scenario most of the growth is due to commercial development and in the "Eco-Environmental" scenario there are lower rates of both settlement and commercial growth compared to the other scenarios.

Figure 9.9 shows the simulation results in greater detail for the urban, inner and outer peri-urban and rural areas of Leipzig-Halle in 2025 (Petrov et al. 2009). Comparing the urban, inner and outer peri-urban and rural areas, the following observations can be made:

1. For all parts of the region, the model assumes an increase of impervious surface due to settlement and commercial growth in the Hypertech and BAU scenarios and a stable development in the Eco-Environmental scenario.
2. There is an increase of urban green space in the Eco-Environmental scenario.
3. Compared to the BAU scenario, commercial growth dominates the Hypertech scenario in the periurban space.
4. In the rural areas a very moderate increase of surfaced areas in addition to a small increase of wetlands or natural green spaces (Fig. 9.9).

In addition, the scenarios lead to more-or-less scattered land use patterns. The Hypertech scenario produces more dispersed land uses in the two cities, leading to large urban brownfields and a subsequent perforation of densely built-up areas

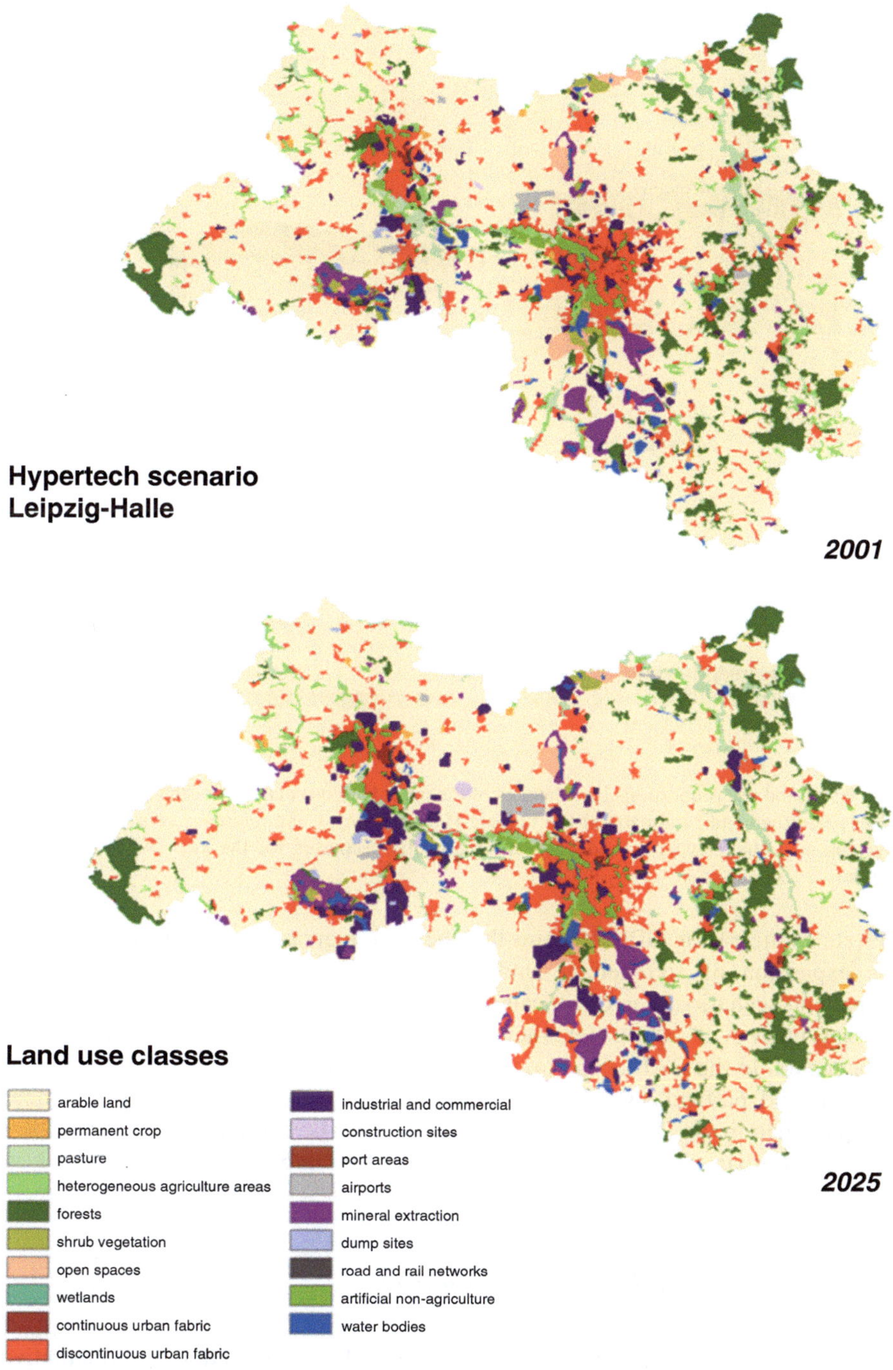


**Fig. 9.8** Output map out of the MOLAND model for the Hypertech scenario set to the year 2025. All land use classes processed in the model are depicted. For visualisation purposes they were amalgamated into six main land use classes and the "rest" in other sections of this report (Petrov et al. 2009)

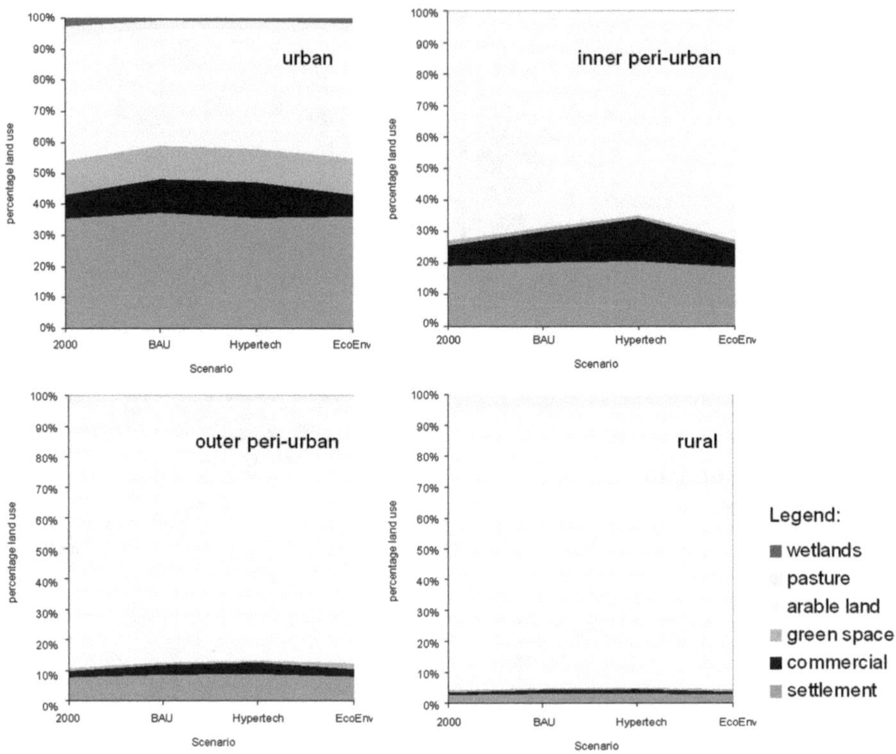


**Fig. 9.9** Land use proportions and changes of the main land use classes, for the urban, inner and outer peri-urban and rural parts of the Leipzig-Halle region for the reference year 2000, and the Business-as-Usual, Hypertech and Eco-environmental scenarios for the year 2025 (layout: D. Haase)

despite an increasing GDP. It seems that the GDP increase tends to feed sprawl and the expansion of commercial land (EEA 2006a; Kasanko et al. 2006). The Hypertech scenario assumes no balance between growth and decline of the urban fabric (Petrov et al. 2009). Here the highest degree of fragmentation of the built-up area occurs. In the BAU scenario, GDP growth leads to an increase of the area of built-up land, particularly in the urban and inner peri-urban zones and is, thus, more compact than in the Hypertech scenario. In the Eco-Environmental scenario, there is the least urban growth but more perforation of the urban areas.

## *9.5.2 Qualitative Land Use Scenarios*

In order to communicate the quantitative approaches described above, qualitative land use scenarios for the year 2025 were also developed by some 40 stakeholders and researchers at a Leipzig-based workshop in 2009. Stakeholders came from local and regional spatial planning agencies from both regions of Leipzig (Saxony) and

Halle (Saxony-Anhalt), including representatives of the core cities. The participating local government officers thus represented a spectrum of interests ranging from urban cores to peri-urban and rural areas. In addition, members of local NGOs and researchers from the universities of Leipzig and Halle took part.

The workshop started with a map of the current land use in the Leipzig-Halle region. Based on this map, participants developed scenarios of growth and shrinkage using the assumptions of the Hypertech and Eco-Environmental scenarios (see above) which were translated into storylines. A "Managed Growth" scenario was also introduced, combining assumptions of demographic and economic growth with an influential spatial planning model. The stakeholders were divided into three corresponding scenario groups. During the workshop each scenario group was supplied with a certain quantity of "land uses" according to the respective scenario assumptions. Coloured paper, symbolising land use classes, and pens were used for preparing simple visualisations.

Each group developed their spatial vision and strategic position in three consecutive stages: A land use map for 2025 was created, analysed and then discussed and assessed regarding its social, economic and environmental implications by the stakeholders and researchers, allowing for a discussion on sustainable regional development (Figs. 9.10 and 9.11).

This qualitative, discursive method enabled researchers to understand local implications of land use change better. In the following section, the main lines of the argument of the evaluation of scenarios are outlined for urban, peri-urban and rural areas.

### *9.5.3 Urban Areas*

A scattered urban structure as produced by the Hypertech scenario and especially by the Eco-Environmental scenario is seen in part as an asset by participants of the scenario workshop. The reasons for this are, firstly, improved environmental qualities, such as more accessible green open spaces; secondly, benefits for local recreation and related businesses are to be expected. Such advantages, it is argued, have to be actively pursued by municipal actors who are restrained by their budgets. However, there were also dissenting voices among workshop participants, who pointed out that scattered urban cores may result from maintaining successful business locations and established residential areas and dismantling disadvantaged areas. Thus, increased social segregation and a widened gap between "successful" and poor communities would tend to occur.

By contrast, increased urban density, created by the Managed Growth scenario, was unequivocally rejected for its negative effects on quality of life since it would lead to a loss of green open space and reduced accessibility to the remainder. Furthermore, increased density was rejected for social and economic equity reasons. It would require investments to be directed towards urban cores and away from peri-urban municipalities.

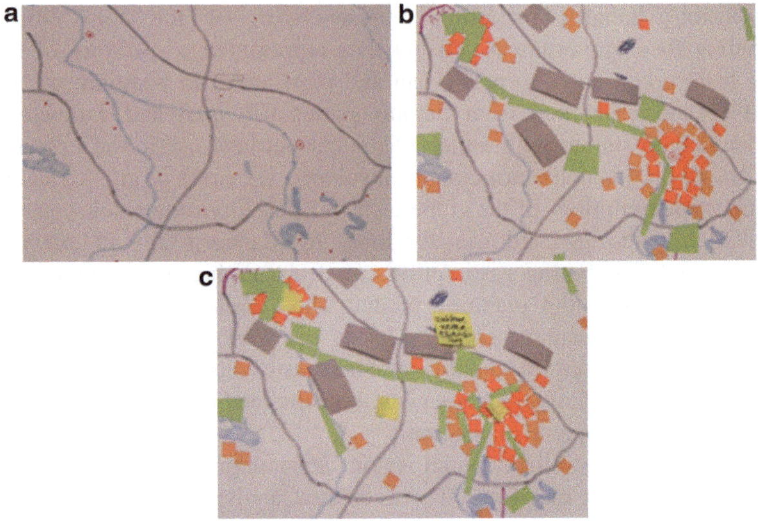


**Fig. 9.10** Different stages in the visualisation of the eco-environmental scenario by participants of the scenario workshop which took place in the Leipzig-Halle region in 2009

**Fig. 9.11** The participants of the regional workshop, composed of local and regional planners from the Leipzig-Halle region, researchers and representatives of local NGOs, at work

## 9.5.4 Peri-Urban and Rural Areas

In the Hypertech scenario in particular but also in the Eco-Environmental scenario, urban growth tends to focus on the inner peri-urban areas as well as along the Leipzig-Halle axis. The main drivers of land use change in the Hypertech scenario, according to workshop participants, would be industrial and commercial investors from both inside and outside the region. Outer peri-urban and rural areas would largely be bypassed for development. Consequently intra-regional disparities would grow and the current configuration of the Leipzig peri-urban area, with leisure in the south and commerce and industry in the north west, would be changed to some extent.

A "laissez-faire" approach to urban sprawl, as anticipated in the Hypertech scenario, was considered beneficial for the regional economy by workshop participants because development options in the former industrial areas of Leipzig-Halle would be fully utilised. The region offers a wide variety of land resources and investors would enjoy an unprecedented degree of freedom regarding location choice. There would be space for creative business start-ups and new job opportunities would arise. Conversely, sprawl would entail follow-up costs and damage to ecological functions of the peri-urban landscapes.

Urban compaction in the managed growth scenario would lead to selective development in peri-urban areas, e.g. along major traffic routes. Such a development would reduce transport costs and lead to a concentration of employment opportunities in the centres or a few commercial areas. Conflicts would probably arise due to the comparatively small scale of developments in peri-urban areas. A development towards urban compaction would require strong and influential regional planning irrespective of current jurisdictions and administrative boundaries.

### *9.5.5 Preliminary Conclusions*

The selected MOLAND scenarios, to different degrees, lead to an increase of built-up areas regardless of population or economic growth and decline. This increase is least pronounced in rural areas and most pronounced in inner peri-urban areas. While built-up areas increase most in the Hypertech scenario, followed by the Business as Usual scenario, the Eco-Environmental scenario leads to the lowest increase of built-up land. There is thus little indication of any distinct land use effects of stagnation. Only the Eco-Environmental scenario, which combines demographic and economic shrinkage with strong spatial planning, creates further urban green open spaces and is distinct in this respect.

Regarding potential future land use configurations, these extreme scenarios lead to scattered (Eco-Environmental) or sprawling (Hypertech) urban forms while the Business as Usual scenario lead to a comparatively compact and dense urban development.

An evaluation by local stakeholders provided insights into their view of future land use. A general preference of scattered urban structures became evident, even though negative social consequences of such a development were identified. By comparison, the peri-urban and rural development options, urban sprawl versus no further urban development, were strongly contested. This was mainly due to opposing evaluations of the likely economic and environmental effects of the same scenarios. Overall, the role of the state in mitigating threats and realising development potential inherent to the three scenarios was emphasised.

## 9.6 Implications for Land Use: New Green-Blue Services?

### *9.6.1 Measuring Ecosystem Services Potential*

Scenarios provide images of future potential versions of the stagnating Leipzig-Halle region. They clearly show differing land use changes and thus present the potential roles for protecting existing and creating new ecosystem services in the region. The concept of ecosystem services translates ecosystem functions into services that are provided for humans (MEA 2003; Bolund and Hunhammar 1999). The maps of future land use configurations produced by the MOLAND model for each of the extreme scenarios were evaluated in terms of their provision of ecosystem services in the case-study region. Furthermore, assumptions regarding the intensity and productivity of agricultural land use were also made, e.g. EU agriculture subsidy programmes and a diversification of energy production, which could lead to positive changes for the ecosystem services of food and energy production (see Chap. 3). Eight ecosystem services were quantified, using the methods described in Chap. 3:

1. Local climate regulation,
2. Global climate regulation,
3. Recreation,
4. Water quantity regulation,
5. Biodiversity,
6. Food provision,
7. Energy provision, and
8. Fresh water provision.

### *9.6.2 Overall Ecosystem Services in the Region*

Evaluation results for these ecosystem services are summarised in a "spidergram" (Fig. 9.12), showing differences across the scenario results for 2025 compared to the year 2000. For local and global climate regulation, biodiversity, water quantity regulation, and fresh water supply, all scenarios were assessed to be slightly lower than for 2000. This is due to the conversion of forested areas to built-up and arable land in the MOLAND model. Conversely, energy and food provision are found to be higher in 2025 than in the year 2000 for all scenarios. The reason is an intensification of agricultural production and the transformation of fallow into productive arable land even though the total area of arable land is declining in all

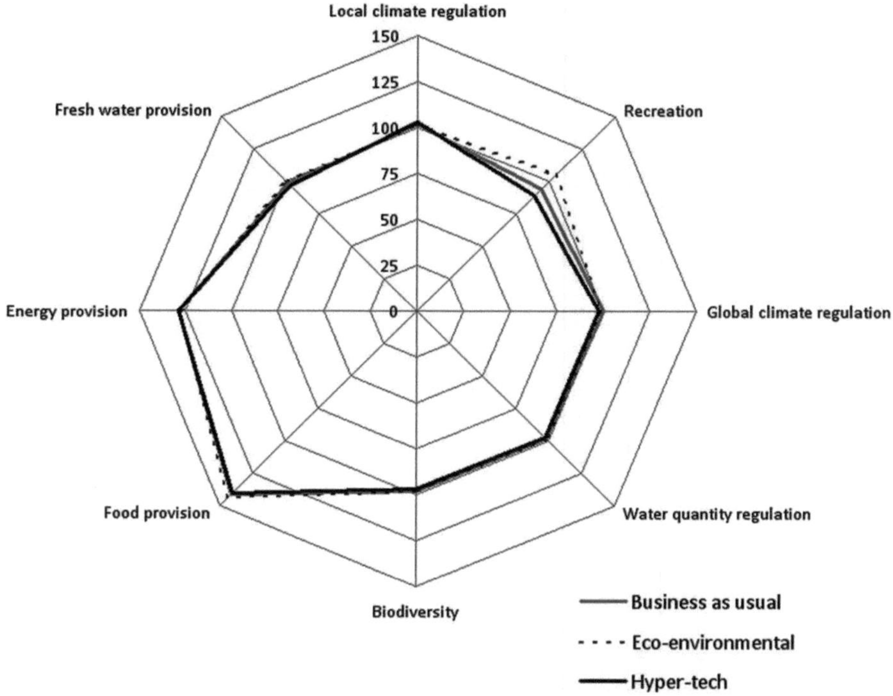


**Fig. 9.12** Spidergram for ecosystem services for all three MOLAND scenarios. The spidergram depicts relative changes of the 2025 scenarios compared to 2000. A value of 100 % indicates no changes from 2000 to 2025, a higher value an increase in the respective ecosystem service, a lower value a decrease (layout: N. Schwarz)

land use scenarios, which increases these two ecosystem services. For recreation, the Eco-Environmental scenario performs better than in 2000, while both Hyper-tech and Business as Usual are worse. This is because the Eco-Environmental scenario is the only one in which urban green space increases. This is a clear signal for a potential to increase recreation by creating new accessible green and blue areas. However, other indicators might also show such tendencies in specific areas, while the effect for the whole region might have been too small or even overshadowed by development somewhere else.

### *9.6.3 Ecosystem Services Potential in Inner and Outer Peri-Urban, Urban, and Rural Parts of the Region*

To assess the potential for ecosystem services in more detail, an analysis distinguishing between inner and outer peri-urban, urban and rural parts of Leipzig-Halle was carried out. Results are summarised in the following Table 9.3.

**Table 9.3** Relative changes of ecosystem services in the three scenarios, split into rural, outer peri-urban, inner peri-urban and urban areas of the region Leipzig-Halle

| | Business as usual scenario | | | | Hypertech scenario | | | | Eco-environmental scenario | | | |
|---|---|---|---|---|---|---|---|---|---|---|---|---|
| Eco-system service | Rural | Outer peri-urban | Inner peri-urban | Urban | Rural | Outer peri-urban | Inner peri-urban | Urban | Rural | Outer peri-urban | Inner peri-urban | Urban |
| Local climate regulation | ↔ | ↔ | ↓ | ↔ | ↔ | ↔ | ↓ | ↔ | ↔ | ↔ | ↔ | ↔ |
| Global climate regulation | ↔ | ↔ | ↔ | ↔ | ↔ | ↔ | ↓ | ↔ | ↔ | ↔ | ↓ | ↔ |
| Recreation | ↔ | ↓ | ↓ | ↔ | ↔ | ↓ | ↓ | ↔ | ↔ | ↔ | ↑ | ↑ |
| Biodiversity | ↔ | ↔ | ↓ | ↔ | ↔ | ↓ | ↓ | ↔ | ↔ | ↔ | ↔ | ↔ |
| Food provision | ↑ | ↑ | ↑ | ↑ | ↑ | ↑ | ↑ | ↑ | ↑ | ↑ | ↑ | ↑ |
| Energy provision | ↑ | ↑ | ↑ | ↔ | ↑ | ↑ | ↑ | ↔ | ↑ | ↑ | ↑ | ↔ |
| Fresh water provision | ↔ | ↔ | ↓ | ↔ | ↔ | ↔ | ↓ | ↔ | ↔ | ↔ | ↓ | ↔ |

Value for ecosystem service is changed by (↑) +5 % or more, (↓) −5 % or more, (↔) less than ±5 %. + always indicates a positive change in terms of more provision of the respective service, − a negative change (layout N. Schwarz)

## 9.7 Lessons Learned

Based on this research, some conclusions for spatial planning practice can be outlined. Firstly, the quantitative analysis of ecosystem services shows a diverse picture of future land use in the Leipzig-Halle region. Indicators may point in different directions for the same scenario (e.g. food and energy provision increases in every scenario, whereas other indicators remain stable or decrease). These trade-offs are due to the fact that, for instance, food and energy provision increase if arable land remains stable in area, whereas other indicators are related to other land uses like forests or water surfaces which change in the modelling process. Consequently, local stakeholders have to prioritise indicators when assessing the outcomes of such a scenario analysis.

Furthermore, no scenario shows positive trends for all indicators. Even in the Eco-Environmental scenario, some indicators (global climate regulation and fresh water provision) decrease, at least in the inner peri-urban areas. A more detailed view of the advantages and disadvantages of different scenario outcomes is therefore needed.

Finally, the spatially explicit investigation of ecosystem services for these scenarios shows possible foci for planning strategies: the inner peri-urban areas especially need more attention in the future, as they are particularly prone to fragmentation and loss of valuable land for most of the ecosystem services. Therefore, planning strategies should

- Focus spatially on these areas and
- Aim at protecting or even enhancing green and blue areas.

Furthermore, the ecosystem service potential of urban green open spaces, as anticipated in the Eco-Environmental scenario, could be realised by planning for such areas.

Such spatial planning strategies already exist in the Leipzig-Halle case study and improving their effectiveness appears to be pivotal. Since the German planning system concentrates many responsibilities at the municipal level, enhancing and protecting green-blue areas first and foremost requires municipal support. However, inter-municipal cooperation in the Leipzig-Halle region is impeded by administrative fragmentation and deeply entrenched conflicts of interest. One way to overcome these obstacles is by strengthening multi-level governance. The analysis of selected nature protection strategies at different levels—the Green Corridors, the Green Ring and the Parthe Floodplain cooperation—suggests that these strategies are more effective when carried out together. Their implementation principles, statutory instruments and financial support as well as voluntary commitment, are complementary and mutually reinforcing. In order to maintain or even increase such effects, an extension of current regional development funding should be considered. Furthermore, the Parthe Floodplain Cooperation with its voluntary commitment could set a precedent for inter-municipal landscape planning within the Green Corridor and Green Ring areas.

The aim of protecting landscapes and unsealed surfaces in peri-urban areas could be further supported by the dispersion of development pressure through increased brownfield redevelopment. The "recycling" of such areas is frequently proposed in the literature (e.g. Lorance Rall and Haase 2011). Brownfield redevelopment requires sufficient funds to reclaim these areas, which, without such efforts, are comparatively costly for developers to use. It would therefore benefit from increased state funding for the clearance of derelict areas and the demolition of surplus infrastructure. Such funding was available for urban areas until 2009 as part of the state's urban renewal programme. The research results suggest that a continuation of the programme and its extension to inner peri-urban areas should be considered.

## 9.8 Conclusions

Stagnation, the coincidence of more or less equivalent rates of population growth and shrinkage, can lead to positive effects for ecosystem services. This hypothesis was examined in the Leipzig-Halle region by focusing on (1) past and recent land use development and regional governance, (2) potential future land use scenarios, (3) their impacts on ecosystem services and (4) corresponding guidance options. By addressing the potentials of stagnation, this research aims to contribute a distinct perspective to the discussion of shrinking cities, which is often concerned with the risks of shrinkage rather than the possible benefits.

In the recent past, stagnation and economic decline in the Leipzig-Halle region coincided with urban spatial growth. The principal population loss occurred between 1990 and 2000 and since then, a gentle population increase has taken place in the city of Leipzig while other parts of the region are still shrinking. In the 1990s, major land consumption took place in the peri-urban area due to commercial and residential expansion and mineral extraction. Recently, this process has been reversed and the opencast mines are being converted into lakes, arable land or forest. Apart from the restoration of former mining areas, there is so far very little evidence of ecosystem service potential arising from stagnation. Administrative fragmentation and deeply entrenched conflicts of interest between municipalities weakens the influence of spatial planning on the control of urban growth. Consequently, much of the past urban growth took place in an uncoordinated manner, leading to an oversupply of commercial and residential areas.

Both models applied in the Leipzig-Halle case study, RUG and MOLAND, lead to urban growth, although this is low compared to other case studies (see Chap. 6 for example). The selected MOLAND scenarios, to different degrees, each lead to a potential increase of built-up areas regardless of population and economic growth or decline. This increase is least pronounced in rural areas and most pronounced in inner peri-urban areas. While built-up areas increase most in the Hypertech scenario, followed by the Business as Usual scenario, the Eco-Environmental scenario creates the least increase of built-up land. There is thus, given the selection of scenarios in this research, little indication of distinct land use effects of stagnation. Only the Eco-Environmental scenario, which combines demographic and economic shrinkage with strong spatial planning, creates further urban green open spaces and is distinct in this respect. Regarding potential future land use configurations, the selected MOLAND scenarios potentially lead to scattered, perforated (Eco-Environmental) or sprawling (Hypertech) urban forms while the Business as Usual scenario lead to a comparatively compact and dense urban development.

These potential future land use configurations were found to be controversial by stakeholders and researchers during the Leipzig-based workshop. Only the scenarios leading to urban density were rejected unequivocally. A perforated or sprawling urban development was seen to benefit local recreation and the regional economy while furthering social segregation and damaging green-blue areas. A compact urban development, with development preferably in selected peri-urban locations, was considered unattractive for peri-urban municipalities but would create advantages for the region as a whole, e.g. through reduced transport costs. The government's role in reducing risks and realising potentials related to land use change were emphasised by stakeholders.

A quantification of the impacts of scenarios shows that, firstly, none of the scenarios yields entirely positive results for ecosystem services. All scenarios led to a conversion of forested areas to built-up land. Consequently, the value derived from the land use in the development scenarios is slightly less than that for many ecosystem services. Only certain services, such as energy and food provision, were found to increase, primarily due to an intensification of agricultural production. The spatially explicit evaluation of scenarios shows that the ecosystem service "recreation" increases in the Eco-Environmental scenario, as this is the only one in which urban green space expands. This is a clear indication of a potential to increase recreation by creating new green-blue areas. Furthermore, the inner peri-urban areas, especially, need more attention in the future, as they are particularly prone to fragmentation and loss of valuable land for most ecosystem services. Discussions with local stakeholders suggest that social and economic consequences of land use change do matter (see above section). This is an indication for future research to take these aspects further into account when evaluating land use change in a stagnating region.

Implications for spatial planning arising from the research results suggest that in the case of Leipzig-Halle, stagnation does not generally decrease residential and commercial development. Nevertheless, some ecosystem service potentials could emerge mainly due to an increase of urban green space. Realising such potentials through spatial planning requires an awareness of them, which is where the PLUREL research results should be of use. In addition, the results, in particular the evaluation of scenarios, call attention to green-blue areas at risk. Therefore, further efforts to keep such areas free of development should be taken, e.g. by fully implementing existing nature protection strategies.

## References

Bauer A (2010) Assessment report on local and regional strategies. Internal report. http://www.plurel.net. Accessed 07 July 2010

Baugesetzbuch (BauGB, 2005) 37. Edition, Deutscher Taschenbuch Verlag, München

Bolund P, Hunhammar S (1999) Ecosystem services in urban areas. Ecol Econ 29:293–301

Bundesnaturschutzgesetz (BnatschG 2009) http://www.bundesrecht.juris.de. Accessed 23 Aug 2010

City of Halle (2010) http://www.halle.de. Accessed 15 Nov 2010

City of Leipzig (2007) Bevölkerungsvorausschätzung für die Stadt Leipzig 2005. Municipality of Leipzig, Leipzig

Eggers O, Schüttlohr K (2005) Öffentliches Baurecht. Bayerische Verwaltungsfachschule, München

EEA (European Environmental Agency) (2006a) Urban sprawl in Europe – the ignored challenge. EEA Report Oct 2006, Copenhagen

EEA (European Environment Agency) (2006b) Corine land cover data download. http://www.eea.europa.eu/themes/landuse/interactive/clc-download. Accessed 21 July 2010

Green Ring (2010) http://www.gruener-ring-leipzig.de. Accessed 21 July 2010

Haase D, Nuissl H (2007) Does urban sprawl drive changes in the water balance and policy? The case of Leipzig (Germany) 1870–2003. Landscape Urban Plann 80:1–13

Haase D, Nuissl H (2010) Assessing the impacts of land use change on transforming regions. Editorial. Land Use Sci 5(2):67–721

Haase D, Seppelt R, Haase A (2007) Land use impacts of demographic change—lessons from eastern German urban regions. In: Petrosillo I, Müller F, Jones KB, Zurlini G, Krauze K, Victorov S, Li B-L, Kepner WG (eds) Use of landscape sciences for the assessment of environmental security. Springer, Dordrecht, pp 329–344

Haase D, Bauer A, Reuther I, Schiffers B, Schwarz N, Seppelt R, Weinert J (2009) PLUREL regional workshop, Spatial development in the Leipzig-Halle region. Helmholtz Centre for Environmental Research, Leipzig

Regional Planning Association Halle (RPA Halle) (2009) Regional plan. http://www.regionale-planung.de/halle/html/frames/rplan.html. Accessed 19 Nov 2009

Herfert G (2007) Regionale Polarisierung der demographischen Entwicklung in Ostdeutschland: Gleichwertigkeit der Lebensverhältnisse? Raumforschung und Raumordnung 65:435–455

Institut für Länderkunde (IfL) (ed) (2001) Exkursionsführer Mitteldeutschland. Westermann, Braunschweig

Kasanko M, Barredo JI, Lavalle C, McCormick N, Demicheli L, Sagris V, Brezger A (2006) Are European cities becoming dispersed? Landscape Urban Plann 77(1–2):111–130

Loibl W, Köstl M (2008) Report on a methodology to delineate RUR-sub regions. Internal report. http://www.plurel.net. Accessed 21 July 2010

Lorance Rall ED, Haase D (2011) Creative intervention in a dynamic city: a sustainability assessment of an interim use strategy for Brownfields in Leipzig, Germany. Landscape Urban Plann 100:189–201

MEA—Millennium Ecosystem Assessment (2003) Ecosystems and human well-being: a framework for assessment. Island Press, Washington, DC

Nuissl H, Rink D (2005) The 'production' of urban sprawl in eastern Germany as a phenomenon of post-socialist transformation. Cities 22(2):123–134

Parthe Floodplain Cooperation (2009) http://www.zv-parthenaue.de. Accessed 7 Dec 2009

Petrov L, Haase D, Lavalle C (2009) MOLAND Model results for the Leipzig region. PLUREL Newsl 5:5

Pütz T (2010) Pendlerverflechtungen im Großraum Leipzig. http://www.bbr.bund.de. Accessed 3 Nov 2010

Raumordnungsgesetz (ROG, 2005) 37. Edition, Deutscher Taschenbuch Verlag, München

Saxon Ministry of the Interior (SMI, 2004) Raumordnung und Landesentwicklung in Sachsen. http://www.landesentwicklung.sachsen.de/index.html. Accessed 24 Oct 2009

Saxon Office for Statistics (2010a) Population statistics. http://www.statistik.sachsen.de. Accessed 20 July 2010

Saxon Office for Statistics (2010b) Territory, administrative structure. http://www.statistik.sachsen.de. Accessed July 2010

Saxon Office for Statistics (2010c) Employment and labour market. http://www.statistik.sachsen.de. Accessed 20 June 2010

Saxony-Anhalt Office for Statistics (2010a) Population statistics. http://www.statistik.sachsen-anhalt.de. Accessed 20 July 2010

Saxony-Anhalt Office for Statistics (2010b) Territory, administrative structure. http://www.statistik.sachsen-anhalt.de. Accessed July 2010

Schwarz N, Bauer A, Haase D (2010) Assessing climate impacts of planning policies—an estimation for the urban region of Leipzig (Germany). Environ Impact Assess. doi:10.1016/j.eiar.2010.02.002

Sinn A, Haase D, Walde A (2008) Analysis of regional spatial planning and decision-making strategies and their impact on land use in the urban fringe. Helmholtz Centre for Environmental Research—UFZ, Leipzig

Sinning H (2002) Leistungsfähigkeit und Grenzen kommunikativer Planungsinstrumente – am Beispiel nachhaltiger Freiraumentwicklung in Stadtregionen. Dissertation, RWTH Aachen

Tosics I, Szemö H, Illés D, Gertheis A, Lalenis K, Kalergis D (2010) National spatial planning policies and governance typology. Internal report. http://www.plurel.net. Accessed 21 July 2010

Turowski G (2002) Spatial planning in Germany: structures and concepts. Akademie für Raumforschung und Landesplanung, Hannover

Regional Planning Association Western Saxony (RPA Western Saxony, 2008) Regional plan. http://www.rpv-westsachsen.de. Accessed 4 Sep 2009

# Chapter 10
# The *Montpellier Agglomération* New Approaches for Territorial Coordination in the Periurban

**Françoise Jarrige, Jean-Pierre Chery, Jennifer Buyck, and Jean Paul Gambier**

## 10.1 Introduction

Due to the attractiveness of its Mediterranean coastal location, the city region of Montpellier in Languedoc-Roussillon in the south of France is a dynamic area with a positive migration balance and a population which keeps on growing (Fig. 10.1). Urban sprawl caused significant changes in the peri-urban landscape during recent decades, with individual housing plots spreading out from the cores of all the villages. New building also took place in former vineyards which used to be the characteristic land-use as result of the historical specialisation in the mass-production of table-wine (Fig. 10.2). This sector has become increasingly weakened by economic problems, as have most of the other traditional agricultural sectors. The local economy now largely depends on tertiary activities, such as tourism, higher education, research, medicine, and new technologies.

These driving forces led to a very diverse population mix and socio-spatial segregation, uncoordinated development and changing peri-urban landscapes—with an increasing *per capita* rate of space consumption—until the creation of a

F. Jarrige (✉)
Montpellier Supagro - UMR Innovation Bâtiment 27, 2, Place Viala, 34060 Montpellier cedex 1, France
e-mail: jarrige@supagro.inra.fr

J.-P. Chery
AgroParisTech - UMR TETIS Territoires, Environnement, Télédetection et Information Spatiale, Maison de la télédétection, 500 rue Jean-François Breton, Domaine de Lavalette, 34093 Montpellier, France

J. Buyck
UFR Institut d'urbanisme de Grenoble, 14, avenue Marie Reynoard, 38100 Grenoble, France

J.P. Gambier
Montpellier Agglomération Land Service, Montpellier Agglomération, 50 Place Zeus, 34000 Montpellier, France

K. Nilsson et al. (eds.), *Peri-urban futures: Scenarios and models for land use change in Europe*, DOI 10.1007/978-3-642-30529-0_10,

**Fig. 10.1** The road of lattes, between lagoon and recent houses

**Fig. 10.2** Old vines and a water container in the east of Montpellier

new form of local authority, the *Montpellier Agglomération*, at the end of 2001. This local government, currently combining 31 municipalities, is now responsible for several major public policies. Among those formerly implemented by municipalities are spatial planning at a regional scale, public transport, water management and housing. The creation of *Montpellier Agglomération* brought about significant changes in local governance and planning practices.

How does this new authority address the challenges of sustainable development at regional scale? What are the governance issues, and what development strategies have been established? What spatial planning tools are used? What contrasting scenarios can be drawn for future development? And what are the impacts of these scenarios when assessed against an integrated analysis in terms of sustainable development? These are the questions for this chapter to answer.

## 10.2 A Growing City-Region on the Mediterranean Seaside

### *10.2.1 Land Use Dynamics and Patterns in Montpellier City-Region*

Montpellier first appears in the historical record in 985 A.D., and through the mediaeval period was well known for medicine (the University was founded in 1220) and the linen trade. At the end of the eighteenth century, this Mediterranean city in the south of France could count 32,000 inhabitants. With the arrival of the railway in 1839 Montpellier region, which enjoyed no comparative advantage for industrial development and relatively poor agricultural soils, specialised in the mass production of table wine which could be easily distributed using the railway network. Even though the wine industry faced major crises throughout its history, it allowed the rural population to maintain a relatively high standard of living and prevented the widespread rural exodus to the cities experienced by many other French regions.

During the first part of the twentieth century, Montpellier was a quiet capital of a region of vineyards and experienced few urban changes. The population rose from 76,000 in 1901 to 91,000 by 1936 but stagnated until 1954 at around 96,000 inhabitants. After the First World War, the first low-cost housing appeared, and some public facilities were provided (schools, university facilities and a big hospital). At the same time, military land was opened for civil use, and built-up areas expanded.

With the creation of 22 administrative regions in France in 1956, Montpellier was promoted to be the capital of Languedoc-Roussillon and the 1960s were years of demographic, economic and spatial growth. Thanks to public and private investments, news roads and new peripheral residential districts were constructed. The sudden arrival of large numbers of repatriated people from North Africa (following the independence of Algeria in 1962) marked the start of the period of uninterrupted demographic growth of the Montpellier region which still continues today. With 120,000 inhabitants in 1962, the Montpellier population rose to

**Table 10.1** Population growth and urban sprawl in Montpellier city-region 1960–2004

| 1960 | 1980 | 2000 | 2004 |
|---|---|---|---|
| 145,000 inhabitants | 280,000 inhabitants | 375,000 inhabitants | 395,000 inhabitants |
| Urban area: 1,000 ha | Urban area: 4,000 ha | Urban area: 9,000 ha | Urban area: 10,000 ha |
| Urban density: 145 inhabitants/ha | Urban density: 70 inhabitants/ha | Urban density: 41 inhabitants/ha | Urban density: 39.5 inhabitants/ha |

Source: Montpellier Agglomération

200,000 inhabitants by 1982. During this period, the city opened up to a wider economic environment with the arrival of "new economy" actors (such as IBM in 1965), tourism development (new seaside resorts) and transport infrastructure (airport, 1965 and motorway from the north, 1967).

In 1984, the reform of the EU common wine policy significantly reduced the public support for table wine production, leaving local wine producers facing an uncertain future and an unfavourable market (a decrease of table wine consumption combined with aggressive international competition). Some growers succeeded in rebuilding a new wine economy by producing wine to meet high quality standards but many of them gave up and the area of the region managed under vineyards was massively reduced. At the same time, on-going economic changes (technocity project since the 1980s) and social evolution (white collar workers looking for quiet residential locations in nearby villages and embracing the car-borne way of life), led to Montpellier city-region becoming a place characterised by a young population (students, young adults working in new economy sectors), major demographic growth and high pressure on housing stocks, with an extension of the urban area in response. The combined effects of the land release triggered by the wine crisis and population growth led to an accelerated consumption of rural land in the peri-urban fringes of Montpellier. In 1999, the agglomeration of Montpellier had 290,000 inhabitants and together with rural neighbourhoods associated with the urban area had 460,000 inhabitants. Table 10.1 shows the relationship between population growth, urban area and population density. According to the typology for rural–urban regions developed in PLUREL (see Chap. 3), Montpellier city-region belongs to the NUTS 3 Département de l'Hérault and is a "monocentric large" type.

### *10.2.2 Land Use Planning, Territorial Governance and the Management of Periurban Areas*

The French definition of peri-urban municipalities—as used for the population census by INSEE (*Institut National de la Statistique et des Études Économiques*)—is based on morphological and socio-economic functional criteria:

- Morphology: urban spaces present agglomerated populations in densely built-up areas (less than 200 m between buildings) and rural areas show open spaces with

sparse population; in between, peri-urban areas are characterized by alternate built-up and open spaces.
- Socio-economic functionality: more than 40 % of the working population living in peri-urban areas commute to the central city or to a surrounding municipality of the first ring of the urban pole.

Apart from their residential function, peri-urban areas also have different kinds of use. This multi-functionality includes:

- Productive agriculture;
- Outdoor recreational uses such as walking, cycling, horse riding, nature watching, hunting and dirt biking;
- Environmental services such as waste treatment, water quality protection, flood expansion areas, biodiversity conservation.

Peri-urban spaces also matter because of the landscape they provide, and in a more symbolic way, they play a role in forming a sense of local identity. In Montpellier city-region all these are found.

France was historically characterised as a highly centralised state but during recent decades a more decentralised territorial administration has been set up. In 1981–1982, the first decentralisation laws implemented the devolution of administrative competences from the central state to local governments:

- Municipalities were delegated the power to deliver legal agreements for development applications (development and land planning); municipalities can belong to one or several inter-municipal cooperation organisations (EPCI: *établissement public de coopération intercommunale*). In 1999 a new law was enacted to promote and simplify inter-municipal cooperation (three main types of EPCI according to different population thresholds, with financial incentives to promote the creation of these new EPCI);
- The *Conseil Général* (*Département*, NUTS 3 level) plays a major role in the field of social support, social solidarity, local transport infrastructure and services, rural development and land policy;
- The *Conseil Régional* (NUTS 2 level) mostly plays a role for future planning and coordination of public policies and develops contracting policies with other local governments, national government and the EU; it elaborates planning schemes at the regional scale (services schemes, SRADDT: *Schéma Régional d'Aménagement et de Développement Durable du Territoire*, Regional Scheme for Sustainable Territorial Development) but these documents only have forecasting/future planning aims and are used as supports for elaborating public policies; they carry no formal legal weight.

Since 2009, major debates about decentralisation have taken place among the French political class because of the reform of taxation on economic activities and the potentially deleterious consequences it might have on the fiscal resources of local governments. The official national government discourse is about more efficiency of public services and fewer taxes on economic activities. However, local governments are being assigned the responsibility for more and more public services without, as they see it, the appropriate resource transfers from central government. This debate illustrates what some authors call the "distant governing state" in the French decentralisation process of territorial government: spatial disengagement of (central) state together with reinforcement of centralised control tools (Le Galès 2004; Epstein 2009; Négrier 2010).

Concerning spatial planning, the legal framework is set out at the national level by the "Urbanism code". It provides common rules for urban design, infrastructure provision, nature and natural resource protection. At a local level, urban design is applied through legal planning documents: a local urbanism plan (PLU: *Plan Local d'Urbanisme*) must be established by each municipality with more than 2,500 inhabitants. Zoning different types of development or non-development areas is no longer enough since the law of urban solidarity and renewal (*loi "Solidarité et Renouvellement Urbain"*) in 2001. A "sustainable development project" (PADD: *Projet d'Aménagement et de Développement Durable*) must demonstrate the political project that justifies spatial planning design and development choices for the following decade, according to demographic trends, housing and infrastructure needs, economic activities and jobs creation planned in the municipality.

This law also introduced spatial planning at a regional scale. Some city-regions had already implemented inter-municipality planning earlier, but it has now become a statutory requirement to draw up a territorial coherence scheme (SCoT: *Schéma de Cohérence Territoriale*) at regional scale in all French city-regions. Like the PLU, the SCoT presents both a spatial planning design and a PADD. Other planning documents—PDU (Plan of Urban Mobility: *Plan de Déplacement Urbain*), PLH: (*Plan Local de l'Habitat*, Local Housing Plan)—are combined with the SCoT to draw and implement urban design policies at the scale of functional urban regions. The SCoT and the PLU should be compatible and participative methods are also supposed to be implemented in local planning process.

The main policy directions concerning the peri-urban aim to protect space by continuous urbanisation, to avoid new isolated urban nuclei to arise from the break-up of agricultural units. Special protection rules apply to sensitive areas (mountains, coast). Specific legal tools are dedicated to protecting peri-urban farmland from being developed: protected agricultural zones (ZAP: *Zone Agricole Protégée*) and protection perimeters for natural and agricultural periurban areas (PAEN: *Périmètres de Protection et de mise en valeur des espaces Agricoles Et Naturels périurbains*). It can be seen that these legal tools have hardly been implemented at the local level, since they have been conceived at the national level. There are two explanations for this: Firstly, local elected representatives are reluctant to remove the right of landowners to develop their land secondly, designating farmland as "protected" in some areas might encourage people to think that development will be authorised sooner or later in "non-protected" areas, and this leads to speculation in peri-urban areas instead of putting an end to it. That was the point of view of urban designers who drew up the SCoT of the *Montpellier Agglomération* and tried to find alternative solutions to this problem.

In a comparative perspective it is necessary to underline the fact that the French legal system has for a long time concentrated on planning rules and zoning to try and counter the effects of development land rent on the loss of land (open space) in peri-urban areas. Only very recently has the loss of farmland—reaching 70,000 ha per year in France—alerted public authorities to the scale of the problem so that it has been decided in the recent *loi d'orientation agricole* (Farming Bill, 2010) to create farmland observatories and to vote a new tax—of 5–10 %—on the value added when farmland is sold to be developed, thus changing its status. This new fiscal resource, implemented at national level under the authority of the Ministry of Agriculture, is supposed to be used to support new farmers to establish themselves.

Recent striking facts regarding territorial governance in the Montpellier city-region are the creation of *Monpellier Agglomération*, as a new local government form, and the drawing up of its territorial coherence scheme, as briefly mentioned in the introduction to this chapter. The city-region of Montpellier has had one of the highest demographic growth rates in France in recent decades but this growth was barely managed at the municipality level, with no planning scheme at the scale of the functional urban area (*aire urbaine de Montpellier*, 93 municipalities, INSEE 2010). Therefore, the *Montpellier Agglomération* (initially combining 38 municipalities with more than 400,000 inhabitants) was created at the end of 2001 as a response to local challenges, facilitated by national legal and financial incentives. *Montpellier Agglomération* brought major changes in local governance and planning practices, thanks to an enlarged holding of competences (transport, housing, planning, water and waste management, economic development, cultural policy) and a strong political leadership with the prominent personality of Montpellier's mayor, who became president of *Montpellier Agglomération*. Sharing the power of decision in the community council was a disputed issue: with nearly 250,000 inhabitants, Montpellier city representatives have a 50 % voting power, which led to the elected representatives of other municipalities to fear a community policy agenda excessively weighted towards the interests of the central city (and those of its mayor, president of the community council). It ended in local political conflict, with 7 of the surrounding municipalities leaving *Montpellier Agglomération*, which since 2005 has a membership of 31 communes. Some disputes about tax sharing occurred at the creation of *Montpellier Agglomération*, with new forms of competition among municipalities according to the gains and losses produced by the association regarding economic activities and infrastructures: what will be the territorial solidarity among members of *Montpellier Agglomération*?

The SCoT was developed from 2002 to the end of 2005 and different experts contributed to the analysis and to the final planning scheme. Different state officials were also involved, with an advisory role. After the "public utility enquiry", the SCoT was approved by the Community council in February 2006. It sets the main planning orientations for *Montpellier Agglomération* for the 15 years from 2006. As *Montpellier Agglomération* only comprises 31 communes now, it does not cover the whole territory of the functional urban area (93 municipalities in the east of the *Département* of Hérault). It should be noted therefore that other inter-municipal associations have also been established in the functional urban area of Montpellier, and that their planning processes are underway, but with low inter-SCoT coordination.

### *10.2.3 Key Rural/Urban Issues and Stakeholders for Montpellier City-Region*

The following table (Table 10.2) summarizes the key rural–urban issues in Montpellier city-region and identifies the main stakeholders. Even if strategic issues had to be prioritised, all six issues identified in the PLUREL project strongly affect land management in the peri-urban area of Montpellier: (1) land pressure due to housing, (2) agriculture under pressure, (3) tourism integration, (4) traffic pressure, (5) water management and flood prevention, (6) high value nature at risk.

The strategy of *Montpellier Agglomération* in the field of peri-urban land use can be explained through its SCoT which concentrates on two main substantive issues:

- Land pressure due to housing development in the urban fringe: Montpellier region is very attractive. A lot of people really want to live there due to the pleasant climate, special geographical situation and regional dynamism, leading to a real “sun immigration”. With strong residential preference for individual housing, the urban fringes here are very attractive. This situation leads to significant space consumption, aggravated by large areas dedicated to transport infrastructure. The price of land is rising, which leads to land discrimination.
- Agricultural land in the urban fringe under pressure: in situations of land price increases, farmers are always tempted to sell their land, especially as farming systems are no longer economically competitive in a context of globalisation and urbanisation.

## 10.3 The Territorial Coherence Scheme: A Strategy for Sustainable Periurban Land Use

The territorial coherence scheme (SCoT) of *Montpellier Agglomération* draws the main lines of the territorial development project at inter-municipal scale (a new planning guidance document derived from the law *Solidarité et Renouvellement Urbains*, 2001). Before the creation of *Montpellier Agglomération* at the end of 2001, the city-region lacked political coordination in spatial planning and suffered decades of urban sprawl as a result. The SCoT aims to bring an end to this unsustainable form of land consumption by renewed and intensified urban development on one hand, and protection of natural and agricultural land on the other.

The boundary of the SCoT was approved in 2006 and required much negotiation and procedural adjustment in order to bring together the views and interests of all stakeholders successfully. Development of the SCoT was based on expert studies (2002–2003), meetings bringing together the elected representatives and the stakeholders from municipalities of the same sector, in order to discuss local development issues (2004–2005) and communication and final debates at the end of 2005, including a public enquiry as required by the planning law. Some

**Table 10.2** Key rural–urban issues and stakeholders in Montpellier city-region

| Actors | Local governments and State officers | Private and public companies | Non-profit formal/collective actions (NGO, associations) | Civil initiatives informal/individual |
|---|---|---|---|---|
| *Land use issues* | | | | |
| 1. Land pressure due to housing | Local governments (multi tiers) | Developers | *Non au béton* (Concrete No) (NIMBY association) | Land owners<br>*Cabaniers* (illegal housing in natural areas) |
| 2. Agriculture under pressure | Local governments (multi tiers) | Agriculture firms (farmers) | *Jardins de Villeneuve* (social integration association)<br>Farmers organisations SAFER[b] (farmland market regulation) | Horses keepers |
| 3. Tourism integration | Local governments (multi tiers) | Tourism sector firms | | *Cabaniers* (illegal housing in natural areas) |
| 4. Traffic pressure | Local governments | *Autoroutes du Sud de la France* (motorways) | | |
| | State (*Direction départementale de l'équipement*) | *Réseau ferré de France*[a] (railways) Airport logistics firms | | |
| 5. Water management, flood prevention | Local governments (multi tiers) | *Bas Rhône Languedoc*[a] (agricultural irrigation) | Fédération de pêche (fishing association) | |
| 6. High value nature at risk | Local governments | *Sita Sud* (waste treatment company) | *Ecologistes de l'Euzières* (environmental association) | *Cabaniers* (illegal housing in natural areas) |
| | DIREN (Ministry of Ecology) | *Areva* (local plant, Suez) | *Gardiens de la Gardiole* (NIMBY association) | |
| | *Conservatoire du Littoral* | | | |

[a]RFF and BRL: operational companies funded by public institutions

[b]SAFER (*Société d'Aménagement foncier et d'Etablissement Rural*) was established in 1962. It is a "parapublic" body, managed by representatives of the farmers under the control of representatives of the state and of local governments. This institution has a mission of public interest—farmland market regulation—but also runs private business, as consultant in land matters for local governments

individual citizens or groups expressed their disagreement with the idea and some institutional representatives even expressed criticism about the project.

Nevertheless, the SCoT was declared to be legal and unanimously approved by the community council in February 2006. The SCoT report includes three parts: the *presentation report* with inventory of site and issues, the *sustainable development project* (PADD: *Projet d'Aménagement et de Déveoppement Durable)* with main political orientations, and the *general orientation document* (DOG: *Document d'Orientation Générale*), stating legal prescriptions and showing maps of future development.

The next step of inter-municipal cooperation for regional planning is the connection between regional and local planning. This is tested for legal compliance (conformity) between the SCoT—at the regional scale—and the PLU (*Plan Local d'Urbanisme*: local urbanism plan)—at the municipal scale: all municipalities of *Montpellier Agglomération* should therefore ensure that their PLU is in legal compliance with the SCoT. Apart from these formal aspects concerning official planning documents, the way they are drawn up and implemented could also be good indicators of how inter-municipal cooperation works in *Montpellier Agglomération*. In order to help municipalities to realise their new urban extensions following the objectives—and the "spirit"—of the SCoT, some special incentives and supporting tools, as well as a dedicated coordination framework, have been set up by the technical services department in charge of development and land issues of *Montpellier Agglomération*:

- A *recommendations handbook* (*cahier de recommandations*, a kind of "good practices handbook") has been written to help the municipalities of *Montpellier Agglomération* in implementing their development projects according to the SCoT;
- Before implementing a new urban extension, municipalities are encouraged to prepare (or have prepared for them) preliminary *urban studies* in order to improve spatial planning, especially when it involves new urban forms and the management of an urban/rural limit; *Montpellier Agglomération has* dedicated a special fund to help municipalities pay for these studies (50 % of the cost), and the community development team may also intervene to support the local team to work out the project;
- Consistent with the ambitious objectives of the SCoT, *Montpellier Agglomération* has developed a land policy and set up a *land management service*, which ensures—directly or indirectly, in cooperation with the municipalities—regulation of the land market.

## *10.3.1 Zoning Land Use at Regional Scale*

One of the main goals of the *Montpellier Agglomération* SCoT is to bring an end to urban sprawl. This involves a combination of urban renewal, urban intensification, protection of open space and the management of the peri-urban fringe as follows:

**Fig. 10.3** New urban extensions in the Scot of Montpellier Agglomeration, example in the south-west side (Source: Montpellier Agglomération 2006)

- *Three minimum levels of housing density* for new urban extensions have been defined according to the local environment and the quality of public transport accessibility (Fig. 10.3). The better the accessibility, the higher can be the housing density.
- *A framework of natural and agricultural spaces* has been defined, with very limited development possibilities (Fig. 10.4); an "agricultural hamlet model" has been developed for farmers to be used in areas dedicated to new farm buildings and housing for farmers, for collective projects only, with possible public support and incentive.
- *Management of the urban fringe*: three types of limit have been defined in the Scot (Fig. 10.5) to limit future urban developments: (1) based on geography, (2) to be strengthened around urban extension, (3) to be strengthened around existing built up area.

This precise zoning in the SCoT gives an overview of possible future urban development for a period of 15 years. The entire consumption of land for development during this period should not exceed 3,000 ha. This represents great progress compared to the sprawling trend of recent decades. The first forecasts, made at the beginning of the SCoT process, were that at least of 6,000 ha would be required for development needs for the same period. This new planning strategy should stop the ongoing despoilation of the landscape.

## *10.3.2 Tools to Implement New Developments at Local Level*

In order to implement this goal of "smart growth", municipalities of *Montpellier Agglomération* first have to design their local urbanism plan (PLU) in accordance

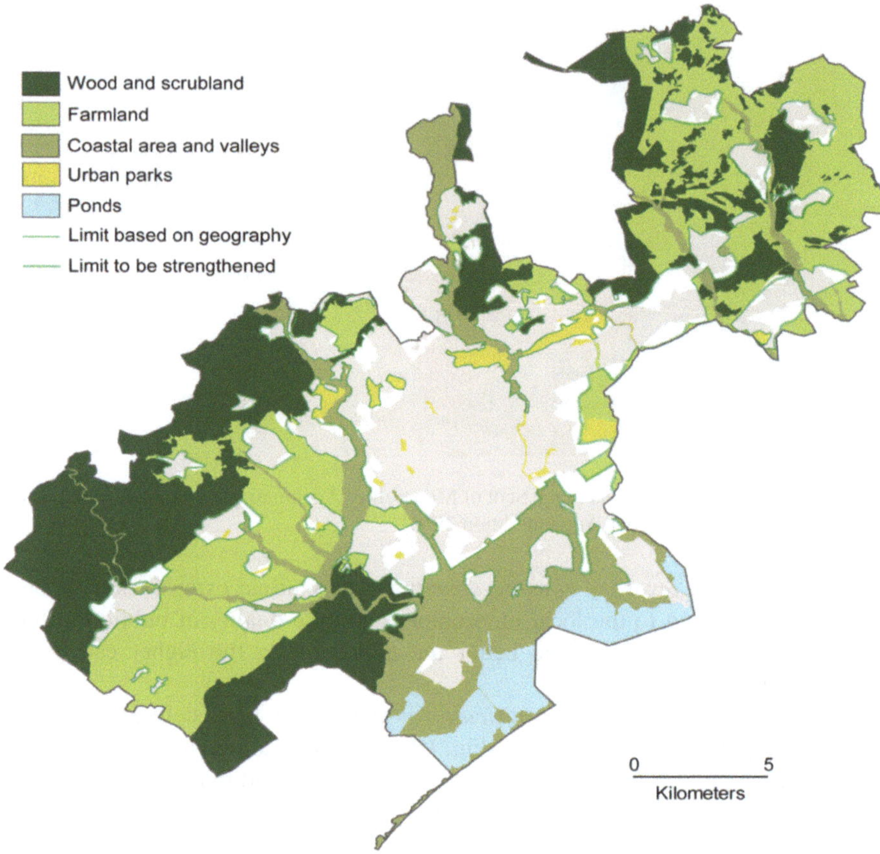


**Fig. 10.4** Framework of natural and agricultural spaces (Source: Montpellier Agglomération 2006)

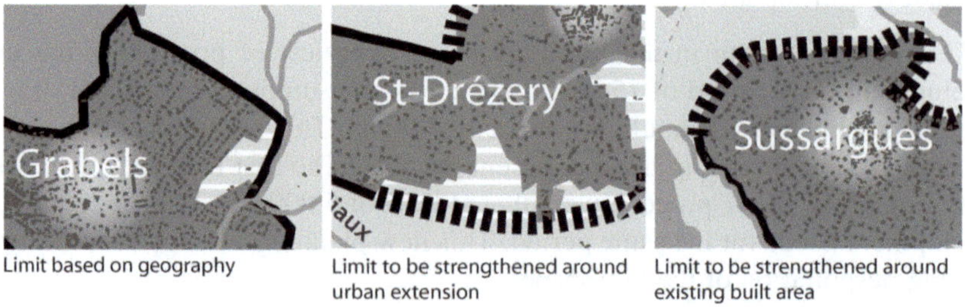


**Fig. 10.5** Three types of limits for urban fringe in the SCoT (Source: Montpellier Agglomération 2006)

with the SCoT. Moreover, different options are proposed for public intervention in or regulation of new urban development and the protection of natural and agricultural land.

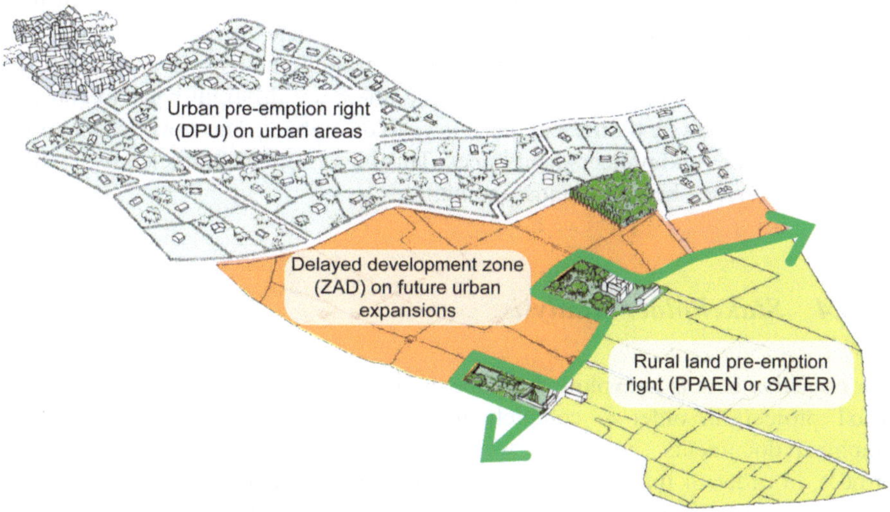


**Fig. 10.6** Pre-emption rights for public land market regulation (Source: Montpellier Agglomération 2006)

- *Public land market regulation* is possible as a result of different public rights that can be used according to the type of zone and land concerned: different types of land pre-emption rights (urban, natural, rural, delayed development zone, for example) can be implemented if necessary by municipalities in urban fringes (see Fig. 10.6);
- *Enhancing the quality of new urban developments*, of higher density, on the urban side of the limit: urban studies are supported by *Montpellier Agglomération* to plan new developments according to the recommendation of the SCoT regarding density, urban design, landscape, green spaces and transport.

## *10.3.3 Landscape as a Vector for the Metropolitan Project*

The SCoT of *Montpellier Agglomération* shows an innovative point of view on landscape in order to transform the city and its periurban areas. Instead of viewing the area as victim of urban pressures, the idea was to focus instead on the qualities of the city-region, and landscape was at the top of the list. Thus, landscape quality was regarded as an objective for the SCoT, not only in terms of protection, but also as an asset to be strengthened. Following this, the SCoT concept was to base territorial planning on landscape and natural and agricultural spaces, a new idea in French thinking on the subject. However, what is hidden beneath this view of landscape? Is it really a vector for a sustainable metropolitan project dealing with peri-urban areas?

Some recurrent urban concepts are found in SCoTs: mobility, proximity of different functions and sustainability, which seem to meet general approval in urban strategies. Landscape is taken into account whatever the type of territory covered by a SCoT, whether rural villages or metropolitan areas. Even if landscape is often associated with rural areas in France, there are different ways to consider landscape in SCoTs, and different discourses to describe it. The conventional point of view evaluates and protects landscapes. This idea is often completed by a will to organise coherence between landscape units.

### *10.3.4 Stakeholder Analysis*

Our objective here is to explore the value of landscape trough the conception of the SCoT project. In order to complete information contained in the written and graphical documents, interviews were conducted with the main actors of the project, in order to understand their various representations of landscape. Seven actors were selected to represent the diversity of points of view. The semi-structured interviews were based on a discussion started by the question: *From your point of view, what is the project of the SCoT?* A list of the main issues to discuss was established, using the final documents of the SCoT and preliminary interviews. These were: increasing housing and transport possibilities, supporting the development of economic activities and protecting and emphasising landscapes.

The analysis of interviews demonstrates first the idea of landscape as an integrative element. It leads to the strategies of all actors converging on the idea of preservation. Landscape is also presented in all discourses as a link between the many heterogeneous parts of the urban and peri-urban territory. The notion of landscape also seems to establish a relationship between different scales of the area, from global to local. Specific representations of landscape proved to be very varied and a large number of elements were identified, ranging from a simple "tree" to the "Mediterranean sea" and including diverse scales such the "square", the "river" or the "beach". Thus, landscape appears as an element of multiple connection possibilities: an element of connection between actors and territories, a link between different scales.

The formulation, and eventually the rephrasing, of the SCoT's objectives based on the discussions on landscape was also analysed. The general objectives of the project expressed by the actors imply some points of view on landscape. The space for new inhabitants, a key objective, implies that landscape, here represented by the rural zone around the urban core, is pushed outwards by the growing city. The objective of developing mobility, especially by public transport, implies a fragmentation of landscape, which is a victim of the urge to connect everything. With increasing accessibility as an excuse, landscape coherence is crossed, divided and thus weakened. In addition, the development of economic activities also leads to the idea of land banking associated with landscape, which is then excluded from any other function of economic development. Landscape can also be thought of as a park in which farmers are the gardeners. The objective of embellishing the housing environment implies that the city is an aggressive milieu opposed to the well-being of the landscape.

These points of view of landscape were contested by the planners and landscape designers in charge of the SCoT project. Indeed, since the beginning of the project, the SCoT appeared to the architect and planner B. Reichen as a "windfall" thanks to which it was possible to make a project that was a "creation". Thus a new landscape image emerged and grew upon stakeholders in the different workshops held during the SCoT process. The pedagogical value of this work was also emphasised by all the actors interviewed. A real process of negotiation, a bottom-up approach, took place during the process. At the end of these discussions, landscape appeared as a substance that we need neither to protect nor to fill. Green spaces around cities could no more be considered to be spaces free for housing and landscape no longer has to be considered as a victim of urban expansion.

Thus, a new definition of landscape evolved. It permitted a new formulation of the general objectives of the SCoT. Landscape emerged as a guarantor of urban quality and of territorial coherence. This implied a change of priorities from the city of mobility to the city of proximity and local identity. Landscape becomes a special economic asset, an active one, establishing relations of interdependence with the urban milieu. The necessity of development intensification, translated into the principle of space saving, then becomes obvious. Landscape is considered as a multi-scale system, always changing, potentially a carrier of latent projects. The planning strategy was also conceived so as to adapt to a changing environment. The objectives of preserving the natural heritage, building a "city of proximity" and intensifying development were introduced, justified and organised.

In fact, the discussion about landscape enabled actors and territories to come together, given the possibility of various individual claims. Therefore landscape was simultaneously an element of consensus and diverse positions. The point of view on landscape developed by the SCoT designers built on this multiple identity. It became the filter through which the general objectives of the actors were rephrased and harmonised in order to build a sustainable strategy in peri-urban areas.

### *10.3.5 From Spatial Planning at a Regional Scale to Development at a Local Scale*

The territorial coherence scheme of *Montpellier Agglomération* aims at ending unsustainable land consumption by intensifying development on the one hand and protecting natural and agricultural land on the other. A major issue for the SCoT is for it to be successfully implemented by municipalities, which remain the key authorities for development regulation. Can the strategy of *Montpellier Agglomération* be considered to be successful in the fields of spatial planning and inter-municipality cooperation? The next section will address the three following issues of (1) territorial governance, (2) intensified urban development and (3) farming as a tool to manage urban/rural limit.

### 10.3.6 Territorial Governance: Cooperation Among Local Governments in the Field of Spatial Planning

Municipalities have a legal commitment to implement the SCoT objectives but what really happens when it comes to local development? How do elected representatives adapt the objectives of the SCoT to their local level? Is the cooperation between municipalities and the inter-municipality authority successful?

The first step of the process of regional planning and inter-municipality cooperation is generally considered to be a success because *Montpellier Agglomération* completed its SCoT very quickly (in less than 4 years). As the first inter-municipality planning document for the city region and one of the first SCoTs in France, it is acknowledged as innovative in the field of regional planning. The fact that the *Montpellier Agglomération* SCoT was unanimously approved by the community council shows that political agreement has been achieved among local politicians. Notwithstanding that, preliminary to the construction of the consensus, there was hard bargaining to persuade local politicians to accept the constraints of the SCoT.

Concerning the next step—the connection between regional and local planning—it should be noted that some municipalities approved new plans in a rush before the SCoT process started, so as to ensure that their local objectives would be met. Others are now carrying out *a posteriori* adjustments, where necessary, so that their PLU is compatible with the SCoT. Whatever the case, most conflicting—or potentially conflicting—issues have been dealt with and solved through political negotiations during the completion of the SCoT. According to the available information, there are only a few very local problems left, for example a NIMBY conflict about the waste storage area project in Fabrègues and the controversial location of new urban extensions in Murviel. Most of the time local elected representatives have clearly understood the objectives of the SCoT and are willing to implement them through their local development projects. That is what was observed in Castries, Fabrègues and Pérols in 2008. These three municipalities are planning new urban extensions and in 2008 they undertook urban studies with the support of *Montpellier Agglomération*.

The observations of the process of developing the SCoT revealed tremendous progress in terms of a vision for territorial development. It also resulted in an adjustment of understanding by local politicians, especially those of small municipalities, from their original point of view based on the issues of villages of fewer than 1,000 inhabitants, to the perspective of a city-region of more than 400,000 inhabitants. The regional planning process thus also appears to be a learning process as well as a tool to strengthen common knowledge and a common vision for the territorial governance of Montpellier city-region (at least for politicians, if not yet for every citizen). These collective territorial advances—made thanks to the SCoT—lead to regrets over the "very suboptimal" perimeter of *Montpellier Agglomération*, being much smaller, with 31 municipalities than the true functional area of the city-region (93 municipalities).

### *10.3.7 Intensifying Urban Development: A Successful Policy*

A major objective of the SCoT is to contain urban sprawl, in order to protect natural areas and landscapes. Since population growth remains the main driving force of the local economy, in order that housing supply should not be reduced, this requires new forms of urban development, intensified and more efficient regarding shared infrastructure (transport, water management, etc.). Do municipalities respect these prescriptions of the SCoT? Are new urban extensions located in the proper zones and do they achieve the objective of intensified development? The strategy of intensified urban development implemented by *Montpellier Agglomération* through its SCoT can be considered to be successful according to several indicators:

– A large degree of public control over development land

Three years after the SCoT was approved, it appears that more than 2,000 ha of land dedicated to future urban extensions are under ZAD zoning (*zone d'aménagement différé*), either at the initiative of municipalities or of *Montpellier Agglomération*. In these areas land price has been under control for 14 years (since the advent of the SCoT) and the public authority has first refusal for land acquisition in order to be able to implement development projects of general interest. This legal tool—the ZAD—allows public authorities to prevent land speculation in strategic areas for future urban developments.

– Managing new urban developments at local level: the success of *urban studies*

Most municipalities of *Montpellier Agglomération* that intend to open land for new development follow the recommendations of the SCoT. Before they revise their PLU, municipalities commission a preliminary urban study in order to ensure new urban extensions are optimised. In 2008 the cases of Castries, Fabrègues and Pérols were studied. These three municipalities had each undertaken urban studies with different consultants (urban designers, architects). In each case, members of *Montpellier Agglomération* urban design services were closely involved.

According to preliminary observations, these *urban studies* took opportunities to develop innovative forms of collective housing, respecting both local landscape and newcomers' residential expectations. Several reasons explain the success of the strategy of intensified development. Firstly, putting an end to urban sprawl and land consumption was a shared objective among the local politicians and the population. Zoning, either at local or regional level, is often considered as an insufficient means to control urban extension and to promote new housing forms. In the case of *Montpellier Agglomération* it seems that statutory tools of public control or public regulation of the land market are strong enough to meet the objective of higher density and efficient new urban developments. This can only be possible because these tools are supported by political will and implemented as part of a proper policy. It appears that the local political consensus was strong enough on the point of intensifying urban developments.

However, none of this would work without efficient technical tools. Drawing precise limits for areas dedicated to future development at regional scale gives developers an overview of their possibilities and allows them to develop projects more likely to be approved. The exercise of local political power to regulate the land market in the urban fringe (land observatory, ZAD) is another decisive point to prevent speculation. Finally, investing public resources—human or financial—in urban studies gives the opportunity to design ambitious and innovative urban developments. Municipalities "play the game" consistently and efficiently with *Montpellier Agglomération* on this point. Here again, it is a pity that the political agreement for a coordinated planning policy could not have been built at a larger scale for the territorial coherence of the development of the city-region. However, it may also be the case that the limited number of municipalities within *Montpellier Agglomération* also helps to explain the efficiency of the political consensus building.

### 10.3.8 Farming as a "Natural Way" to Manage Urban/Rural Limit in Urban Fringes?

Among other objectives, the SCoT was meant to communicate the new rules of spatial planning and to give strong signals to land owners on future zoning, in order to prevent widespread land rent expectation and speculation: has the message been heard and are its objectives achieved? The assumption of the SCoT designers was that drawing clear development limits would secure land tenure and consequently allow new farming projects to be realised in the mid- or long-term in the urban fringe. Farmers would, by the same token, fulfil urban demands for "landcare" and landscape. Can these intended effects of the SCoT be observed? Is a secured spatial planning at regional scale enough of a tool to "boost" farming in the urban fringe?

A long term guarantee (15 years) for the agricultural status of land is a necessary but inadequate condition to maintain, let alone to boost farming dynamics in the urban fringe. Observations in the three municipalities—Castries, Fabrègues and Pérols—tend to confirm the following facts:

- There is an ongoing fragmentation of agrarian property in urban fringes;
- There is an ongoing conversion from perennial crops to short-cycle crops in the urban fringe;

- There is an ongoing de-capitalisation and fragmentation of peri-urban farming structures.

The changing of land cover in Fabrègues west fringe from 1971 to 2008 (Delay 2008) shows an illustration of the generalised process of dismantling peri-urban structures in urban fringes. The traditional perennial crop—vines—almost disappeared during the period. The total cropping surface appears to remain but annual crops are taking over from perennial ones, while scattered housing developed largely in urban fringes.

Land owners' strategies have not yet changed, even with the SCoT, and development anticipations remain. The land owners' strategy has been analysed in Castries (Montfraix 2008). Results show the impact of former urban zoning on land owners' strategy: the proximity of developed land leads to expectation of a change in land status. The agricultural use then changes; for example, vines are uprooted and annual crops are grown instead.

General observations in the situations studied show that agricultural status is not enough to stop development expectations in urban fringes. Another criterion appears to be necessary, one that landowners consider to be legally stronger and more credible in the long term to guarantee non-building land status: for example the protection of a historic monument (Castries's castle), or flooding risk zoning.

Securing development zoning in the SCoT is not enough to:

- Reverse landowners' strategies, because they might not know of the SCoT (which is a recent document), have such a long experience of changing urban limits and may even believe in their own power to have them changed by lobbying local decision-makers about future development zoning;
- Help farmers (grape growers) facing the crisis in the wine sector (selling part of their land to be developed is sometimes a means to safeguard the financial position of their farm, and in doing so providing funds for their agricultural activity, which is a form of de-capitalisation);
- Allow or encourage new farmers to settle and provide a diversified agricultural supply structure: landowners are still reluctant to commit to long term land-rent contracts; as a consequence, access to farm land by would-be farmers remains the limiting factor to them settling in urban fringe areas.

Facing these difficulties and the limited impact of the SCoT, *Montpellier Agglomération* has used other tools and recently established a new policy in order to support the farm sector. Different types of actions are carried out in order to provide complementary support to farmers in periurban fringes:

- Special actions were dedicated to support grape growers and the wine industry from the very start of *Montpellier Agglomération*, even before the SCoT was achieved: a wine route, a wine fair and support to new wine cellars are examples.

- More recently, as a response to the limits of the SCoT in sensitive areas, the creation of "agriparks" is being considered in order to combine farming with landscape protection, outdoor activities, gardening and outdoor markets for agricultural products, for example. Two projects are already planned in cooperation with municipalities in high value places close to Montpellier. Public acquisition of land is being explored, in addition to other supports for the creation of these multifunctional areas where farming will be part of the land uses (Montpellier Agglomération 2010).
- Support to short supply-chains for agricultural products, from farmers to consumers: the thought process on that point is still at an exploratory stage. It requires for *Montpellier Agglomération* to get in touch with new partners from the farm sector, because most professional institutions in this field are still largely concentrated on wine sector problems and not very open to innovative forms of farming.

The heritage of the mass production of table wine and vine monoculture, as well as the land tenure system with farmers as major landowners, make it complicated to develop innovative forms of peri-urban farming. Nevertheless, the SCoT can be considered to be the first step of a working process to "embed" farming in urban development planning. The decision makers of *Montpellier Agglomération* have already gone further with special support to wine industry and the current project of agriparks.

Eventually, compared to the situation of other large French city-regions, the planning strategy of *Montpellier Agglomération* appears to be successful, and its SCoT is widely known for its strong recommendations to promote "smart growth". However, as far as cooperation between the farm sector and the city is concerned, the situation still needs to be improved.

## 10.4 Scenarios Exploring Possible Futures for *Montpellier Agglomération*

### *10.4.1 Four Scenarios for the Future of Montpellier City-Region*

Four scenarios have been developed to explore future land use patterns for *Montpellier Agglomération* with a special concern to urban/rural relationships and sustainability of the urban system. These scenarios are locally interpreted versions of a selection from the four scenarios used throughout PLUREL and described in Chap. 2. In addition, the text box below represents the results from the RUG model which allow comparison of the Montpellier case with all other European case studies of the PLUREL project (see Chap. 4 for details on the methods of the different approaches to scenario modelling) (Box 10.1).

**Box 10.1 Modelling Built-Up Areas in Montpellier: RUG Results**

The four scenarios differ in the intensity and patterns of change in artificial surfaces (Fig. 10.7).

In the *Hyper-tech* scenario, rapid technological change, new transport technologies and few planning constraints lead to counter-urbanisation. Urban growth is also high in city centres, probably because the peri-urban and rural areas are unable to absorb all of the high projected population increase.

In the *Extreme water* scenario, the rural, hilly areas north and west of Montpellier suffer the most from extreme events such as drought or landslides brought on by storms, which makes them less attractive to potential new residents. Increases in artificial surfaces are therefore found mostly in urban and peri-urban zones.

In the *Peak oil* scenario, high fuel costs and strict planning policies concentrate growth in the urban cores of cities. There is little increase in the peri-urban zone, even along the main transport axis between Montpellier and Nîmes, and next to none in the rural areas.

In the *Fragmentation* scenario, the new artificial surfaces spread into the peri-urban zone more than in *Peak oil*. This spread is not uniform but consists of clustered communities of different age groups, ethnicities, etc. Green enclaves form in rural areas as older native people move out of the socially and ethnically diverse cities.

The first two scenarios present two radically different paths for local governance.

1. The "Hypertech Metropolis" shows the reinforcement of the power of the local government, renewing and enlarging the vision of local development coordinated at the scale of the entire city-region, socially and spatially integrated, and strengthened around Montpellier as the centre. This scenario follows through the virtuous choices initiated by *Montpellier Agglomération* with the SCoT.
2. "Back to Old Business as Usual" includes a weakening of local governance that will lead to a reduction of urban sprawl control and an increase in socio-spatial segregation. It also means freed up space for market forces and no or little public control of land use, as experienced during three decades of demographic growth and urban sprawl before the creation of *Montpellier Agglomération* and the drawing of the SCoT. This matches with the Hyper-Tech scenario of the RUG model.

The comparison of these two scenarios can be considered as an assessment of the relevance and robustness of the tool of strategic planning of *Montpellier Agglomération*, the SCoT.

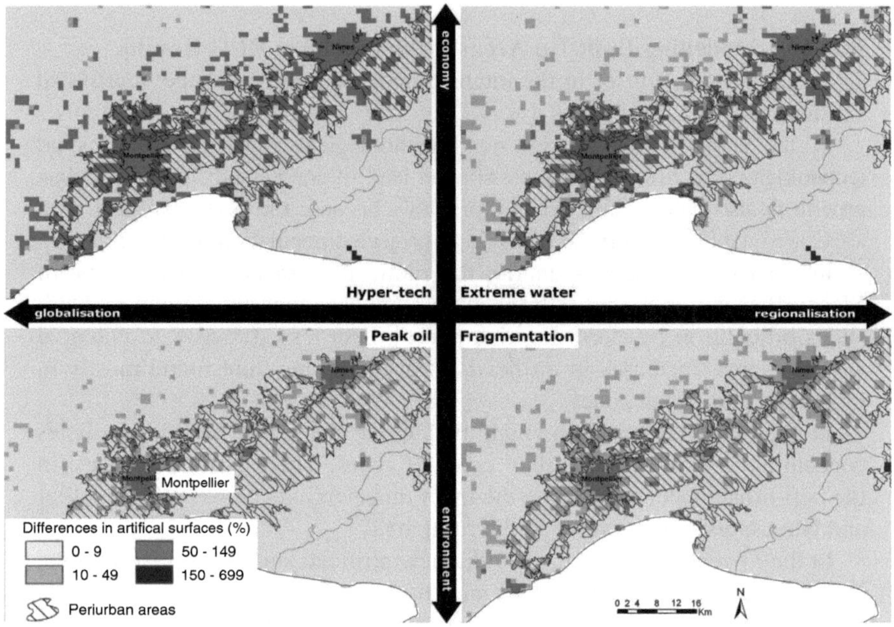


**Fig. 10.7** Development of artificial surfaces in the Montpellier region according to the RUG (regional urban growth model) model for the year 2025 (Source: University of Edinburgh)

External driving forces are decisive in the two other scenarios:

3. "Peak oil and the decline of the Technological Metropolis", where there is long term economic decline and a demise of private motor transport.
4. "Extreme Water": Montpellier city-region as a victim of the Mediterranean Sea.

These two scenarios have been drawn as comparisons with other PLUREL case studies. These two last scenarios are less sensitive to local changes in governance. Whatever the future development strategies may be, external drivers have major impacts. Thus these two scenarios test the development decisions and choices made nowadays in the SCoT.

These four scenarios are presented following the same list of items: policy, economy, urban planning, infrastructures, agriculture and climate (Table 10.3).

**Table 10.3** The four scenarios used in the Montpellier case study

| Scenario | The "long March" or the advent of the hypertech metropolis | Back to old "business as usual" | Peak oil and the technopolitan model decline | Extreme water |
|---|---|---|---|---|
| Politics | The same party wins regional elections in 2010 and keeps the power in *Montpellier Agglomération*. Directly or indirectly, the same party controls regional public policies at all levels. This coordination is accelerated by the merger in a unique regional authority (nuts (2) of the 5 General Councils (Departements, nuts 3), after their bankruptcy following the reform of local governments and their taxation system (decrease of their tax resources). *Montpellier Agglomération* becomes an Urban Community, with enlarged competences, and covers an extended metropolitan area from Sète in the south west to Nîmes and Ales in the north east. Unified political management allows achieving a balance between economic, social and environmental issues at the scale of the city-region, which | There is a change in power at regional elections of 2010. At national level, the same party wins the presidential elections in 2012, and locally, the political majority of the city changes. Economic liberalism, either chosen or imposed because of lack of public resources following financial crisis, drives public policies. State disengagement carries on and social inequality rises. The poorest have little access to quality of life and live mostly in cities. The spirit of gated communities now also applies to inter-municipalities. *Montpellier Agglomération* becomes an Urban Community competing with neighbouring territories: Communities of Thau lagoon, Pic Saint Loup and Pays de Lunel. Contrary to the unified and centralized situation at the scale of the city-region in scenario 1, in this case there is | The time of finite urban world started with the disappearance of oil. A law now prohibits any new urban development without integrated solution of collective transports. It marks the end of the period known as peri-urban which was born, in France, in the 1960s with the explosion of car market. The Minister of Social Cohesion and Territorial Solidarity launches a program to support new neighbourhoods of large peripheral housing estates inherited of the old urban sprawl and penalized by their isolation. | The politics is not in the heart of this scenario. Whatever the political options may be, they have little impact compared to global/natural/external factors which play a decisive role. |

(continued)

**Table 10.3** (continued)

| Scenario | The "long March" or the advent of the hypertech metropolis | Back to old "business as usual" | Peak oil and the technopolitan model decline | Extreme water |
|---|---|---|---|---|
| | now fits with the functional urban area. At the end of 2025, the capacity of local actors to come into negotiation with their neighbours holding different, thus complementary, resources, becomes essential. An integrated regional governance system is achieved. | a "balkanization of territories". | | |
| Demography | There is a record population growth: +1.7 % to +2 % per year. Through Local Housing Program, the population increase is absorbed without problems thanks to social and territorial solidarity in housing policy at the scale of the great metropolitan area. Long-distance commuters living part of the week in residential areas, dwellers in small towns and villages adjacent to the new centres, rural metropolitans: all newcomers have ways to take advantage of the new inter-territoriality. | Demographic growth is still high (1.3 % per year), but spatially segregated with "social sorting". Social barriers and local identity are at the heart of conflicts between local "ghettoized" populations. | The high cost of fossil fuels imposes a halt to population growth (0.1 %). Peri-urban housing, which implies individual commuting, has become a major trap for people who have no access to employment. High social tension is transferred to the city centre. Municipalities are trying to support their citizens with difficulty. | There is a very moderate demographic growth (0.3 % per year) or even a population decline. This is due to two factors: the sea level rises of 1 or 2 m (less room for new residents or for re-housing climate refugees) and Cevennes rain episodes are now very common (natural hazards also reduce land capacity for urban fabric). |

| | | | | |
|---|---|---|---|---|
| Economy | The sector of personal services knows a record development in a "French California" type new economic metropolitan model (*Sud de France* University, coupled with Research and Development of local enterprises). New technologies are booming, from firms-nurseries like the public Montpellier International Business Incubator (MIBI). | The technopolitan model is questioned. Main economic activities are in the fields of personal services and residential economy. Most investments made in local economy come from offshore funds, in real estate and high-end services. | The announcement that the forecasts of oil reserve stocks were completely overestimated has the effect of a global tsunami. Transport becomes the largest household budget item. Local economy has to be completely reviewed: it is necessary to resolve the decline of the technopolitan economic model. Logistics hubs of the languedocian corridor close one after the other. There is a widespread conversion with great difficulty. Only new technologies make the most of this delicate situation. Alternative energies are developing (solar cells, wind mills). | Local economy follows the model of Agenda 21. All urban planning schemes are reviewed to put people out of risk. Tourism activities lose the seaside component since beaches have vanished, and most direct and indirect jobs linked to tourism disappeared. On the other hand, global warming allows taking advantage of hot winds that provide good generation power in addition to mass production of solar cells. |
| Urban planning | The SCoT gets into version 2: the perimeter of the SCoT is extended to the whole corridor of Languedoc (from Sète to Nîmes). High building density and precise urban limits are generalised urban planning rules. The establishment of a Local Public Urban Planning Society, in charge of urban planning, "in house" operator | Urban planning is going out the window: the SCoT is put on ice. "Land hunting" is reactivated. Public planning tools are given up (deletion of land pre-emption rights and of landowning public establishments) due to litigation according to European law of free market in urban planning. This gives | Public transports, which pool the costs and are affordable, are at the heart of this scenario: the long phase of housing redistribution within sprawled urban areas seems to have stabilized along transport infrastructure and near services. Urbanisation refocuses in a sense of strong polarization on Regional | New urban developments have to respect enlarged corridors for possible flooding. The SCoT is reviewed in this perspective. Concentration of risk-free housing is the priority. It leads to the expansion of urban areas to accommodate climate refugees who fled away from littoral municipalities. Land |

(continued)

**Table 10.3** (continued)

| Scenario | The "long March" or the advent of the hypertech metropolis | Back to old "business as usual" | Peak oil and the technopolitan model decline | Extreme water |
|---|---|---|---|---|
| | of municipalities, enables the implementation of public policies without the hazards of setting competition between developers. Tested in Montpellier, it has now extended jurisdiction over the entire metropolitan area as quasi-monopoly public service. | full place to private real estate monopolistic developer. Urbanization is opened to provide access to sites with high landscape value, dedicated to high income executives and retirees. City centres are impoverished: there is no more social housing programs. Socio-spatial segregation is reinforced with new gated communities. | Express Train (TER) and tramway. Through this prism, spatial segregation is reinforced. A "gerontocratic" atmosphere prevails in peri-urban villages where high income European pensioners, released from commuting, are concentrated. | conflicts are exacerbated on the fringes of remaining spaces. The airport disappears under water. The topping out of High Speed Train Line is the frontline of the fight against the Mediterranean Sea. |
| Infrastructures | The Urban Transport Plan is completed in the whole new area of the great SCoT. Tramway network reaches Mèze in 2015. All major transport infra-structures planned are now built: highway A9bis, Nîmes-Montpellier High Speed Rail Line bypass, new TGV Rail station in 2020, single airport for the metropolitain area, Sète competitive harbour. Like new TGV station district, mobility hubs gained strategic values that help structuring new metropolitan urbanity. | Public transport infrastructures like tramways are given up because of lack of public funding. The A9 highway is doubled south of the *Montpellier Agglomération* area. High Speed Railway (LGV) is built and the operating of the new TGV rail station is licensed in 2020 to a private company. | Public transport is strengthened but not enlarged: the peri-urban is neglected. Regional train network is maintained for the benefit of central cities. Municipalities have no means to pay for transport extensions such as new airport. A9bis highway bypasses and High Speed Train Line (LGV) project is abandoned. There are no water projects for agriculture. | The dream of the President of Region Languedoc-Roussillon and former mayor of Montpellier materializes, but it is not "Montpellier which goes to the sea", it is the reverse. . . Montpellier harbour has to be built as Sète became an island, and its harbour has been overwhelmed. The High Speed Train Line is along the coast. The building of the TGV rail station in the south-east of the city is compromised. |

| | | | | |
|---|---|---|---|---|
| Agriculture | Since the demise of General Councils in France, the legal competence on natural areas has been devoluted to Regional Councils, and rural land development to urban authorities. Public supports contribute to farming activities, more and more considered as common heritage of new urban territories. In the Languedocian Metropolis, agriculture is now based on high-tech and high quality farming systems: organic crops, greenhouses,. . . with increasing local sourcing for food products. Some high tech vineyard remains, successfully achieving competitiveness in a globalized wine economy. Besides vineyard, multifunctional agricultural areas are dedicated to both production and recreational uses. | Horsification expands at the expense of farmland and vineyards. Owners of farmland massively turn to production equipment of solar energy. Accelerated disintegration of farming goes on because of globalized competition and the end of public support (from EU or local governments). | Although people from countryside feel they are the losers of development and spatial planning policies, "return to land" is a necessity. There is a boom in family food gardens, and for the movement "back to the land". The concept of agripark is developed. A problem remains for local food system if no solution is found to provide water for agricultural diversification. | Agriculture is delocalized in the northern rural districts of the region. Municipalities have to care and provide open space freed up to maintain public access to nature, for people in situation of ecological stress. There is a revival of some Mediterranean productions like sheep. But warming raises a problem with no possibility of irrigation: the issue of water and water for agriculture is deteriorating. |

(continued)

**Table 10.3** (continued)

| Scenario | The "long March" or the advent of the hypertech metropolis | Back to old "business as usual" | Peak oil and the technopolitan model decline | Extreme water |
|---|---|---|---|---|
| Climate | The climate is not decisive in this scenario. | The climate has no major impact in this scenario. | This scenario is sensitive to climate: warming limits the production capacity in the farm-to-fork system, imports have become expensive because of transport costs. | Sea level rises of 1 or 2 m and the Cevennes rain episodes are more frequent and devastating. Temperatures are rising and with them the risks of sun exposure for people. The attraction of the seaside disappears, it is swarming with jellyfish and the beaches no longer exist: seaside tourism collapsed. |

### *10.4.2 Land Use Changes According to MOLAND Scenarios*

The scenarios developed for Montpellier case study have been tested using the Moland land use model (see Chap. 4). The model produces simulations of land use evolution under the constraints of the local context with the inclusion of different elements: the situation of land use in the years 1990 and 2000, the existing and in progress or programmed transport networks into the simulation period (until 2025), the zoning maps already established locally, including those related to Montpellier's SCoT since 2006, and the suitability of the land cover types, considered as potentially subject to conversion to artificial surfaces: arable land, vineyards, pasture, forest, heterogeneous agricultural areas, shrub, sparsely vegetated areas etc. The model simulates the consumption and changes in land, considering possible extensions of built land uses: continuous urban fabric, discontinuous urban fabric, industrial and commercial areas, construction sites, ports and airport (see Chap. 4 for more details of the method) (Table 10.3). Comparison of changes of land use is presented for three of the scenarios: "Back to the Old Business As Usual" (BAU), "The advent of the Hypertech Metropolis" (HT) and "Peak oil and the decline of the Technological Metropolis" (PO). This comparison is based on the spatially explicit changes, particularly in the continuous and discontinuous urban fabric. The Moland model shows maps of these changes, with a resolution cell of 1 ha (Fig. 10.8). These results therefore allow comparison of the impact of the different scenarios on these different areas by 2025.

The geographic extent chosen for this modelling is the functional urban area of Montpellier (93 municipalities), according to the criteria of the French National Institute for Statistics and Economical Studies (INSEE) for 1999, supplemented by the portions of local government territories that straddle the functional area and surround the *Montpellier Agglomération* territory (134 municipalities in all). Thus, it is an extended area beyond the limits of the Montpellier SCoT, that allows an analysis of the various different impacts of the scenarios, depending on the location, whether inside the perimeter of the *Montpellier Agglomération* and its spatial planning tools or outside, with the potential effect of borders.

According to the three maps of Fig. 10.9, the spatial impacts of the scenarios are quite different, the main ones being the changes of the discontinuous urban fabric, which is the main characteristic of the urban sprawl.

A more detailed analysis of results from area changes between 2000 and 2025 helps to provide a better understanding of spatial impacts of each of the different scenarios. The study region covers an area of 219,417 ha and of this the artificial types of land use covered 21,087 ha in 2000, i.e. 9.6 % of this region. By 2025, the scenarios provide contrasting artificial growth patterns (Fig. 10.10) both in terms of quantities and in their locations across the peri-urban and rural areas.

For scenario "Old Business as Usual" (Old-BAU), the growth is of 6,787 ha, with the largest increase of all scenarios in the area of peri-urban type of the Rural–urban Region (RUR) PLUREL typology (4,960 ha) (see Chap. 3). Figure 10.9 locates this growth in two spatial types: one is urban sprawl on the fringes of existing artificial

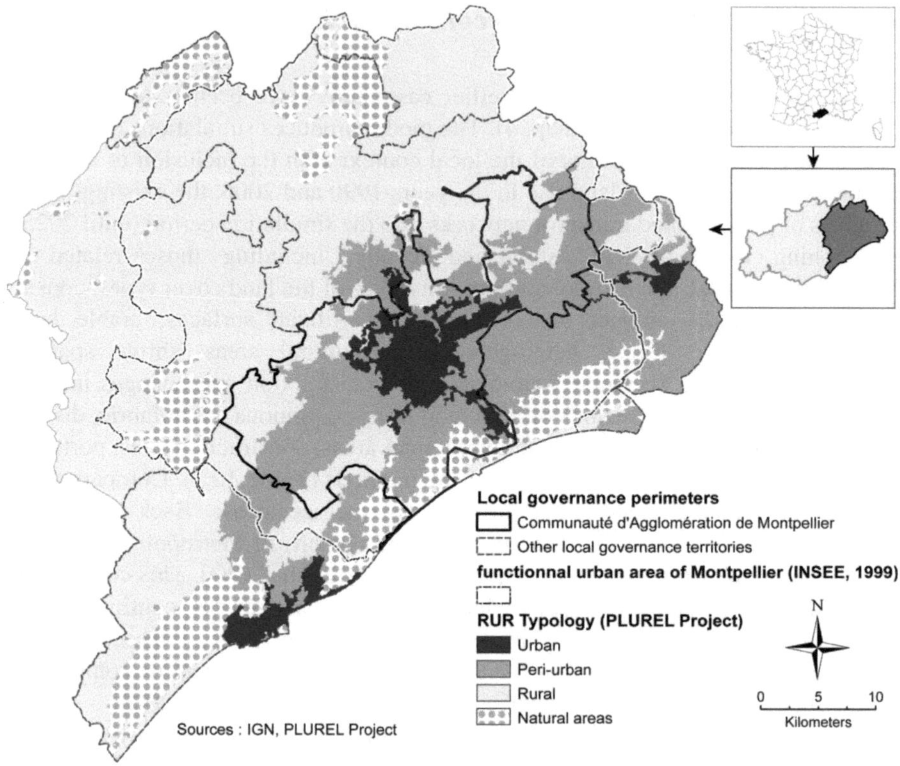


**Fig. 10.8** Area used for the MOLAND application of Montpellier scenarios

areas, with a process of neighbourhood expansion and with an affinity for major transport infrastructure; the other is sparse urban development reaching significant areas previously unaffected. This second type corresponds to urban development considered to have the greatest negative impact on the quality of landscape. This phenomenon is very strong in the eastern region of Montpellier, towards Lunel and beyond Nîmes, along the major axis of transport routes in the plain.

The scenario of the Hypertech Metropolis (HT) shows an increase in artificial areas (Fig. 10.10) which is the most significant of all scenarios: 7,168 ha over the entire study area. However, this growth has spatial arrangements that appear consistent with the logic of the Montpellier SCoT. This scenario takes into account the different housing densities of the SCoT of Montpellier (see Fig. 10.3) in the Moland model simulations. It is established by considering the A and B densities levels (greater than 50 and 30 houses per ha) integrated in the type of continuous urban fabric land use, while the C level (between 20 and 30 houses/ha) belongs to the discontinuous urban fabric type of land use. Indeed, areas consumed in the peri-urban areas are less than in the previous scenario (339 ha less), while the artificial surface in the rural area is higher, with 2,557 ha (as against 1,844 for the

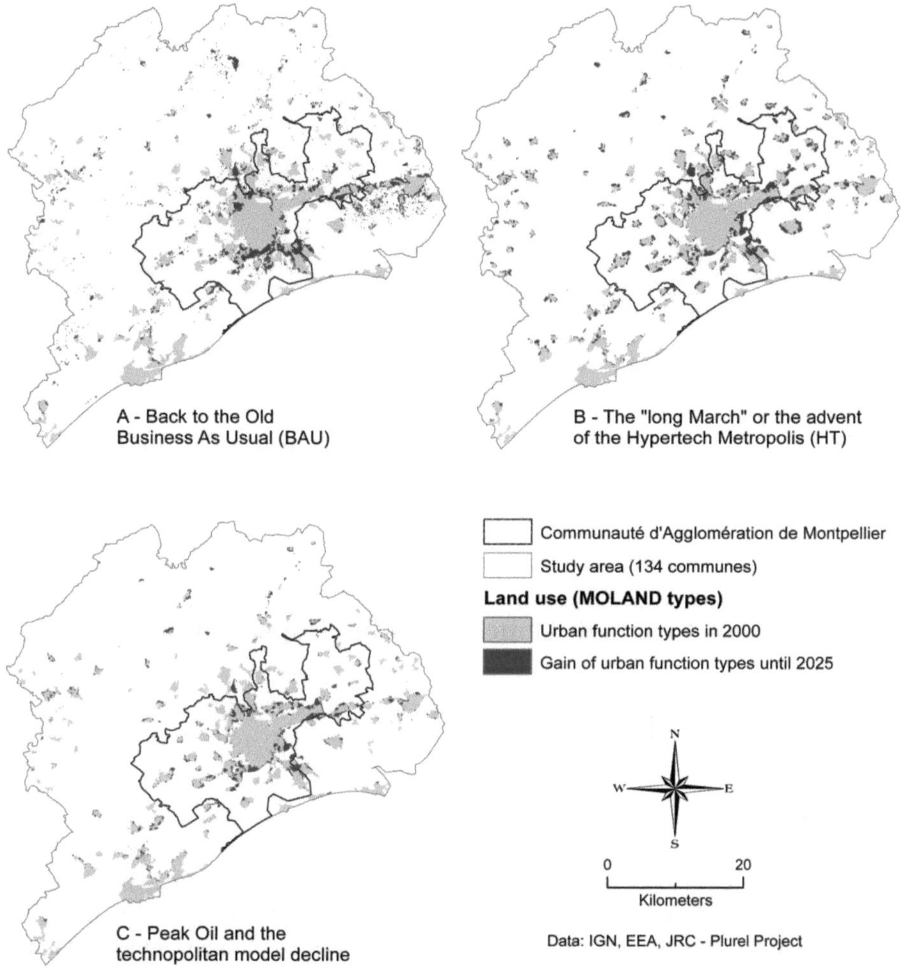


**Fig. 10.9** Growth of discontinuous urban fabric areas in three scenarios with MOLAND simulations

"Old-BAU" scenario). This is because the difference is due to the higher demographic growth in the HT scenario, requiring larger areas dedicated to housing. However, the simulation result is interesting in its spatial expression: urban development is made in a single neighbourhood process on the fringe of existing urban areas. This feature corresponds to the logic of the SCoT for the preservation of landscape, and greatly limits urban sprawl. Thus, areas consumed affect all villages surrounding the centre of Montpellier urban pole, thereby limiting the dispersion of local housing. The eastern plain, between Montpellier and Lunel, is particularly preserved in comparison to the previous scenario.

Finally, the scenario "Peak oil and the decline of the Technological Metropolis" (PO) appears as the one with the smallest increase in artificial surfaces: 3,763 ha

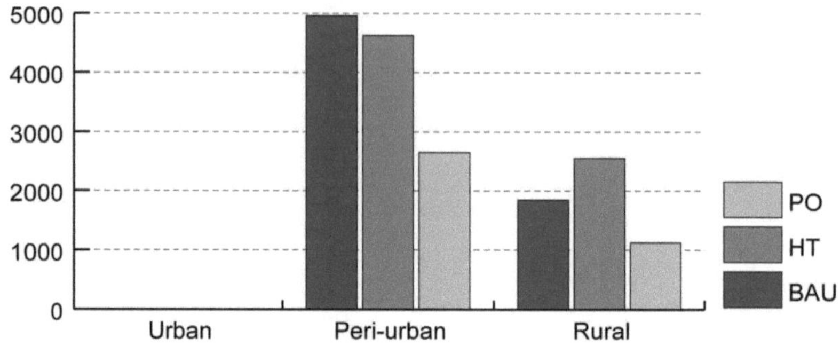


**Fig. 10.10** Comparison of area changes in urban land use (2000–2025) in PLUREL Rural–urban types, according to the different scenarios

(half of the HT scenario) (Fig. 10.10). The spatial extent in 2025 affects fewer sectors within the study area, and many isolated villages away from Montpellier know virtually no growth.

## 10.5 Lessons Learned: Conclusions for Improved Strategies

In the French context of decentralisation and spatial planning reform, and the regional context of high demographic growth of the city-region of Montpellier, some conclusions can be drawn from this case study.

- The institution of a new local government tier—*Montpellier Agglomération*—empowered to conceive and implement coordinated development policies at an inter-municipal scale, especially in the field of spatial planning, has with no doubt permitted the emergence of a strong territorial project (Volle et al. 2010);
- Whether one agrees with the choices made and the governmental style of the first president of Montpellier Agglomération or not, a strong political leadership, an "iron will" development policy and high levels of technical competence helped drawing up and implementing an ambitious territorial project translated into the SCoT; the territorial development project and the political leadership have been strong enough to join uncoordinated, even formerly competing, 31 municipalities. After decades of chaotic development, a consensus has finally been achieved;
- Nevertheless, the efficiency of the planning policy—thanks to the SCoT—within the perimeter of *Montpellier Agglomération* leads to many observers regretting the fact that the perimeter of the territorial authority is much smaller than that of the functional city-region; the leadership of the President of *Montpellier Agglomération* (who died in October 2010) was double-edged: a strong political project based on a true territorial prospective vision on one hand, a strong non-consensual personality failing to ensure broad political adherence on the other; the project of enlarging the perimeter of *Montpellier Agglomération* from Nîmes-Alès in the east to Sète in the west could not have been realized during his lifetime;

- An innovative, detailed, and prescriptive spatial planning document has been drawn at the inter-municipalities scale: the strengths of the SCoT are its ambitious objectives and strict development rules for saving space;
- As a counterbalance to the previous points, and maybe as a result of the emerging character of the SCoT, the weaknesses are:
  - Lack of public consultation, lack of involvement of some major stakeholders, inadequate support to peri-urban farming, a key sector for the future of peri-urban areas
  - Issue of urban sprawl jumping over the frontier of the 31 municipalities of *Montpellier Agglomération*, because of lack of inter-territorial coordination within the area of socio-economic impact of Montpellier city-region (the larger urban area of 93 municipalities)
- Scenarios for future land uses in periurban zones confirm the efficiency of the SCoT's model of *Montpellier Agglomération* as spatial planning tool. If implemented in the whole functional area of Montpellier city-region (Hypertech scenario) it proves to be an efficient tool to combine population growth, controlled urban development and preservation of peri-urban landscape and open spaces;
- Recommendations for improved strategies are:
  - Strengthening the support to the SCoT's objectives by improving a shared vision of regional sustainable development
  - Increasing the political commitment for enlarging the perimeter of coordinated development policy at the scale of the functional city-region (moving from 31 to 93 municipalities)
  - Improving the reflection on the place and role of farming in peri-urban sustainable land use relationships, and consequently improving the support to farming as a sustainable peri-urban land use, and consistently increasing farmers consultation and involvement in policy making regarding the management of urban/rural relationships.

## References

Delay C (2008) Le maintien de l'agriculture dans les franges urbaine: enjeux et outils pour Montpellier Agglomération. Exemples de Pérols et Fabrègues. Montpellier Agglomération. Mémoire d'ingénieur Supagro Montpellier, TERPPA. 40 p. + ann

Epstein R (2009) Après la territorialisation: le gouvernement à distance. In: Vanier M (ed) Territoires, Territorialité, Territorialisation. PUR, Rennes, pp 131–139

INSEE (Institut National de la Statistique et des Études Économiques) (2010) les chiffres-clés de Montpellier Agglomération, p 16

Le Galès P (2004) La restructuration de l'État en Grande-Bretagne. Les instruments de contrôle et de surveillance. In: Lascoumes P, Le Galès P (eds) Gouverner par les instruments. Presses de Science-Po, Paris

Montfraix P (2008) Evaluation du devenir de l'agriculture dans les franges périurbaines de la Communauté d'Agglomération de Montpellier. Application au cas de Castries. Montpellier Agglomération. Mémoire de CESA Supagro Montpellier, p 110

Montpellier Agglomération (2006) Schéma de cohérence territoriale de l'Agglomération de Montpellier: Rapport de présentation. Projet d'aménagement et de développement durable. Document d'orientations générales. Plans de secteurs, p 211 + ann

Montpellier Agglomération (2010) Agriparcs, cultivons nos agricultures de proximités. La lettre de l'Agenda 21 no. 3. septembre 2010, p 3

Mubareka S, Baranzelli C, Lavalle C (2010) Land use projections based on Moland output. Montpellier test case. Delivrable 2.4.2. PLUREL project, IES—DG JRC, unpublished report, p 45

Négrier E (2010) La différenciation positive. In: Nemery JC (ed) Quelle nouvelle réforme pour les collectivités territoriales françaises? L'Harmattan, Paris, pp 315–325

Volle JP, Viala L, Négrier E, Bernié-Boissard C (2010) La ville inventée. Collection la ville en train de se faire. Editions Parenthèses, p 261

# Chapter 11
# Koper: Beyond the Rural and Urban Paradigm

**Marina Pintar, Anton Perpar, Andrej Udovč, Marko Zupan, Majda Černič-Istenič, Vesna Miličić, Tjaša Babič, Davor Deranja, Georgi Bangiev, and Andrej Mlakar**

## 11.1 Introduction

Koper is a historical mediaeval city dating from Venetian times and also a major port on the Adriatic Sea in southern Slovenia in the region known as Istria. It can be described as a national nerve centre due to its port and logistic role, which in turn also makes it one of the eight functional regional centres in Slovenia. As a regional nucleus, it influences economic, social and spatial development and is in turn perceived as a cultural centre and a focal point for the region. It is the smallest of the case studies by far, having an urban population of only 24,000 people (more in the municipality as a whole), although it presents some unique aspects not found in the other case studies.

There is high pressure for urban growth in the Municipality of Koper for two main reasons. Firstly, the attractiveness of the coastal area, which is already very developed for tourism, is still one of the main driving forces of demand for new housing. Secondly, the port, which is one of the largest in the Adriatic Sea, is constantly requiring more space to increase its activity–it is already as big as the town itself in area. In this case study the main challenge for researchers and practitioners is how to seek a better balance between urban expansion and the protection of valuable agricultural land and green areas.

The chapter firstly describes the main features of the town and its hinterland, then examines the system of land use planning before addressing three strategies,

M. Pintar (✉) • A. Perpar • A. Udovč • M. Zupan • M. Černič-Istenič • V. Miličić
University of Ljubljana, Biotechnical Faculty, Agronomy Dept., Jamnikarjeva 101, 1000 Ljubljana, Si
e-mail: marina.pintar @bf.uni-lj.si.

T. Babič • D. Deranja • G. Bangiev
Municipality of Koper, verdijeva 10, 6000 Koper, Si

A. Mlakar
Studio mediterana, Pittinijeva 9, 6310 Izola, Si

K. Nilsson et al. (eds.), *Peri-urban futures: Scenarios and models for land use change in Europe*, DOI 10.1007/978-3-642-30529-0_11,

after which the local scenarios are explored and the land use change modelling demonstrated. The three strategies that were evaluated were: (1) land use efficiency and the protection of the best agricultural land, (2) the Rural Development Plan 2007–2013 and (3) the use of green and recreational areas to increase quality of life.

Based on these key issues and strategies, two locally interpreted scenarios (Hyper-Tech and Peak-Oil), derived the original PLUREL scenarios as discussed in Chap. 1, were defined with assistance from the local stakeholders. The third scenario, Business-as-Usual, is the baseline for comparison of these scenarios.

### *11.1.1 Context: Sunny, Warm and Relaxed*

The Municipality of Koper and the town of Koper, its centre in both functional terms and on a symbolic level, has an important role for a wider area of Slovenia. Koper is a logistical nerve centre whose port serves the movement of goods of many kinds into and out of central Europe as well as Slovenia itself. As a regional nucleus, it influences economic, social and spatial development and in turn is also a cultural centre and a focal point for the region (Regional Development Centre of Koper 2002). It has an interesting history, once being an island town like a miniature Venice (it was founded by that city and was once part of the territory of the Venetian Republic) and between the wars was also part of Italy when the salt pans that lay between the island town centre and mainland were filled in to joint it to the mainland. This character remains strong and the narrow paved winding streets of the old town remind the visitor of Venice to this day. Formerly known as Capodistria, the area occupies part of the short stretch of Slovenian coastline squeezed between Italy and Croatia and is very close to its rival port in Italy, Trieste. It has a Mediterranean feeling (see Fig. 11.1) different from the more Austrian character of Ljubljana, the capital.

The municipality is used as the unit of analysis in this case study, which does not necessarily equate exactly to the rural–urban regional definitions as discussed in Chap. 2. According to the definitions in that chapter Koper is actually classified as a rural area, although it obviously has an urban core: this is owing to its small size when the urban areas of Europe as a whole are analysed. Using only the indicator of population density per settlement, according to OECD, for delineation between urban and rural areas and peri-urban (Fig. 11.2) as a transition zone between them, the results for Koper are shown in Fig. 11.3.

Koper is integrated into many diverse networks and systems, such as the Slovene coastal area, the Južnoprimorska statistical region, the functional region and urban network of the border region between Italy, Slovenia and Croatia, the national urban network of Slovenia, the functional region and urban network of the North Adriatic, the intersection between European macro-regions (Alps, Central European, Adriatic, Danubian and South-eastern Europe), the Trans-European Networks (TEN) and the European urban network (Regional Development Centre of Koper 2002).

Koper municipality belongs to the Južnoprimorska region in southwest Slovenia, has 17.6 km of coastline and is today the sixth largest urban conurbation in Slovenia

**Fig. 11.1** Koper—the historic city centre and the harbour (Photo: S. Klemenc)

**Fig. 11.2** Urban development on the fringe of the city of Koper (Photo: A. Perpar)

and second in terms of per capita GDP (17,807 EUR) in 2009 (Statistical Office of the Republic of Slovenia 2010).

The basic advantages of its economy are diversity and versatility. In addition to tourism, the most important activities are crafts, industry, entrepreneurship,

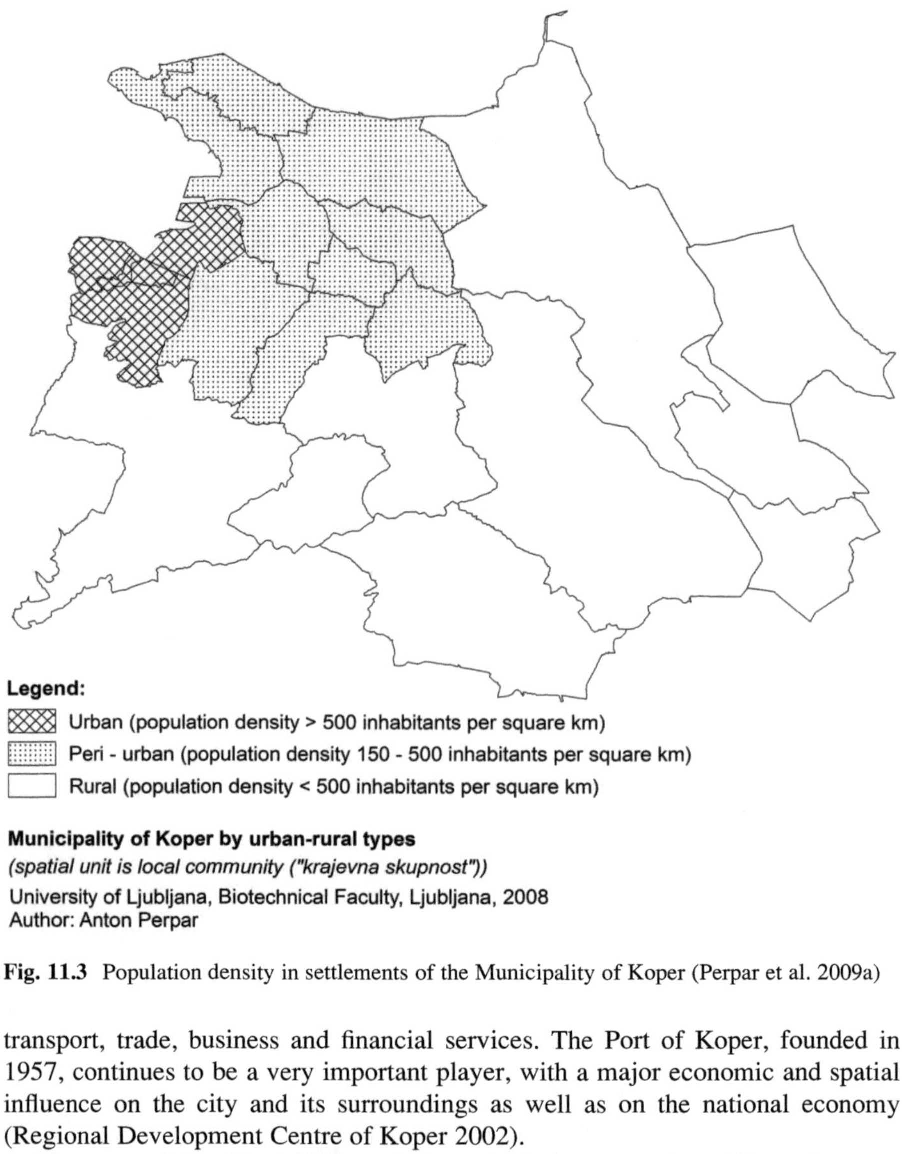


**Fig. 11.3** Population density in settlements of the Municipality of Koper (Perpar et al. 2009a)

transport, trade, business and financial services. The Port of Koper, founded in 1957, continues to be a very important player, with a major economic and spatial influence on the city and its surroundings as well as on the national economy (Regional Development Centre of Koper 2002).

The area of the Municipality of Koper includes more than 100 settlements. According to the latest population census (Statistical Office of the Republic of Slovenia 2002), the population of the municipality was 47,539 in 2002 and rose to 51,915 by June 2009 (Statistical Office of the Republic of Slovenia 2009). It is characterised by a sub-Mediterranean climate, with long hot summers, mild rainy winters with occasional strong winds. The administrative centre of the municipality is the town of Koper, with approximately 24,000 inhabitants, where most of the administrative, economic and cultural activities are concentrated.

The spatial development of the municipality has been shaped by geographical factors, of which its coastal position has the major influence. This affects many different aspects of spatial development, from the purely physical attributes of space, climate and geological structure to demographic, economic and transport processes, which in turn act on the social structure, ecological status and material flows. The most direct and easily recognisable consequence of the coastal location is the Port of Koper, which is the single most important economic and transport facility in the area.

### *11.1.2 Land Use Patterns*

The present pattern of land use is a result of society's current needs, as well as a reflection of historical factors. One of the effects of the fast economic development after the Second World War and of the attractiveness of the coastal area is the present high level of urbanisation of the municipality. In the last two decades sub-urbanisation is most evident in the urban fringe and in settlements close to Koper. While the increase in population is evident in the coastal area, almost all settlements in the rural hinterland, by contrast, have until recently been demographically endangered, with a shrinking population, although the situation has stabilised today. The lack of sufficient housing capacity in terms of numbers and sizes of houses, a higher standard of living and a changing socio-economic situation in the urban zone, have encouraged people to move to the urban fringe and the rural hinterland (reversing the local flow of people from villages, for example) and has boosted the rate of individual housing construction as well as the enlargement of existing houses, in order to accommodate the increased number of families.

The main land use interests today are the development of land for housing, the expansion of economic activities (Koper harbour, industrial zones, trade centres, tourism etc.), infrastructure (recent construction of a new highway through the area, anticipated construction of a second railway track from Koper to Divača with a branch to Trieste) and agriculture, as well as nature reserves and other protected areas.

Almost half (49 %) of the total area of the municipality is covered by forests and their proportion is increasing, because 16 % of the former agricultural land is abandoned and is becoming colonised by secondary forest, mostly in the remoter rural hinterland areas, mainly due to difficult or uneconomic production conditions such as steep slopes. Agricultural land represents 39 % of the total area of the municipality, with grassland, mostly extensive, occupying 48 % of this. Vineyards cover 13 % of agricultural land, olive plantations 5 % and orchards around 2 %. Built-up areas occupy the remaining 7 % of the total area of the municipality (Fig. 11.4).

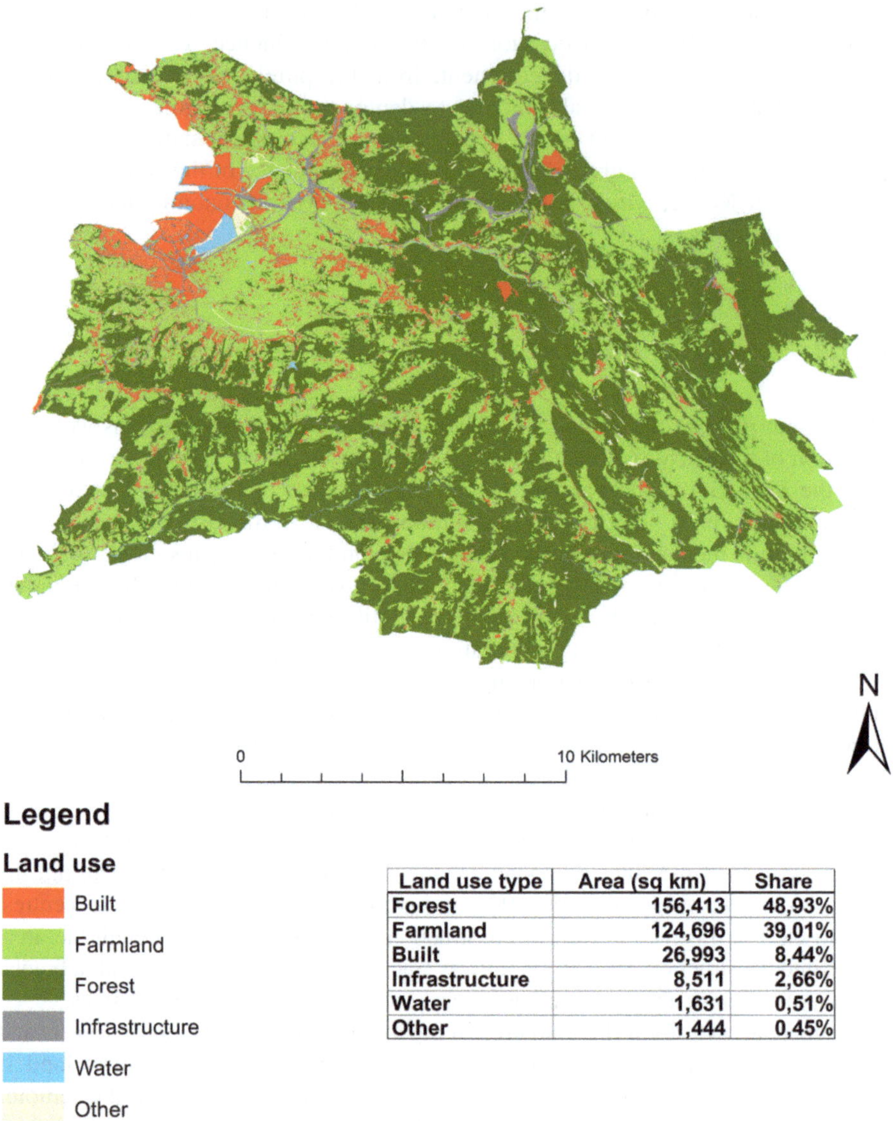


| Land use type | Area (sq km) | Share |
|---|---|---|
| Forest | 156,413 | 48,93% |
| Farmland | 124,696 | 39,01% |
| Built | 26,993 | 8,44% |
| Infrastructure | 8,511 | 2,66% |
| Water | 1,631 | 0,51% |
| Other | 1,444 | 0,45% |

**Fig. 11.4** Land use in the Municipality of Koper (Perpar et al. 2009a)

## *11.1.3 Important Land Use Issues for Koper*

The case study described in this chapter deals with three important issues for the area as identified by the stakeholders from the Municipality who were involved in the PLUREL project: (1) land pressure due to housing and industrial development,

(2) agriculture under pressure from built development, and (3) high value nature at risk from destruction.

1. The housing and industrial development pressure along the coast is due to it being a very attractive area to live or to have a holiday house, not only for people from the region but also for people from all over Slovenia and, since EU accession in 2004, for foreigners, especially Italians. There is therefore a major interest and competition for available building parcels along the coast and in the urban fringe generally.
   The Port of Koper (see Fig. 11.1) is an important actor in present and future land use of the municipal area. The port's real estate amounts to 400 ha. The port is state owned and is controlled by the Koper Port Authority.
2. Agriculture is characterised by a traditional use of terraces on steep slopes (dating back many centuries), which is very well adapted to the environment, since it reduces erosion, increases the cultivated surface and plays an exceptionally valuable role in water retention as well as being an essential part of what gives the area its landscape character. Agricultural ownership structures in the Koper area (and in Slovenia in general), are very specific: compared to the EU average farm size, land parcels in the area are very small and they have been fragmented (as a result of heredity laws and customs) down to 0.5 ha plots on average; with only about 10 % of farms being larger than 5 ha.
3. The Municipality of Koper has rich biodiversity because of the favourable natural conditions, such as the sub-Mediterranean climate and the various ecosystems and landscapes resulting from different rock and soil types. Due to this rich biodiversity, the municipality has several ecologically important areas (e.g., the Karst, Dragonja river valley, Rižana river valley, Škocjanski zatok Nature Reserve) (Fig. 11.5, Debeli rtič and the coastline). Natura 2000 areas, for example, cover 44 % of the municipal territory.

## 11.2 The Slovenian System of Land Use Planning

Slovenia has two official levels of planning: national and municipality (local) with no legislated regional level until now. At the state level the main spatial planning system is based on the Strategic National Spatial Plan, which represents the basis for the preparation of all spatial planning documents at national, regional and local level. At the municipal level, the main strategic planning document is the Municipal Spatial Plan, which deals with all the spatial issues of a municipality and is the basis for preparing projects leading to the granting of building permits. The 2007 Spatial Planning Act of Slovenia also makes it possible for a municipality (if necessary) to adopt the strategic part of the spatial plan as an independent document, the Municipal Strategic Spatial Plan. In addition to the Municipal Spatial Plan, municipalities can prepare Detailed Municipal Spatial Plans for specific spatial arrangements such as restoration of areas of dispersed settlement. The 2007 Act also allows for an inter-municipality spatial document, a Regional Spatial Plan.

**Fig. 11.5** Aerial view of Škocjanski zatok nature reserve (DOPPS-BirdLife Slovenia 2010)

This approach tries to compensate for the missing regional administrative level in Slovenia by using a "bottom up" approach to regional spatial planning. Regional Spatial Plans are based on consensus among or between interested municipalities and cover spatial arrangements that impact on the territory of multiple municipalities (Perpar et al. 2009a).

After 1990 (the independence of Slovenia), changes in social and economic conditions, such as land ownership, privatisation, the changed position of municipalities and the conversion from the planned to the market economy, also influenced the urbanisation process. Due to the new spatial planning legislation from 2002, the spatial development potential of Koper had to be re-evaluated.

A strategy to ensure sustainable development and the most reasonable land use in the municipality had not yet been officially adopted at the time of writing, although there was a draft proposal of a new Municipal Strategic Spatial Plan. The general idea of the plan is to divide the territory of the municipality into three semi-circular belts, in line with the existing situation (settlements, traffic connections, coast, the importance of the Port of Koper), the natural conditions (agricultural land, forests etc.) and to organise activities in each belt in a balanced way, taking into account the needs, situation and limitations that each of them possess (Perpar et al. 2009a).

Studies and analysis of the built-up space show a very clear concentric (semi-circular) hierarchical division (Municipal Spatial Plan of Koper–strategic part 2008). Three of them are very characteristic:

- The first belt represents the central zone–the town of Koper, its historic centre and immediate vicinity, including the coast (the urban belt),
- The second belt represents the peri-urban area, with a medium density network of settlements,
- The third belt represents settlements in the rural hinterland with lower density.

The general concept of how to achieve the most reasonable and efficient sustainable land use in the municipality is to re-arrange the existing spatial situation and to direct development (settlement, economic activities, traffic, agriculture etc.) into the three belts as appropriate. The suggested directions are as follows:

- *Central belt*: To maintain or develop only urban activities or those connected with or dependent on the sea (economic, recreation etc.),
- *Peri-urban belt*: To move and concentrate other economic activities here; the problem is that the highest quality agricultural land is also located in this belt and is strongly protected by legislation; some green and recreational areas are also envisaged for the second belt and agriculture will remain here wherever possible,
- *Rural belt*: This is intended as the main place for agriculture, forestry and also for the development of service activities and of existing settlements, adjusted to reflect the rural characteristics (Municipal Spatial Plan of Koper—strategic part 2008).

In view of the recent spatial development trends in the municipality, the most important policy option that has been chosen and adopted is to "hold the line" (i.e. try to retain construction within the existing developed areas, avoiding the spread of urban areas and stimulating the renovation and reactivation of existing urban areas), which has been implemented in the most urbanised or intensively used parts of the coastal zone. Only those parts of the coast that are either nature reserves or uninhabited areas are subject to natural processes and are left to natural dynamics, with the soft policy option of "limited intervention". In the area of the port, the preferred policy option is to "move seawards". In general, the approach to solving coastline erosion problems is local, using the right technical solutions and appropriate land use techniques such as green cover in vineyards, green cover in winter period, etc. (Municipal Spatial Plan of Koper—strategic part 2008).

According to Slovene legislation (Agricultural Land Act 2003), agricultural land in Koper is divided into two categories: best quality and other agricultural land. However, recent studies have shown (Perpar et al. 2009a) that soil characteristics are not the only parameter that determines the production potential and, consequently, the economic potential of agricultural land. The suitability (e.g. slope, exposure) of the land for particular agricultural crops, the availability of agricultural infrastructure, the possibility of carrying out economically viable production, the accessibility of plots and isolation from pollution sources are also very important factors (Perpar et al. 2009a; Ministry of Agriculture, Forestry and Food 2010).

Thus, because of the factors described above, the existing agricultural land categorisation should be amended, which is also envisaged in new soil/land legislation, currently in preparation in Slovenia (Ministry of Agriculture, Forestry and Food 2010). Soil quality, the specific suitability of agricultural land for a particular agricultural category, the possibility of introducing new production types (e.g. sustainable and organic production) and certain other parameters that significantly influence agricultural productivity, should be taken into account. This much more sophisticated approach should also be diversified by spatial criteria and, in cases of special agricultural categories, even into specific production areas, for example districts or other defined spatial units (Perpar et al. 2009a; Ministry of Agriculture, Forestry and Food 2010). Prescriptions for the restoration or substitution of agricultural land if the best land is converted to another use, are already incorporated in existing (Regulation on Criteria for Planning 2008) as well as in planned legislation (Ministry of Agriculture, Forestry and Food 2010) on agricultural land.

## 11.3 Policy Arrangements

The opinions of stakeholders at a local PLUREL workshop in May 2009 were that the Municipality undoubtedly plays a key role in the planning and developing of the spatial structure but that future changes may affect its current dominance. On the one hand, pressure from private investors is constantly increasing and, on the other, physical space is a limited resource which is also constantly increasing in value, making the municipality an important actor only as long as it maintains control of the physical space (Perpar et al. 2009b).

The municipality is losing its developmental role in affecting spatial changes, becoming an intermediary between national central government and local private initiatives for building construction. Since the Port of Koper is already one of the largest in the northern Mediterranean, the municipality, the port and national authorities have come together to form an informal coalition and all share an interest in increasing the port capacity, which also implies further spatial requirements. Some additional space demands for new infrastructure also arise from the development of tourism and lead to urban sprawl outside the city. Conversely, the Port of Koper has supported the restoration of Škocjanski zatok Nature Reserve, as a complement to these investments.

Environmental groups active in the Koper area very often deal with both the environmental protection and social aspects of rural development. An example of this is Škocjanski zatok Nature Reserve, of which Društvo za opazovanje ptic Slovenije (DOPPS, i.e.–Slovene Bird Life) initiated the restoration and now manages the area. In the hinterland of Koper, the revitalisation of Šmarje village was initiated by the local association named Pangea. In contrast to these good examples, the governmental Institute of the Republic of Slovenia for Nature Conservation has encountered difficulties in persuading politicians to protect environmentally important areas, although it has the mandate to care for the most valuable natural areas in the country. This is clearly because of the weak enforcement guidelines for nature protection in the face of the considerable economic power of development interests. It is up to the politicians on the municipal council to reconcile social/cultural and environmental interests with those of commercial actors, which puts a certain pressure on them.

Environmental NGOs and EU environmental policies play an important role in raising environmental awareness among the local population. However, these remain within closed sectoral (e.g. agri-environmental) boundaries. In contrast, development interests pursue a very different discursive path of providing economic prosperity through infrastructure and housing expansion. Local actors consider social issues related to sustainability to be less important and they are least emphasised in debates as well as in the outcomes of strategies relating to the peri-urban areas in the municipality.

The majority of the land in the municipality is privately owned, with a relatively small part therefore being publicly owned either by the state or the municipality, while this ratio increases a little for agricultural land, mainly because of the Farmland and Forest Fund of the Republic of Slovenia. Speculation on land value increases by means of land designation change is present in the area, mainly for plots with a sea view. Some companies and individuals are land-banking agricultural properties in anticipation of a designation change, thus creating certain direct and indirect pressures on the planning authorities.

In terms of formal regulation, procedures to protect the best agricultural land are embedded in spatial plans (Municipal Spatial Plan of Koper–strategic part 2008), although it is difficult to predict their effects. Knowledge, in addition to legislation, is the main influence on decision making in favour of the preservation of the best agricultural land. In the past, the procedure of land use change enabled not only legal but also serious illegal financial transactions by investors (Perpar et al. 2009b). There are no coalitions of interest in relation to this. The strategy of the preservation of the best agricultural land is already established in the legislation but has very often been abused. Land use efficiency is the main aim in the municipality and could be promoted through debate and discussion but this is not happening at present.

Despite the small average size of farms, agriculture in the municipality remains important. The size of farms or, more accurately, of individual plots, is a vital factor for the multi-functional role of agriculture, including an 'identity factor' in terms of landscape attractiveness, ecosystem functions and maintenance of local agricultural products. Varied and relatively well preserved nature is still one of the characteristics of the Koper area.

The development of Koper port, urbanisation and industrialisation of the region and the development of tourism are the predominant national interests in terms of the development of Slovenia. Such needs influence spatial development in the municipality. There are some important discrepancies between the spatial development dynamics of the newer parts of the town and the historical core, driven by the intensive development dictated (also) by national interests. These discrepancies are especially accentuated from the point of view of the standard of living and social activities.

The centre of development has shifted from the historical core towards its outskirts, with the space in between being forced to adapt quickly to the fast-changing environment. Building on vacant land has proven to be less financially demanding than revitalising the historical core, mainly because of the changing structural needs of new activities, which the old buildings are incapable of accommodating. The driving force behind such changes has been the intense social and economical development, which found the coastal city's spatial capacities insufficient. The problem of stagnation of the city centre arose when activities that could still benefit from the spatial setting of the city started to vacate the historical core (Perpar et al. 2009b).

## 11.4 Strategies for the Municipality of Koper

To cope with the three main land use issues that are most pronounced in the peri-urban area of the Koper, three strategies were defined at the workshop with the stakeholders that took place in September 2008. These were: (1) Land use efficiency and protection of the best agricultural land, (2) the Rural Development Plan 2007–2013, and (3) Green and recreational areas to increase the quality of life. Since only spatially explicit strategies can be considered in the MOLAND modelling programme (as indicated by the black arrow in Fig. 11.6), in the Koper case study only the first strategy can be modelled.

### *11.4.1 Strategy 1: Land Use Efficiency and Protection of the Best Agricultural Land*

The pressures of urbanisation have led to increased surface sealing and loss of high quality agricultural land in the peri-urban area (Fig. 11.7). Suburbanisation and individual house construction along main roads have also reduced land use efficiency and thus led to lower agricultural production. A clear fact-based policy on spatial development and land use can provide a better tool for the preservation of valuable agricultural land and production capabilities based on scientific criteria, as well as defining rules for resolving disputes among different land use needs.

At the time of writing the strategy still has to be approved by the municipal council. Private construction companies have an influence on politics through lobbying and they almost never work directly. Private investors and companies can therefore propose alternative initiatives for new spatial interventions.

The national legislative dimension of the strategy is expected to provide order in terms of land use efficiency. This means preservation of the important natural resource, the prime agricultural land. A strengthening effect on agriculture and the empowerment of farmers is expected to result. The strategy is incorporated into the municipality's spatial plan.

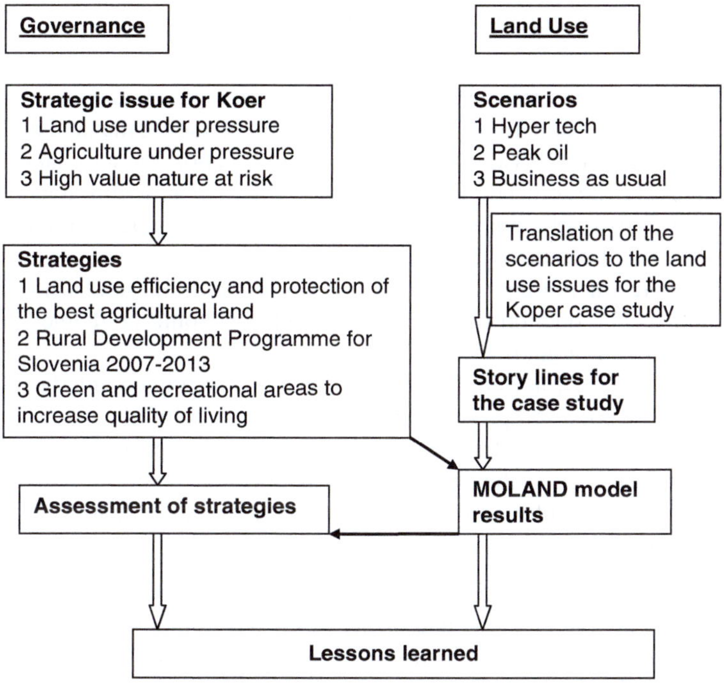


**Fig. 11.6** Position of the strategies and other process elements (e.g., strategic issues, story lines, lessons learned, etc.) in the PLUREL project

**Fig. 11.7** Urbanisation on valuable farmland in the hinterland of Koper (Photo: A. Perpar)

### 11.4.2 Main Objectives of Strategy 1

- For the best agricultural land, the following principle should be applied: the higher quality the land, the stricter the criteria for changing the land designation, because the economic value of agricultural production on such land is higher than on less fertile soil. The same should be applied to agricultural land in larger complexes, especially those in which the basic infrastructure required for intensive agricultural production is already in place.
- After prior reconciliation with other interests, agricultural land with very good and good soil production potential for agriculture should be defined as the top quality agricultural land.
- Measures to upgrade agricultural production could be implemented in areas with very good and good soil production potential for agricultural land use and where there are no special restrictions regarding groundwater protection. Existing agricultural structure, the field road network, the spatial orientation (e.g. directions of field cultivation), ecological corridors and the logic of the landscape pattern should be taken into consideration.

The first strategy aims to prepare a tool to contribute to optimal and sustainable land use planning in the municipality and to enable the development of the municipality in a sustainable way. It will first focus on the primary soil quality and other qualities important for individual sectors (forestry, water management, environmental protection etc.) and subsequently on inter-sectoral cooperation in order to find optimal land use solutions in the new municipal spatial plan.

The aim of the strategy is to find solutions for all three main land use issues: land pressure for housing and the preservation of agriculture and the protection of biodiversity in areas with the highest pressure in the peri-urban area. This will help spatial planners to avoid using the best quality land for construction purposes, so that it will be retained for agricultural production and important environmental areas will also be preserved.

Biodiversity will be maintained with a mosaic of activities across space, with due consideration of environmental demands and the maintenance of natural resources, as well as appropriate management of Natura 2000 and protected areas. Areas planned for future settlement, economic and other development, will be directed away from areas important because of their primary soil quality, environmental importance or high agricultural productivity, and will be redirected to lower quality land.

Semi-circular belts, aiming to divert development centres away from the congested coastal belt are planned in the wider context of this strategy and decentralisation is to be re-orientated towards peri-urban and rural areas. It is also planned to round-off new construction in the town, while renovation and reconstruction in both the town and rural areas will be promoted. It is planned for the hinterland to be connected to the city by radial lines of infrastructure.

The spatial concept of the three concentric belts and polycentric development are used as spatial development models but do not seem (yet) to be actively promoted via a discourse/branding that can unite actors into a coalition to support the plan.

### 11.4.3 Strategy 2: Rural Development Plan 2007–2013

The second strategy will try to stimulate the implementation of rural development measures from the Rural Development Plan for Slovenia 2007–2013. Available measures and support mechanisms from this programme can steer development of the Istrian countryside so that the area will remain settled, abandoned houses will be renovated, overgrown terraces will be re-cultivated, new infrastructure will be established in such a way as to enable a contemporary way of life and, together with the development of tourism and other environmentally friendly activities, residents will be able to earn additional income.

Tourism and recreation associated with agriculture should be stimulated because of the traditional quality and attractiveness of the cultural landscape, which is a result of the mainly agricultural land use and settlements in rural areas in the past. For this purpose, maintaining the traditional land use, as well as promoting alternative agriculture, will be stimulated, especially in areas of exceptional landscape quality, across the whole hinterland and in other areas already protected or designated for protection for their natural attributes. Within this, the principles of sustainable management of renewable natural resources, maintaining the cultural landscape and environmental protection should be stressed. In rural settlements, agro-tourism, possibilities for farm enlargement to provide overnight accommodation and other supplemental farm activities in connection with tourism will be assured.

### 11.4.4 Main Objectives of Strategy 2

- To develop further the abundant potential of the natural and cultural heritage, bearing in mind that a spatial and cultural continuum enables the building up of a friendly environment with abundant job possibilities.
- Maintaining settlement and the revitalisation of villages.
- Concern for environmental protection and revitalisation of the cultural landscape.
- Maintaining agricultural production, the development of sustainable agriculture, and assurance of adequate income for farmers.
- Preservation and renovation of the cultural heritage and its inclusion in the market offer.
- Development of rural tourism.
- Development of supplementary enterprise and handicraft activities on farms.
- Management of village centres and renovation of common infrastructure and buildings.

In terms of formal regulations, this strategy is not directly connected with the land use plan. It is up to farmers to join the programme. In the case of measures to renovate a village, this could also be an initiative on the part of either a municipality or a village. In terms of financial resources, incentives from the EU (and partly from the national budget −15 % to 50 % participation) are intended to cover the costs of measures that are in accordance with the Rural Development Plan. The Ministry of

Agriculture, Forestry and Food, together with the Municipal Office of the Economy are the main actors engaged in the development and promotion of the plan. Farmers have an interest in obtaining these incentives. The Farm Extension Services play an important role, as does also the Agency for Agricultural Markets and Rural Development, because this is the funding agency and it implements the measures of the Rural Development Plan. There are some actors at the national level (e.g., the Ministry of Agriculture, Forestry and Food) and some local companies (e.g. Agraria–for vegetables, Vina Koper–for wine) or growers associations (e.g. the Association of Olive Producers), who promote local agriculture. The official agricultural policy contributes to preserving high biodiversity on traditional farmland. It contributes towards promoting farming related to the identity of the region and promoting regionally specific food and products. This is also part of the policy of protection of the best agricultural land.

In view of the development problems of the Istrian countryside (lack of employment opportunities and low incomes, depopulation, abandonment of farmland, inadequate infrastructure) the policy makers of the municipality wish to provide an impetus to tourism and recreational development connected to agriculture in the region. Obtaining resources through the measures of all four axes of the Rural Development Plan 2007–2013 seems a suitable approach for stimulating these activities. It is expected that resources obtained in this way could enable the maintenance of traditional land use, as well as stimulating alternative agriculture and preservation of the cultural landscape and environment.

### *11.4.5 Strategy 3: Green and Recreational Areas to Increase the Quality of Life*

Urban sprawl and unorganized spatial development have led to a state in which green spaces and recreational areas are scarce and do not provide for an adequate quality of life in the peri-urban area of Koper. Roadside construction and lack of organized settlement centres make it difficult to plan and allocate green spaces, thus diminishing the quality of life because not everyone has easy access to adequate areas for recreation. Recognizing the importance of such areas and their integration in the spatial planning process is essential to guaranteeing a higher living standard of the population in the peri-urban area.

Overall, this third strategy should help to increase the recognition of the importance of green and recreational areas and their integration in the spatial planning process. They are essential for guaranteeing a higher living standard for the population in all areas but especially in the peri-urban (Box 11.1). The aim is also to create a policy on green area allocation.

**Box 11.1 Quality of Life and Land Use Change in Koper**

As part of the quality of life indicator work (see Chap. 3), a limited sample of people were asked about their perceptions of different factors of the environment in relation to their perceived quality of life. Although the results are not statistically valid owing to the smaller than desirable sample size, they offer some insights into the factors which people view as important. Safety and security is the top concern followed closely by air quality, housing quality and nose problems as the factors which affect quality of life related to land use. Adequacy of waste collection, followed by accessibility of green space are the next in the order of priority, quite a bit lower in importance for the population as a whole than the first group. These are followed by the numbers of shops and then by transport convenience.

The variation in comparison with the other case studies is greatest with the transport convenience being lowest–probably because of the small size and population of Koper compared with Warsaw or Greater Manchester. The air quality issue may result from the presence of the port, as may some noise aspects (Fig. 11.8).

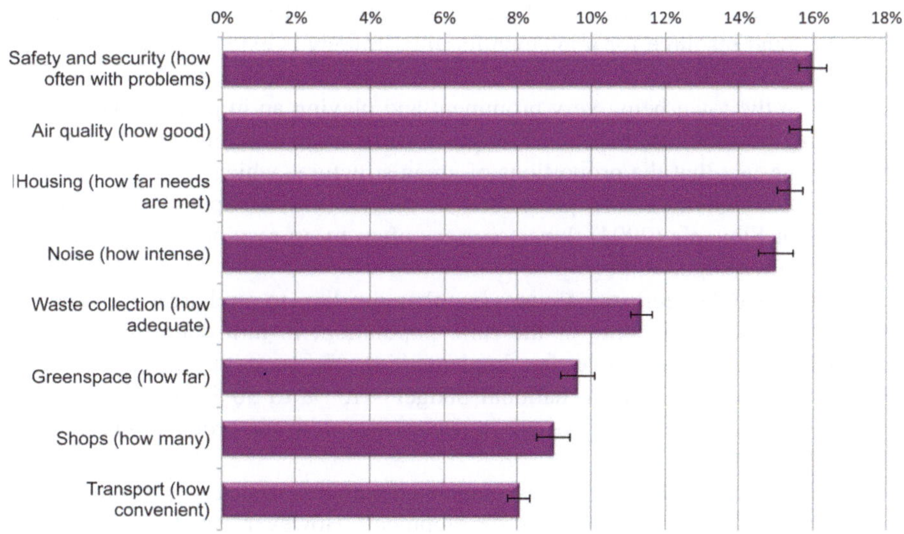


**Fig. 11.8** Ranking of the indicators of quality of life for Koper (Source: Openspace Research Centre, Edinburgh College of Art)

### *11.4.6 Main Objectives of Strategy 3*

- A green system for settlements should be developed, by which green spaces within a settlement, water areas and waterside areas, agricultural land and forest areas would all be included as constituent elements. Formal, ecological and social functions should be defined for each part of this green system. Their connection internally and with other systems beyond the urban area is also essential in order to create a balanced spatial structure.
- Planned conservation of green areas and management of recreational areas maintains the ratio between urban and non-urban areas. In the city's hinterland the possibility of the simultaneous use of agricultural and forest areas for recreation should be turned to advantage.
- Green and open areas, crucial for the green system of the city in ecological and social terms, should be protected from land use change.
- A balance between open and built up areas is important, especially in urban areas. Using vegetation, especially forest, agricultural and other green areas, to separate built-up areas should be used to prevent the fusion of settlements. Artificial or natural settlement borders should be restored and protected.

A major factor, which can significantly influence this strategy, is land ownership, divided among farmers, house owners, and the Farmland and Forest Fund in the rural areas, and house owners and the municipality in the urban area. On the one hand, the land resource for implementing this strategy is extensive, since 44 % of the municipality is Natura 2000 areas, with Škocjanski zatok natural reserve at the entrance to the town being very prominent and playing an important role. On the other hand, however, there is a lack of green structure in residential areas, notwithstanding the fact that the proportions of green structures which should be retained within residential areas are prescribed in the Decree on Spatial Order of the Republic of Slovenia (2004). Investors very often try to manipulate the Decree to their own advantage.

Tourism and the recreational visits by urban dwellers to rural areas are expected to contribute to the income of the rural population, in turn contributing to the financial resources available for implementing the strategy. Incentives from the EU (and partly from the national budget −15 % to 50 % of the cost) will cover the costs of measures that are in accordance with the protection regimes of Natura 2000 sites.

There are no coalitions of interest associated with this strategy. The municipality is the main actor involved in its development. However, small-scale farming practices, which are characteristic of the area, are very positive for biodiversity. In ecological terms, the approach of the strategy has benefits in producing higher biodiversity and better living conditions. The green fingers in the future urban region of Koper are expected to contribute to a higher extent of available green space for recreation for peri-urban residents. Polycentric development is expected to combine social and environmental benefits with the predominantly economics-based priorities of those in power (e.g., mayors, politicians). Natura 2000 status is also expected to protect important parts of the rural areas.

## 11.5 Scenarios Exploring Possible Futures for Koper

Out of the four SRES scenarios developed for PLUREL (see Chaps. 1, 3 and Box 11.2) two were selected by the stakeholders and locally adapted for the Koper situation: Hyper-Tech and Peak-Oil. The third scenario is a baseline for comparison of these scenarios and is called Business-as-Usual (Table 11.1).

**Box 11.2 RUG Model: The Impact of Four Scenarios on Artificial Surfaces in the Municipality of Koper**

The four scenarios (Hyper-Tech, Extreme-Water, Peak-Oil, and Fragmentation) differ in the intensity and patterns of change in artificial surfaces. In the Hyper-Tech scenario, rapid technological change, new transport technologies and few planning constraints lead to counter-urbanisation, with the highest growth occurring in the peri-urban area around Koper and the rural area above Koper and Trieste.

In the Extreme-Water scenario, the coast holds the risk of storm surges for the city of Koper itself, as well as the port, while extreme drought or storm risks may curb urbanisation in the hilly, rural zones further inland. This leads to homogenous urban growth in all zones, as population growth is high and people must live somewhere.

Urban growth is much lower in Peak-Oil than in other scenarios. High fuel costs and strict planning policies concentrate growth in the urban centre of Koper and along the rail network in smaller settlements. Some of these belong to the commuting area of the capital, Ljubljana (off the map to the north-east).

In the Fragmentation scenario, there are clusters of artificial surfaces spread throughout the urban and peri-urban zones, as communities segregate by age group, ethnicity etc. Green enclaves form in rural areas as older native people move out of the city (Fig. 11.9).

### *11.5.1 Likely Land Use in the Municipality of Koper in 2025*

How land use conflicts in the Municipality of Koper will develop in the future in relation to urban development and the protection of the best agricultural land is a topical question. Possible answers were explored using the MOLAND model. In addition to the Business-as-Usual scenario, the Hyper-Tech, and the Peak-Oil scenarios were tested for the period 2007–2025. Data on soil quality (soil number from 7 to 100 is an indicator for soil quality–a higher number means better soil (Mubareka 2009)) was included in the modelling, since such data are among the most important for determining the best agricultural land protection strategy. The final results show a significant decrease in vacant urban zones for all scenarios; growth in all economic sectors and in residential areas. The latter is especially true for the Hyper-Tech scenario.

**Table 11.1** Description of the scenarios and their adaptation (story lines) for the Koper case study

| Scenario | Storyline |
|---|---|
| Hyper-tech | Economic sector of concentration is tourism |
| | Increased pressure on the coastal areas |
| | Specific transportation plans to be determined |
| | Specific building plans to be determined |
| | Building unrestricted in terms of housing and economic sectors |
| | Port expanded and density increases |
| Peak-oil | Building restricted |
| | Relieve pressure on coastline by encouraging building in hinterland |
| | Shift of economic activity towards tourism and agriculture |
| | Tourism sector emphasized natural attractions |
| | Port density increases, but not expanded |
| Business-as-usual | Preservation of natural and agricultural land is a priority, but based on a more detailed and sound division of agricultural land, based on production capability. It's senseless to preserve land as agricultural if it is not suitable for agriculture. So preservation yes, but for the land that really "deserves" it. |
| | Strong pressure on land—high demand for stand alone housing |
| | Coastline somewhat protected |
| | Tourism increasing |
| | Competition for coastal lands between four sectors: tourism, agriculture, port, and housing. The coast (sensu strictu) is highly attractive for individual housing (the view of the sea) |

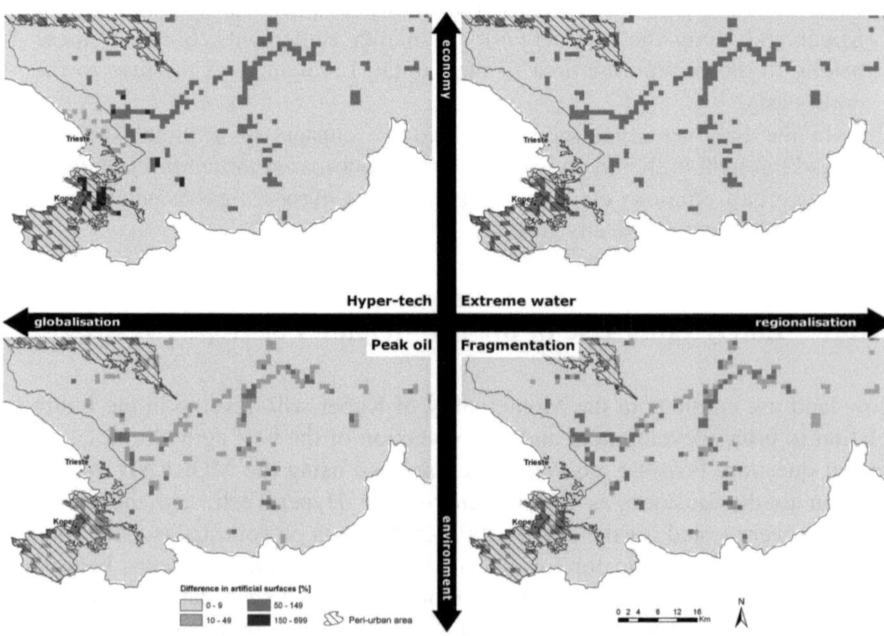


**Fig. 11.9** Regional urban growth (RUG) results for the Municipality of Koper (Source: University of Edinburgh)

**Table 11.2** Trends in residential, economic and natural and semi-natural land use classes in the three Moland scenarios for the period 2007–2025

| Land use | Residential | Economic | Natural and seminatural |
|---|---|---|---|
| Soil quality | ha | ha | ha |
| | Business as usual (BAU) | | |
| Poor soil | −5.0 | −5.5 | 5.8 |
| Moderately poor | −12.5 | −39.5 | 41.5 |
| Moderate | −25.3 | −265.5 | 312.8 |
| Moderately rich | −14.5 | −150.0 | 151.8 |
| Rich soil | 90.5 | 16.3 | 31.0 |
| Total change | 33.3 | −444.3 | 542.8 |
| | High tech (HT) | | |
| Poorest soil | −5.0 | −5.5 | 6.0 |
| Moderately Poor | −11.8 | −40.5 | 41.8 |
| Moderate | −18.5 | −306.3 | 351.3 |
| Moderately Rich | −8.3 | −175.0 | 177.5 |
| Richest soil | 90.5 | −78.5 | 134.8 |
| Total change | 47.0 | −605.8 | 711.3 |
| | Peak oil (PO) | | |
| Poorest soil | 0.0 | 4.3 | −4.3 |
| Moderately Poor | 6.0 | 25.0 | −24.3 |
| Moderate | 37.8 | 381.3 | −344.5 |
| Moderately Rich | −6.8 | 332.3 | −326.5 |
| Richest soil | −5.8 | 687.0 | −636.0 |
| Total change | 31.3 | 1429.8 | −1335.5 |

### 11.5.2 The Possible Future Fate of the Best Agricultural Land

The results of MOLAND show that housing would tend to occupy the best land in the Business-as-Usual (Table 11.2) and Hyper-Tech scenarios, whereas the Peak-Oil scenario would result in housing mostly occupying moderate quality land. The incentives for agricultural exploitation and strong zoning impediments in the Peak-Oil scenario would result in a revival in the municipality's hinterland, even in the remotest rural areas.

### 11.5.3 The Possible Fate of Natural and Semi-Natural Areas

MOLAND results for the Business-as-Usual scenario reveal that natural and semi-natural, i.e. forest and pastures (Mubareka 2009) areas are likely to take over as land cover in places where economic and residential areas go into decline by 2025 (Fig. 11.10). In the Hyper-tech scenario, local communities in the south-west of the municipality would experience an increase in natural and semi-natural land uses and a decrease in economic and residential activities, whilst communities on the coast would be affected by the stimulation of the construction of a new marina and sports facilities (Fig. 11.11). The main feature that characterises the Peak-Oil

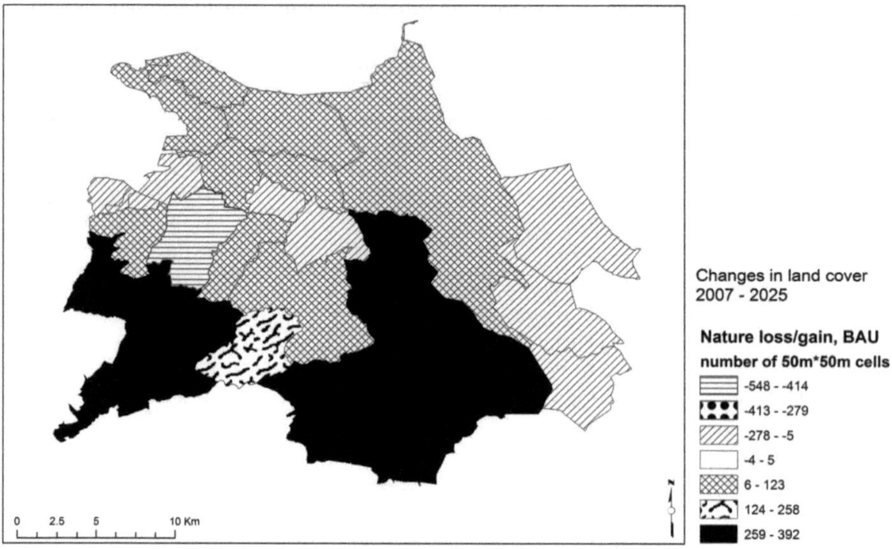


**Fig. 11.10** Trends in natural and semi-natural land use classes for the Business-as-Usual scenario projections for 2025 for the Municipality of Koper. The land use cell area is 50 m × 50 m (Mubareka 2009)

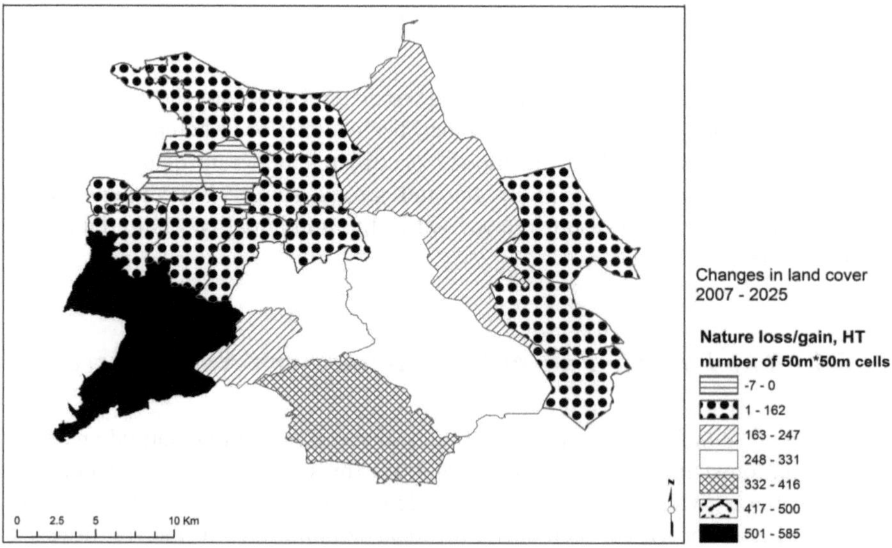


**Fig. 11.11** Trends in natural and semi-natural land use classes for the Hyper-Tech scenario projections for 2025 for the Municipality of Koper. The land use cell area is 50 m × 50 m (Mubareka 2009)

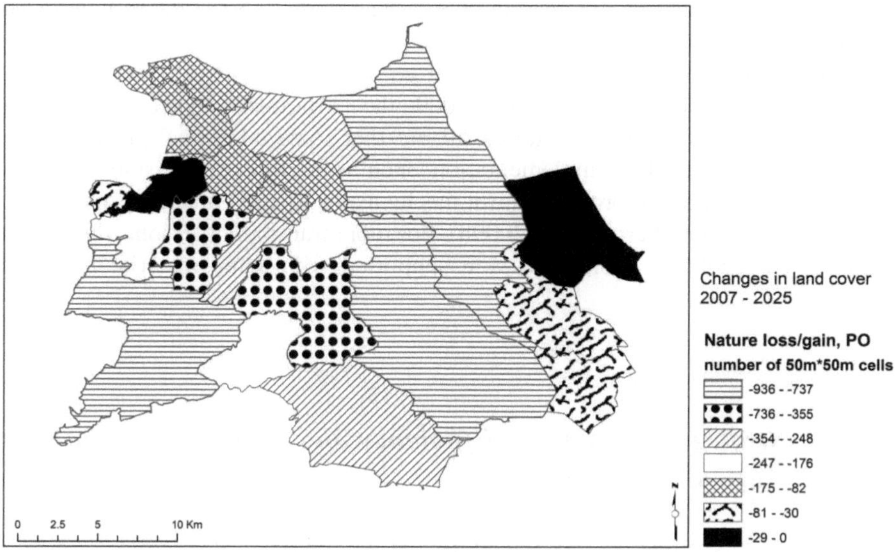


**Fig. 11.12** Trends in natural and semi-natural land use classes for the Peak-Oil scenario projections for 2025 for the Municipality of Koper. The land use cell area is 50 m × 50 m (Mubareka 2009)

scenario is growth in the agricultural sector, predominantly an increase of permanent crops on pasture areas in the hinterland (Fig. 11.12). The remotest parts of the municipality would be significantly impacted by the Peak-Oil scenario, favouring more economic land use types at the expense of their natural and semi-natural cover. Natural and semi-natural areas would be best preserved in the case of the Hyper-Tech scenario in the hinterland, because of the further intensification of activities within the urban and peri-urban areas and abandonment of agricultural activities in rural areas. Natural areas within urban and peri-urban areas would still be under pressure and shrinking.

## 11.6 Lessons Learned from the Koper Case Study

### *11.6.1 Strategies*

The consensus among the various actors in the Municipality of Koper (municipal officials and experts, spatial planners, agricultural and forestry experts, environmentalists etc.) and researchers about the participation approach is that such a tool is necessary to confront the different interests and problems in space and find the optimal solution. However, the final decision is still in the hands of the municipal council, where professional recommendations do not always play the main role, since political and economic power and other vested interests are also important.

At the workshop with stakeholders from Koper and researchers from the University of Ljubljana held in May 2009, the participants mostly agreed that the chosen strategies could be successful but that some additional preconditions will be necessary for their implementation, such as appropriate legislation and its strict implementation, improved control mechanisms, assured financing and public and political support (Perpar et al. 2009b). For more transparent lessons, the results from the stakeholders' workshop were supported with some material from the literature.

### *11.6.2 Strategy 1: Land Use Efficiency and Protection of the Best Agricultural Land*

According to scientific and technical knowledge, the existing system of land evaluation and the existing agricultural land categorisation in Slovenia (agricultural land of the best quality and other agricultural land) is not satisfactory. It should be modified and upgraded. Stricter criteria should be defined for changing the designation of the best agricultural land, especially where the basic infrastructure required for intensive agricultural production is already in place. Furthermore, agricultural land with very good and good soil production potential for agriculture should be defined as the top agricultural land. Measures for modernising and intensifying agricultural production or land recovery also need to be implemented. In the preparations for the Municipal Spatial Plan, these scientific and technical recommendations are considered, as well as other existing legislative documents (e.g., the Spatial Planning Act 2007).

Bearing in mind that scientific and technical knowledge are only one part–the priority issues–the strategy also needs to contain other elements, particularly well-defined procedures and actors involved in its implementation. In this regard, the basic issue is who will exercise power and authority in decision-making and implementing protection of the best agricultural land, and how.

Thus, the improved strategy in Koper needs to include the de facto participation of local actors (of various professional backgrounds) in the planning processes for local land use. They need to have full access to the process in order to express and discuss their ideas, initiatives and concerns with the various professionals employed by the municipality engaged in land planning and conservation and to take an active part in the process of decision-making.

Although this is already being done, since, at least formally, public participation is mandatory in spatial plan preparations, all proposed spatial changes in the area need to be publicly presented and discussed among various interest groups. In order to equip local actors with knowledge about the value of land protection, good information and education programmes also need to be set in place. Then local actors can take full responsibility for land use and also exercise a supervisory function over the formal inspection bodies. Through these activities and learning processes, local actors can gradually gain skills and capacities for taking part in environmental governance.

### *11.6.3 Strategy 2: Rural Development Plan 2007–2013*

Some recent research in Slovenia (Černič Istenič 2010) has shown that the effectiveness of the implementation of the Rural Development Plan is significantly related to the capacities and skills of local people to participate in this programme and to apply for available resources (financial etc.). Since Rural Development Plan measures to promote rural development through the aforementioned activities were introduced for the first time in Slovenia in 2007, the acquisition of such capacities and skills is a challenge for local people and for many other stakeholders. In order to encourage the local population to participate in design and decision making about rural development within the Rural Development Plan, networking in local public-private partnerships–local action groups–is encouraged and financially supported (Ministry of Agriculture, Forestry and Food 2007). As a new institutional form, however, these groups are encountering some problems and inadequacies in their governance.

The short-term interests of individual members, particularly those who are in a position of better access to various kinds of information, have dominated over common strategies and benefits and have led to discrimination among the members in terms of access to project resources. This shortcoming is related to a general lack of a culture of cooperation and feeling of social responsibility in society in general, which leads to a lack of trust and insufficient accumulation of social capital among the members.

An additional reason for the shortcomings of the governance of local action groups is a general lack of sufficient resources available to all members–there are far more ideas among local people than resources (also from the Rural Development Plan) to accomplish them. No private funds, such as exist in many EU countries, have been made available to date for rural development in Slovenia.

Another issue indicated by the research is poor understanding of the actual meaning of state support. There are considerable discrepancies in perceptions among many beneficiaries related to the amount of work associated with projects (preparation of project applications, gathering documents and receipts, writing reports) and the value of this work expressed in payments through the Rural Development Plan. Beneficiaries thus complain that their work is not adequately valued or paid. They do not recognise that state support is only part of the available resources, or that their personal involvement is also required. These misunderstandings have had a negative impact on the willingness of potential applicants, particularly farmers and NGOs, to participate in projects of the Rural Development Plan's Axis 3 more fully.

It is assumed (by some stakeholders) that with the presentation of good practices–successful and satisfied beneficiaries–such attitudes will change and become more realistic. However, at the same time, more attention and support in the future from higher level institutions, e.g., Ministry of Agriculture, Forestry and Food, should be made available (such as more clearly defined procedures of project tenders and less demanding criteria for applicants) in order to stimulate potential applicants who lack sufficient initial capital and project work experience.

Those responsible for the strategy of the Rural Development Plan 2007–2013 in Koper should bear these findings in mind. It is very important that the operational rules of local action groups and the responsibilities of their members are very clearly defined. To evolve a culture of cooperation, more room should be given to negotiations and the reconciliation of different interests among members. Advocates of this particular strategy also need to search for other, not only public, resources to accomplish their ideas. In this regard, they need more knowledge and skills related to rural policy, legislation, administrative procedures and the philosophy of the creation of social capital. An additional element of the strategy could therefore be the organisation and maintenance of high quality educational and informative activities on various topics (including subjects studied within the framework of the PLUREL project) for anyone interested in rural development issues.

### *11.6.4 Strategy 3: Green and Recreational Areas to Increase the Quality of Life*

This strategy has not so far been successfully implemented in Koper. In spite of existing official norms being integrated into the Municipal Spatial Plan, which claim that, within building areas, the preservation of a certain proportion of green space is obligatory, the fulfilment of such requirements, especially with new constructions, has failed. Due to insufficient state control–a lack of inspection efficiency–the interest of construction enterprises to expand their activities on to land earmarked for green space has prevailed. This phenomenon is a consequence of an unsuccessful environmental policy, which is still mainly the domain of the state. Thus, as in the case of the agricultural land protection strategy, a multi-level environmental governance system should also be applied. In this regard, the municipality should be given much greater authority over the control of green space within its territory and set up its own control service. The main issue is allowing the problem to be solved parcel by parcel rather than for the area as a whole.

In addition to official authorities, as in all the other cases, the best resource for ensuring green space protection is the local people, who are aware of the importance of these areas for their quality of life and are willing to participate actively in their maintenance, which is very locally dependent. Today, only a relative small part of the population, including farmers, own or actively manage land. The vast majority of the public are becoming increasingly separated from direct involvement with the land itself. However, most of the population nowadays ‘consume’ land through access to it for active recreation or more passively by simple enjoyment of everyday surroundings (parks, green spaces), if available.

Considering the above findings, the strategy of Koper in relation to green spaces should also contain elements of public involvement. The policy makers of the municipality should encourage local people to participate in caring for green spaces through the activities of civil society associations, clubs, housing councils, schools and kindergartens, enterprises, tourist agencies and the like. Much more attention should be devoted to educating and informing people about the benefits of living near green spaces and raising their awareness of the need to take responsibility themselves for preserving green spaces.

At the same time, local people should be much more involved in the process of land planning and policy making; their wishes and suggestions should be fully considered. As in the case of The Hague School of Painting and its linkage to Green Rings (see Chap. 4), a similar strategy of developing the commitment of local people to green spaces should also be developed in Koper. The reconstruction of the old town centre, in the sense of bringing together culture and landscape, would be a good opportunity for combining people’s ideas and their efforts at obtaining funds from national and international sources. Furthermore, as in the case of Leipzig (see Chap. 8), in Koper too, the availability of green space for its urban dwellers can be extended by the creation of green corridors running between the town and its rural surroundings. In addition to beneficial spatial effects, these actions could give also more opportunity to evolve cooperation of various sorts among urban and rural dwellers of the area.

In relation to the space near Koper port, a strategy could be applied of a green belt combined with building on brownfields, as in the Manchester case (see Chap. 7). In some respects, this has already been put into action by the Koper port authorities but with additional cooperation among the local people and municipal officials these endeavours could be even more effective and satisfactory for all interested parties.

### *11.6.5 Reflection of the MOLAND Modelling Results on the Strategies*

According to the MOLAND results, the Business-as-Usual scenario, if it came to pass, would impact strongly on the best agricultural land, so the strategy “Land use efficiency and protection of the best agricultural land” would not be very effective, unless it was incorporated earlier in the planning system (e.g. as zoning). Because of the decline of residential areas in the hinterland, there would be possibilities for the implementation of the “Green and recreational areas to increase quality of life” strategy for the benefit of the urban inhabitants, if hinterland areas were incorporated in the green system of the town.

The impact of the Hyper-Tech scenario would mainly be on the coastal zone and the hinterland would remain less attractive for many kinds of activities. Thus the Rural Development Plan 2007–2013 would hardly be implemented, since its main idea is to increase environmentally friendly tourism in connection with an extensive farming system in the hinterland. As with the Business-as-Usual scenario, the Hyper-Tech scenario would lead to more naturalised areas at the expense of economic and residential uses in the hinterland. It would be positive for the "Green and recreational areas to increase quality of life" strategy for urban inhabitants if these naturalising areas were also incorporated in the city green system. The results from the MOLAND model are not very optimistic for the "Land use efficiency and protection of the best agricultural land" strategy. The best agricultural land would either be taken for new built development or would become overgrown and transformed to naturalised areas. In the first case, the land would be lost permanently, while the land would not be lost permanently in the second case but would involve a decrease in agriculture which would have a weaker position in negotiating for land.

The results for the Peak-Oil scenario, if they came about, would be very promising for the "Rural Development Plan 2007–2013" strategy, because there are sufficient natural resources for it to be properly implemented. The shift in agricultural production towards it being much more important economic activity than at present would reactivate rural areas and improve the quality of life of farmers. However, in this scenario the "Green and recreational areas to increase quality of living" strategy would be competing for the same natural resources as the Rural Development Plan, at least in the peri-urban zone. In terms of the "Land use efficiency and protection of the best agricultural land" strategy, this scenario shows the most positive results among any of the three scenarios.

### *11.6.6 Comparison Between the Current Land Use Situation, MOLAND Results and Future Plans*

When comparing the current situation in the municipality and its plans (Fig. 11.13) with the MOLAND results, it is most likely that future development in Koper will go in the direction of the Business-as-Usual or Hyper-Tech scenarios. The increase in population density and future land use changes at the community level, as envisaged by the MOLAND model in all three scenarios, can be compared to the actual state in the municipality and its future plans. In reality, the predicted spatial development plans of Koper tend to coincide with the outcomes derived from the MOLAND model.

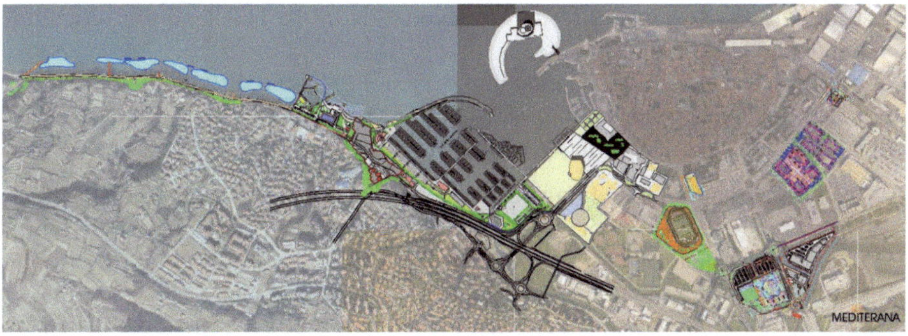


**Fig. 11.13** Aerial photo of the Municipality of Koper with planned interventions (Studio Mediterana 2010)

## 11.7 Suggestions for Better Practice in the Future

Agriculture in the Municipality of Koper plays an important role due to its multifunctionality. Its economic function is of great importance for commercial farms and agricultural enterprises but other functions, such as the social and environmental, are also very important. Agriculture contributes to the attractiveness of the area for tourism by preserving the cultural landscape and producing local products that are seen as an important tourist asset. Moreover, the maintenance and further development of farming activities is a prerequisite for the conservation and sustainable use of agricultural ecosystems. Since the amount of good agricultural land in the municipality is limited, special attention should be devoted in spatial planning to avoiding the use of the highest quality agricultural land when planning interventions in space and maintaining as much of it as possible in the long term.

In addition to agriculture, the area is characterised by an extremely varied and relatively well preserved nature. The relatively small municipal territory has high biodiversity, which is primarily a consequence of different converging climate types, the underlying geological structure and a great variation in altitude and which is also largely connected with the traditional, small scale farming. Since the goal of small farms is often not maximising the profitability of their agricultural production, they tend to use less intensive production techniques and the farmers are often more interested in participating in agri-environmental schemes. Instead of high productive crop varieties, they often maintain locally developed varieties of plants and animals and use extensive production methods instead of intensive ones. These characteristics must be maintained in the future.

The variety of interests in the region leads to many land use conflicts, these being most intensive in the peri-urban zone, which is the most vulnerable to future spatial development changes. The MOLAND results for the Koper suggest a significant decrease in vacant urban land for all scenarios and growth in all economic sectors and in residential areas up to 2025. Housing would mainly tend to occupy the best agricultural land in both the Business-as-Usual and Hyper-Tech scenarios, while the Peak-Oil scenario would result in exploitation of the best land for agriculture and moderate land for housing.

In order to improve the situation and decrease the trend of building on the best agricultural land, the agricultural land use policy should be supplemented and strictly implemented, particularly by the following measures:

- The best agricultural land must be precisely defined legally and spatially;
- New settlement areas should be directed to areas characterised by agricultural land of lesser quality, or to forest areas;
- The best agricultural areas can be used for construction interventions only when other solutions are not available, and then only for the construction of transport, public or social infrastructure;
- A proportion of revenue from taxation of changes in land designation should be directed to funds set aside for other agricultural land quality improvements;
- The quantity of the best agricultural land that is transferred by sale to non-farmers (to limit land speculation) must be restricted;
- Recording of agricultural land transactions must be carried out.

The main purpose of spatial planning is to ensure harmonious spatial development through the coordination of economic, social and environmental protection aspects of development. Such planning is one of the main mechanisms for ensuring sustainable development in space, since it possesses tools that can guarantee the conservation of nature, the protection of natural resources, of the cultural heritage and other high quality features of the natural and cultural environment, by the application of a range of methods.

The current phase of preparation of new spatial documents in Koper is a good chance to analyse past development, present problems, future development possibilities and needs and to evaluate the municipal space. Through the spatial planning process the concept of development in three semi-circular belts, has been confirmed as one of the most appropriate models, due to the spatial characteristics of the area. The coastal belt is already overburdened with existing settlements, with the presence of the Port of Koper as an economic activity of national importance with large spatial needs, activities connected with the sea etc. Because development cannot be halted, its direction into the "second belt" (peri-urban areas) is the only possible and reasonable solution, but planning its development needs special attention to be devoted to the qualities of this area, which must be preserved as far as possible.

Agriculture must also be retained in the second belt, not just because the most favourable conditions for it are there but also because of the "identity" that agriculture gives to the area (i.e., landscape character). Agriculture cannot therefore be preserved only in the "third belt" (rural areas). Bearing in mind the quality of the natural resources (soil quality, water sources, important environmental areas etc.), an inter-sectoral planning approach and adjustments can contribute to more optimal land use.

Spatial planning should, after all, be based on professional, objective background studies and recommendations. Political and economic interests can present a real danger and modify objectively developed plans for the worse. Changes to spatial legislation have been frequent in the recent past in Slovenia and almost every change of government has also been reflected in spatial legislation. Frequent changes are not good, because they are often poorly considered in detail; they cause confusion and, at the same time, provide opportunities for speculators to take advantage of the "unclear situation". Fortunately, the Municipality of Koper is very interested in an inter-sectoral planning approach, which can avoid land use conflicts in the early preparatory phase. Finally, it is to be hoped that the right spatial planning approach and good cooperation amongst all the stakeholders will help the municipality to prepare spatial plans that will ensure the sustainable future development of the area.

Arising from the final results of the PLUREL project, it is recommended that improved strategies in the Municipality of Koper should include: (1) de facto participation of local actors in the local land use planning process, (2) a process of educating and informing people about the benefits of living near green spaces and raising their awareness of the need to take responsibility themselves for preserving green spaces and (3) the organisation and maintenance of high quality educational and informative activities on various topics for all interested in rural development issues.

## References

Agricultural Land Act (2003) Official Journal of the Republic of Slovenia 2003/55. Online from: http://www.uradni-list.si/_pdf/2003/Ur/u2003055.pdf. Accessed 1 Feb 2010

Černič Istenič M (2010) Inter-sectoral coordination of policies affecting diversification and quality of life in rural areas in Slovenia. Deliverable 8.1. Assessing the impacts of rural development policies—project funded by the 7th framework programme for research and technology development of the European Commission. Online from: http://www.rudi-europe.net/79.html. Accessed 23 Mar 2012

Decree on Spatial Order of the Republic of Slovenia (2004) Official Journal of the Republic of Slovenia 2004/122 (= Uredba o prostorskem redu) (in Slovene). Online from http://www.uradni-list.si/1/content?id=51961. Accessed 1 Feb 2010

DOPPS-BirdLife Slovenia (2010) Škocjanski zatok nature reserve—the green heart of the Slovenian Coast. Online from: http://www.skocjanski-zatok.org/index_en.php. Accessed 10 Mar 2010

Ministry of Agriculture, Forestry and Food (2007) Rural development programme of the Republic of Slovenia 2007–2013 (= Program razvoja podeželja v Sloveniji 2007–2013) (in Slovene). Online http://www.mko.gov.si/si/delovna_podrocja/program_razvoja_podezelja_2007_2013/. Accessed 1 Feb 2010

Ministry of Agriculture, Forestry and Food (2010) Proposal of the act of changes and amendments of the act of agricultural land. Online from: http://www.mkgp.gov.si/fileadmin/mkgp.gov.si/pageuploads/Zakon_predlogi/zakoni/osnutek_ZKZ_javna_razprava.doc. Accessed 20 Dec 2010

Mubareka S (2009) Land use projections based on Moland output—Koper test case. Module 2. D. 2.4.2, PLUREL study, unpublished report

Municipal Spatial Plan of the Municipality of Koper—strategic part (2008) (=Občinski prostorski načrt Mestne občine Koper—strateški del) (in Slovene), Municipality of Koper, Office for the Environment and Space, Koper, pp 153

Perpar A, Pintar M, Udovč A, Černič-Istenič M, Zupan M, Prus T, Pirnat J, Hladnik D, Glavan M (2009a) Analysis of regional spatial planning and decision-making strategies and their impact on land use in the urban fringe—KOPER case study. D 3-3-5-Koper, PLUREL Study. University of Ljubljana, Biotechnical Faculty, Agronomy Department, Ljubljana, Slovenia. Unpublished report

Perpar A, Pintar M, Černič-Istenič M, Udovč A, Zupan M (2009b) Assessment of regional strategies—KOPER case study, Slovenia. FIRST DRAFT—for internal use only. University of Ljubljana, Biotechnical Faculty, Agronomy Department, Ljubljana, Slovenia. Unpublished report

Regional Development Centre of Koper (Regionalni razvojni center Koper) (2002) Regionalni razvojni program za Južno Primorsko 2002–2006. Regional development program for the Južno–Primorska Region 2002–2006, in Slovene. Municipality of Koper

Regulation on Criteria for Planning of Spatial Arrangements and Planning Intervention on the Best Agricultural Land Outside Settlement Areas (2008) Official Journal of the Republic of Slovenia 2008/110 (=Pravilnik o kriterijih za načrtovanje prostorskih ureditev in posegov v prostor na najboljših kmetijskih zemljiščih zunaj območij naselij) (in Slovene). Online from http://www.uradni-list.si/1/content?id=89321. Accessed 1 Feb 2010

Statistical Office of the Republic of Slovenia (2002) Si-Stat Data portal. Online from: http://pxweb.stat.si/pxweb/Database/Popis2002/Popis2002.asp. Accessed 1 Feb 2010

Statistical Office of the Republic of Slovenia (2009) Rapid reports: population, 14 Dec 2009, No. 42. Online from: http://www.stat.si/doc/statinf/05-si-007-0902.pdf. Accessed 01 Feb 2010

Statistical Office of the Republic of Slovenia (2010) Slovene regions in figures. Online from: http://www.stat.si/doc/pub/REGIJE-2010.pdf. Accessed 01 Oct 2010

Studio Mediterana (2010) Personal communication

# Chapter 12
# Hangzhou: Fast Urbanisation and High Population Growth

**Martin Spiekermann, Youjoun He, Jianjun Yang, Irene Burkhardt, Fei Yan, Xin Yi, and Stephan Pauleit**

## 12.1 Introduction

Hangzhou is a major Chinese city located at the centre of a prospering urban region in Zhejiang province, part of the rich Yangtze delta region. The municipality covers 16,596 $km^2$ and has over six million residents with an estimated annual population growth of about 100,000 (HSS 2010). Urban growth takes place not only in the peri-urban area but in the urban core itself, which is still significantly expanding, while former rural farmland faces transition into the peri-urban area.

First, historic and current urban development patterns will be reviewed, including an analysis of urbanisation processes. The next section will then introduce the Chinese system of government and planning which is an essential prerequisite for understanding the mechanisms that drive urbanisation as well as the means that can

M. Spiekermann • I. Burkhardt
Irene Burkhardt Landschaftsarchitekten, Fritz-Reuter-Str. 1, 81245 München, Germany

Y. He
The Chinese Academy of Forestry, Research Institute of Forestry Policy and Information, Dong Xiaofu 2, Haidian District, Beijing 100091, P.R. China

J. Yang
Department of Regional and Urban Planning, Zhejiang University, Hangzhou 310058, P.R. China

F. Yan
Urban and Regional Planning Department University of Florida, P.O. Box 115706, Gainesville, FL 32611-5706, USA

X. Yi
Department of Urban Planning, School of Architecture, Southeast University, Sipailou 2#, Nanjing 210096, P.R. CHINA

S. Pauleit (✉)
Chair for Strategic Landscape Planning and Management, Technical University of Munich, Emil-Ramann-Str. 6, 85354 Freising, Germany
e-mail: pauleit@wzw.tum.de

K. Nilsson et al. (eds.), *Peri-urban futures: Scenarios and models for land use change in Europe*, DOI 10.1007/978-3-642-30529-0_12,

be employed to control urban expansion. Three sub-cases were selected to explore specific challenges of peri-urbanisation in more depth and how these are dealt with in planning. The strategies employed were suggested as good practice by the Chinese partners from the Zhejiang University. The chapter will go on to discuss the potentials and limitations of these strategies and to draw conclusions about planning for sustainable development of peri-urban areas in Hangzhou and potentially for other Chinese cities.

In contrast to the European case studies, it was not possible to develop scenarios for future land use with local stakeholders for Hangzhou and implement these into the RUG and MOLAND models nor to carry out any quality of life surveys. Therefore, the Hangzhou case study concentrates on assessment of strategies employed to address current challenges of peri-urbanisation.

## 12.2 Hangzhou City Region

### *12.2.1 Ancient Origins Combined with Modern Dynamism*

The City of Hangzhou was founded about 2,200 years ago during the Qin Dynasty, although the first evidence of human settlement can be traced back more than 5,000 years. The city is recognized as one of the seven ancient capitals of China. In the thirteenth century Marco Polo, the Italian traveller, admired Hangzhou as "the most splendid and luxurious city in the world". Hangzhou has a world famous heritage of historic houses, temples, parks and gardens which mainly lie close to the West Lake, a natural water body which was enlarged and modified over the centuries. The West Lake area is still the basis of Hangzhou's fame as a tourist destination and the city is considered to be the most attractive place to live in China (Fig. 12.1).

After the foundation of the People's Republic of China in 1949, Hangzhou became the capital and also the political, economic, scientific and cultural centre of Zhejiang Province. It is second only to Shanghai in the Yangtze delta in south-eastern China (Fig. 12.2). The whole Yangtze delta is densely populated and is one of China's three economic powerhouses.

The urban centre of Hangzhou is located at the lower reaches of the Qiantang River on Hangzhou Bay in the southern part of the Yangtze River Delta and at the southern end of the historic Beijing-Hangzhou Grand Canal (Fig. 12.2). The rural–urban region is far from being completely urbanised (Fig. 12.3). Topographically, hills and mountains occupy 65.6 % of the total municipal area, the plain 26.4 % and the streams, rivers, lakes and reservoirs 8.0 % (Hangzhou Statistical Yearbook 2005). The north-western and south-western parts of Hangzhou are mountainous regions with woodlands and mainly small-scale agriculture. Here, the peri-urban area can be described as a mosaic of small urban settlements with tourist infrastructure, small scale industry, educational facilities and farmland. The north-eastern and south-eastern parts of Hangzhou are situated in the densely

**Fig. 12.1** West lake scenery

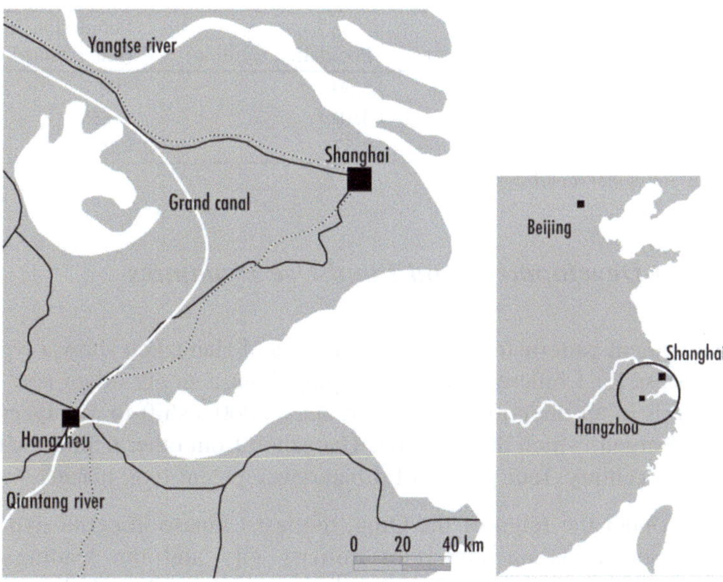


**Fig. 12.2** Location of the city of Hangzhou

populated and fertile north plain of Zhejiang, which is known as a country of fish and the region's breadbasket. It is characterised by a dense network of streams and rivers. Here, manufacturing and high-tech industry has developed in the peri-urban areas. Table 12.1 shows some facts about the area and population of Hangzhou.

**Fig. 12.3** Administrative regions of Hangzhou

**Table 12.1** Facts of the city of Hangzhou

| | Area ($km^2$) | Population (million) | Population density (inhabitants/$km^2$) |
|---|---|---|---|
| Municipality | 16,596 | 6.60 | 398 |
| City proper | 3,068 | 4.10 | 1,335 |
| Hangzhou city (built up area) | 314 | 2.52 | 8,004 |

Source: Hangzhou Statistical Yearbook (2005)

## *12.2.2 Recent Development and Land Use Dynamics*

The early development pattern and land use dynamics of Hangzhou show a typical urbanisation process for Chinese cities, developing from a single urban core and sprawling outwards in all directions. However, since 2000 a shift to a polycentric pattern can be observed, which distinguishes Hangzhou from other Chinese cities.

In the twentieth century, four phases of urban development took place:

1. *Before 1949:* Under the reign of the Qing, the last Chinese imperial dynasty, Hangzhou prospered as an important military city and the leading silk manufacturing centre in China. The city then went through turbulent times after Chiang Kai-shek had conquered Hangzhou during the revolution in 1911. Yet it was only in the 1920s that the city gave up its leading economic role in the Yangtze River delta to Shanghai. Nevertheless, it continued to grow and remained an important industrial town and the transport hub of Zhejiang province.

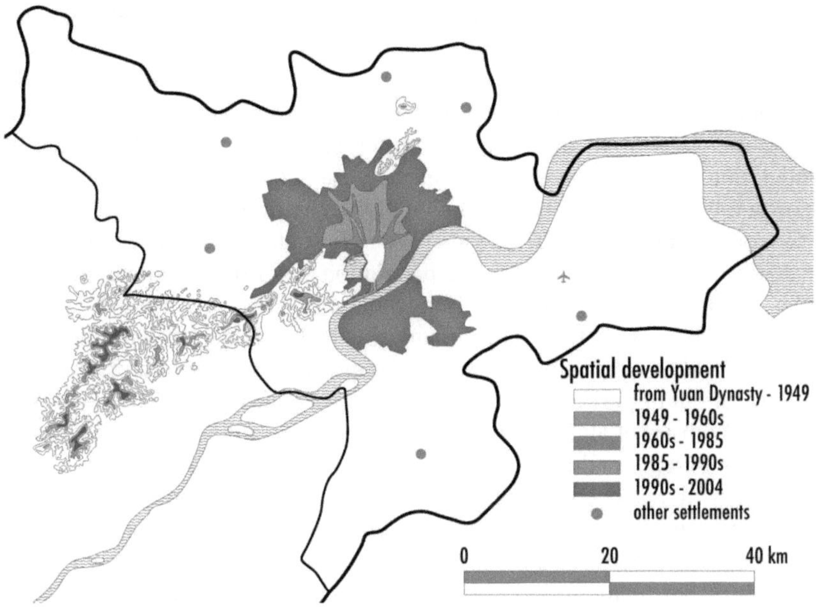


**Fig. 12.4** Development of Hangzhou city from 1949 to 2004

2. *From 1949 to 1978*: After the foundation of the People's Republic of China, civil war came to an end. The social situation and the economy underwent fundamental changes. The scale of industrial production rapidly increased and several large industrial zones were built around the city core of Hangzhou. As the new industrial zones were only a few kilometres away from the urban centre, most of these zones eventually grew together and merged with the city core. During this period, Hangzhou also started to expand along major road, rail and water course axes, especially in the 1950s and the 1960s, resulting in a hand-shaped structure.
3. *From 1978 until 2001*: Economic reform known as "Socialism with Chinese characteristics" was introduced in 1978. While the state maintained ownership of a large proportion of the Chinese economy, the market was also opened to private parties. At that time, Hangzhou's small urban core could not provide enough space for its economic development. Therefore, a number of new factories chose peri-urban locations. In addition, many existing industries relocated from the core city and moved out to cheaper land. Consequently, the spaces between the five fingers of the hand-shape city became filled up. The city began to sprawl outwards on a large scale (Fig. 12.4).
4. *Current urban development situation and post 2001 trends*: One of the characteristics of Hangzhou is the dynamic demographic situation with high levels of in-migration leading to high rates of population growth which drives urban development. Economic development in turn attracts more migrants which puts further pressure on urbanisation.

Between 2001 and 2005, the non-agricultural population in Hangzhou City increased from 1.93 to 2.46 million and its proportion of the total population of the municipality as a whole rose from 50.9 % to 60.0 % (Hangzhou Statistical Yearbook 2005). The population of the entire municipality grew on average by 1.7 % between 1990 and 2000 (Webster et al. 2003, see also Table 12.1). Population increase was strongest in the urban core where the annual average growth rate was 5.2 %. The population of the city proper is currently 4.4 million (Yao, oral comm., 18/08/07, see also Webster et al. 2003). It is expected to peak in 2030 when it will have reached a population of 6.7 million (note: these figures refer to the eight districts of Hangzhou city—the area shown in grey in Fig. 12.4). By contrast, the peri-urban areas had a slower growth rate while the rural areas of the municipality even lost population as people moved from there into the urban or peri-urban areas. Migrants account for 22 % of the population overall (Webster et al. 2003) while migrant-to-local ratios can be even higher in highly-favoured localities.

Urban growth has been driven by strong economic development which has also drawn migrants to the city. Hangzhou's GDP increased in the period of the 10th Five-Year Plan (between 2001 and 2005, more recent figures were not available) by 13.6 % per year. In 2005, the per capita GDP of Hangzhou exceeded US $5,000 (Leman 2005).

### 12.2.3 Peri-Urbanisation

For the case study of Hangzhou researchers developed and applied a methodology for identification and delineation of peri-urban areas based on their spatial character.

Figure 12.5 shows the development of the peri-urban area from 1988 to 2004. The dark grey colour represents the built-up area of the core city while the light grey indicates the peri-urban area. According to the results from this analysis, the peri-urban areas expanded by 471 km$^2$ between 1988 and 2004, corresponding to an increase of 458 %. In 2004 they covered 19.7 % of the area of Hangzhou City. Urban core areas expanded by 158 km$^2$ during the same period, corresponding to an increase of 668 %. The urban core covered 6.1 % of the city's area.

Periurban areas have grown differentially: there is strong growth in the southeast along the Hangzhou—Ningbo corridor whereas the northern area stagnates. The growth in the east is supported by infrastructure development (bridges over the river, etc.). This growth led to a major loss of farmland. The municipality lost 8 % of its total cultivated land in the 10-year period between 1991 and 2001 (Webster et al. 2003). Most of the loss occurred in inner peri-urban areas close to the core city.

In 1988, Hangzhou was a monocentric city. After 1994, the City government decided to move developments across Qiantang River and locate a new centre on the south bank in order to relieve the pressure from the urban core. The beginnings of this shift however date back to the early 1990s, when Binjiang was established as

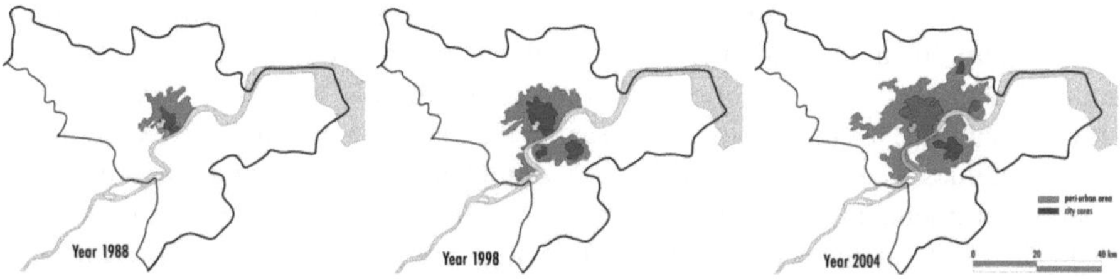


**Fig. 12.5** Land use change in the Hangzhou city between 1988 and 2004 (urban areas are in light grey)

a new urban development zone. Several new bridges enabled new urban development on the south bank of Qiantang River to take place. The image from 1998 clearly illustrates the result of this strategy: two newly built areas appeared on the south bank. In 2001, the reform of administrative divisions, incorporating adjacent towns into Hangzhou city, improved the coordination of urban development. By 2004, separate centres on the south bank joined together to form a larger centre; while on the north bank, two new centres, although in small size, began to take shape. As a result, the city has expanded enormously both north and south of the Qiantang River.

## 12.3 The Government and Urban Planning System as the Main Drivers of Urban Sprawl

### *12.3.1 General Political and Economic Conditions Leading to Urban Sprawl and Excluded Farmers*

China's mainland is divided into 31 administrative units at the province level, 22 provinces, 5 autonomous regions, and 4 cities under the direct jurisdiction of the central government, such as Beijing and Shanghai. The governmental system shows a strong top-down character. Every department of the administration is seen as a branch of the central government and higher level administration always controls that at the lower level.

Since the reform policy started in1978, socialist ideology has been fundamentally weakened and a series of adjustments have been carried out in the transition from a centrally-planned to a market economy. The strong economic growth is a sign of the success of this policy. However, instead of a radical break with the old centrally-planned system, the Chinese government adopted a gradual transformation. Consequently, a dualistic system now exists which leads to tensions between the economic and political systems. Moreover, decision-making has been to a great degree devolved from central government to the local governments which are the main driving force for urban development.

Since a large part of local revenues generated by taxes are handed over to Central Government, local governments need to generate or obtain financial sources for the large investments in new public infrastructure required as a result of the rapid rate of urbanisation. This has led to an increasing gap between revenue and expenditure. Since revenues from land sales stay locally and because they have land as an asset, local governments are willing to sell it, especially undeveloped land in peri-urban areas, in order to raise capital (see Ding and Song 2009 for a more profound discussion of the current problems of the tax system).

Moreover, land in peri-urban areas is often used inefficiently, since the land price is low, which allows buildings to expand horizontally rather than grow vertically. Furthermore, a number of strongly developing Chinese cities suffer from land market speculation by people who bought land without intending to use it at present and which remains unused and unmanaged.

### *12.3.2 Institutional Frameworks in Terms of Management, Land Ownership Not Equipped to Control Sprawl*

The simultaneous existence of the system of central planning and the market economy is reflected in the field of urban planning. In a country under public ownership urban planning is seen more as a management tool for the government than for the regulation of the economic and legal relationships between the public and private sector or among the different segments of the private sector itself. Although the property right has re-emerged with the rise of the market economy and a series of housing and land reforms have been implemented, the old centrally-planned system still has a strong influence as an instrument for the control of resources by the government.

In China, there are only two types of land ownership. One is state-owned land and is administered by public sector organisations, such as city government or some other state-owned enterprises. The other is known as collective land. This is owned and operated by groups of local farmers and can only be used for the purpose of agricultural production and for farmers' houses and other agricultural buildings or facilities. The requisition of collective land for urban development must be implemented through the city government.

For the development of the market economy, land markets have been established in urban areas through legal reform whereby land use rights can be transferred. The land use market is divided into two types. The primary land market refers to the transfer of land use rights from the state to the users. The second land market refers to the transaction of land use rights among users.

In the periurban area specific conditions for the control of land use need to be considered that have a strong impact on their spatial development:

1. Periurban areas are located in between the rural and urban area. Therefore, the two forms of land ownership, collective land for agriculture and land used in the market economy in the designated urban areas, neighbour and intermingle with each other.
2. A considerable amount of land belongs to state-owned enterprises which are authorised to administer it. In turn, they are overseen by their own public sector regulators at a higher administrative level than the local government. As long as the land use is still according to the original purpose, usually industrial production and the associated workers' housing, state-owned enterprises can carry out their own construction programmes and only have to register them in the local planning department. Before the reform policy, many state-owned enterprises were located in the relatively remote peri-urban area. However, land owned by state-owned enterprises has been increasingly integrated into the expanding urban fabric and other economic functions have often been developed on these lands without the permission of the local planning department.
3. Collective land is often illegally converted to economic uses such as industrial production and housing markets may also arise on collective land. Local governments often tolerate such development in order to promote economic growth. Therefore, the simple prohibition of conversion of collective land to economic functions cannot halt urbanisation because of the enormous economic pressure.

Based on the Urban and Rural Planning Law enacted in 2007, a city government is authorised to re-designate some collective land as urban areas. In principle, subsequent construction activities there are based on urban plans regulated by the planning law. However, the problem of self-administration of the public sector land and the illegal conversion of collective land are obstacles for well-ordered urban development. Thus, the lack of a unified administration needs to be considered as the main problem in peri-urban areas. The various forms of control of land use rights are a structural flaw for the spatial development of urban areas.

### *12.3.3 Government-Driven Development and Subordination to Economic Planning*

The city government level is authorised to administer the adjacent rural and peri-urban areas, so that urban development can be implemented as quickly as possible. The large cities of a higher administrative level can regulate the development of the adjacent smaller cities and counties of a lower administrative level through a series of financial and regulatory instruments. Since 2001, the City of Hangzhou has expanded its surface area through a series of administrative reforms. Several adjacent cities and counties have been integrated into the city where they take the role of subordinate organs of Hangzhou.

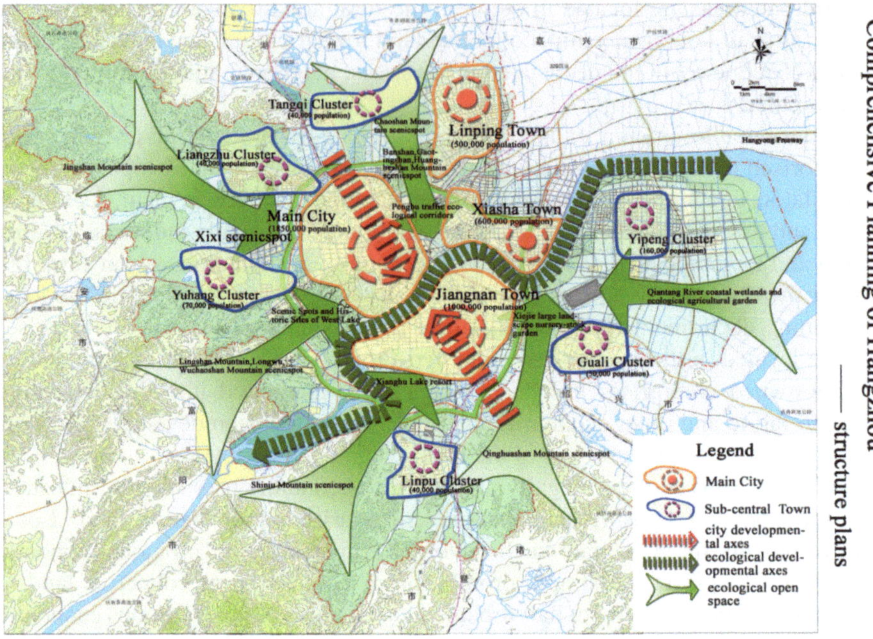


**Fig. 12.6** Hangzhou comprehensive plan (2001–2020)

Urban planning is used as an important instrument to guide urban development. Based on the planning law, legal planning instruments are divided hierarchically into three levels: 'Urban system plan', 'Comprehensive plan' (Master plan) and 'Detailed plans'. The 'Urban system plan' is only used to determine the main aspects of city size and provides only a few details for guiding the following development. By contrast, the 'Comprehensive plan' provides more detailed information for urban development and plays the central *de facto* role in determining the layout of development, functional land use differentiation, the general character of land use, and the transport system. The comprehensive plan of a large city like Hangzhou always includes both the urban area of the city itself and the adjacent cities and towns. Therefore, it influences development at city regional scale.

It is the intention to develop the city of Hangzhou as a polycentric, green city with rapid economic growth and well balanced levels of employment and residency. These goals are reflected in the current Hangzhou Comprehensive Plan (2001–2020) which outlines the spatial strategy as 'one urban centre, three sub centres, six town clusters and six green belts (Fig. 12.6)'. However, from the preceding analysis, intense conflicts in the peri-urban areas are expected to arise because of existing institutional and organisational frameworks. Moreover, the various control forms of land use rights cause a major obstacle for unifying administration and development.

In conclusion, rapid urban development is facilitated by a centralised, top-down system of planning and decision-making. These centralised arrangements stem from the pre-reform period. At that time, city planning was restricted to the implementation of strategic industrial projects. That is also the reason for the existence of so many state-owned enterprises with self-administrative authority. This allocation of authorities still continues to exist after the economic reforms and leads to an uneasy relationship with market orientated policies.

Significantly, urban planning remains a task subordinated to economic planning, which is carried out independently, by the Development and Reform Commission. The basic purpose of the Comprehensive Plan is to support economic planning by providing the corresponding spatial allocation.

Urban development is characterised by a project-orientated mode where the peri-urban area is basically seen as a free space for the allocation of strategic projects and the uptake of land resources. On the other hand, due to their specific character, peri-urban areas represent a more complex situation for planning than urban areas. Economic growth encounters multiple restrictions due to other goals e.g. environmental protection and landscape conservation.

The co-existence of a centrally-planned system and market economy will make communication between stakeholders difficult. Cooperation will be arranged around the goal of economic growth among the government levels, state-owned enterprises and private developers while local inhabitants will be by and large excluded from involvement. Cooperation in other fields will be limited and only feasible when combined with projects for economic growth.

## 12.4 Development Strategies for Peri-Urban Areas: Selection of Case Study Areas

The following three case studies will reveal a series of challenges for the development of peri-urban areas in Hangzhou. Study areas outside the original core city were selected, representing different geographic and socio-economic parts of the peri-urban area and also different urbanisation pressures and development paces. The study areas are shown in Figs. 12.7 and 12.8.

The Zhuantang area in the southwest represents an attempt to combine moderate urban development with tourism and protection of the cultural and ecological heritage. The Xixi wetland area in the west serves as a case where landscape restoration is combined with development of up-market residential areas. Finally, the Binjiang district represents the case of a rapid large scale development from rural to urban structures and from low to highly productive commercial conditions.

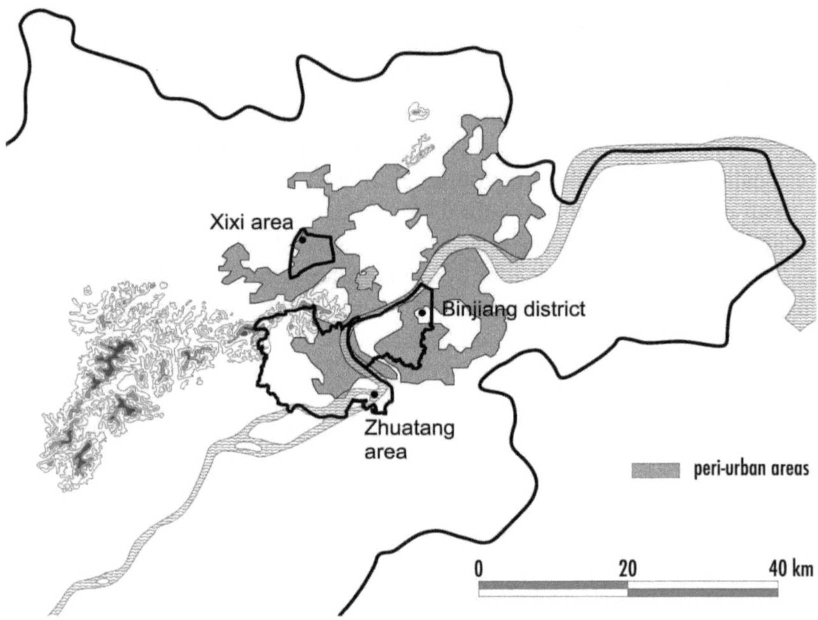


**Fig. 12.7** The three cases study areas: Zhuantang, Xixi and Binjiang

## 12.5 Zhuantang: Strategies to Combine Moderate Urban Development with Tourism and Protection of the Cultural and Ecological Heritage in an Area that is Lagging Behind in Development

The Zhuantang area is designated as a tourism and ecological protection area in Hangzhou but it is lagging behind in terms of urban development. There are some large urban areas with specific functions in this region such as the Zhuantang Science and Technology Economic Zone, Zhijiang campus of Zhejiang Industrial University, military land, the driving school of the municipal police station, a school of the Chinese Academy of Art, Fushan Ecological Resort and the Longwu and Linshan scenic areas. It is also the water supply area for the whole of Hangzhou. In recent years, the regional development strategy and spatial planning process have suggested policy changes. The area can therefore be described as a case where an attempt has been made to combine conservation of ecological and cultural heritage with controlled growth.

### *12.5.1 Main Issues and Strategies for Sustainable Development of Land Use in Zhuantang*

The Zhuantang area covers 156 km$^2$ and had a total population of 126,197 in 2006 (Table 12.2). It consists of three towns, one township and the Zhijiang National

a

b

c

**Fig. 12.8** (**a–c**) Typical views of Xixi wetland park, urban development in Binjiang and tea farming in Zhuantang study area (from *left* to *right*) (Photos: M. Spiekermann 18a&c, S. Pauleit 18b)

**Table 12.2** Population structure in Zhuantang area (2006)

| | | | Part 1 | | Part 2 | |
|---|---|---|---|---|---|---|
| Zhuantang | Area (km$^2$) | Total population | Migrating population (%) | Household population (%) | Urban population (%) | Rural population (%) |
| Total | 156.35 | 126,197 | 15.1 | 84.9 | 55.1 | 44.9 |

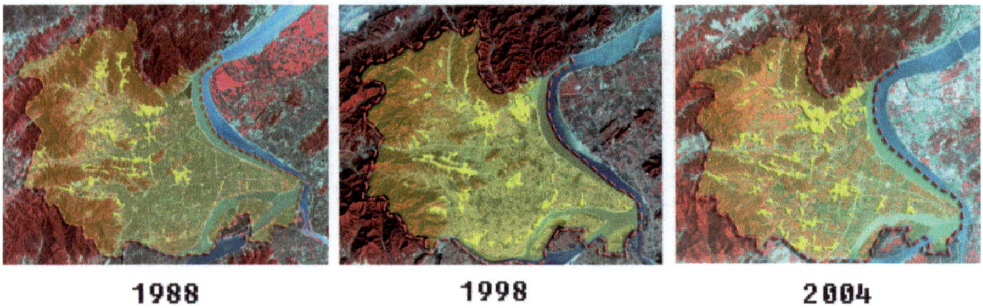


**Fig. 12.9** The development of built-up areas (*yellow*) in Zhuantang

Tourist and Holiday Resort. The population growth in the Zhuantang region has been slow by comparison to other peri-urban areas around Hangzhou and a large part of the population in this area is still rural, i.e. employed in agriculture. The famous Longjing tea is produced in the Zhuantang area.

In the early 1990s, the Zhuantang area became a part of the Hangzhou Scenic and Water Resources Protection Area, primarily to supply the city with water and agricultural products. As a consequence, the social and economic development was restricted in many aspects because of the environmental protection policies within the area. Industrial and residential development projects were unable to expand into the Zhuantang area, and industrial development within towns was also limited. Thus, the pattern of land-use was simple and built-up areas were far fewer than the green and open spaces (Fig. 12.9).

Later in the 1990s there was a significant change in policy for the development of Zhuantang, in order to respond to the need for local economic growth. As an outcome of the new plans and policies, the land designated for construction expanded. By 1998, built-up areas had extended towards the riverside of Qiantangjiang, and large expanses of farmland and water began to disappear in the eastern and southern parts of the area. By 2004 the fast expansion of built-up areas had created several urban cores which accommodated public and private residential estates, often in the upmarket sector, and tourism facilities. All of these developments resulted in the corresponding loss of farmland within the region.

In our study, we identified five main strategies that have been applied in the Zhuantang area to address these challenges.

1. *'Active' protection of the environment and natural resources*: 'Protection' is the most important theme, since Zhuantang has a valuable natural environment, and more importantly, it is the drinking water source for the entire city of Hangzhou as noted above. However, in recent years, the protection strategy has shifted from mere protection to a new concept called "active protection". The rationale of this strategy is that protection of the environment and natural landscapes should still be prioritised but also be combined with development. New developments should concentrate on the expansion of already existing settlements while leaving other areas as green and open space. The strategy also aims to increase the economic viability of traditional farming activities. Organic farming has been supported by strictly limiting the use of pesticide, providing market information and skill assistance to farmers in order to increase the production of quality agricultural products which can be sold at higher prices; public infrastructures (especially water and sewer networks) have been improved; green industries, i.e. industries with a low environmental impact, and services such as education as well as activities associated with the cultural and creative economy have been encouraged; heavily polluting industries are strictly prohibited.
2. *Development of tourism*: Zhuantang has attractive natural landscapes where tourism facilities have been in existence for a long time. In 1992, the Zhijiang National Tourism Holiday Resort was set up. The resort is one of 12 such resorts approved by the State Council. Therefore it enjoyed preferential policies to promote the development of tourism. For example, foreign funded enterprises which invested in tourist projects with at least a 10-year operation are exempt from income tax in the first 2 years. With this and various other preferential policies, the resort has attracted domestic and international investment. However, a government report (HMCDR 2009) pointed out that the tourism development of Zhuantang did not actually benefit the local economy since most tourism development projects are independent of Zhuantang and show little economic connection with it. Furthermore, tourism projects are managed by three different governing bodies at the same level. The local Zhuantang government has little decision making power to exert and gains little share of the profits.
3. *Development of the cultural and creative economy*: As a major element of this strategy, the Hangzhou government cooperated with the China Academy of Art and decided in 2007 to locate the Academy's School of Visual Art in Zhuantang. This school is expected to bring a number of design companies and to attract artists to the area. In addition, a broader plan for a Cultural and Creative Park was started in 2009. It includes redevelopment of villages and abandoned factories, to be dedicated to the cultural and creative industries. Therefore, Zhuantang is envisioned to be 'the town of the arts and design'. Development of rural tea houses is another important part of the cultural economy in Zhuantang. Customers can experience the farmers' life style and enjoy the famous Longjing tea.

4. *Pollution prevention approaches*: From the late 1970s, Hangzhou started to adopt a strict environment pollution prevention strategy to ensure the drinking water safety for the 1.8 million strong population. Since then, various polluting industries have been shut down or moved out of the area. In addition, the Hangzhou government has taken advantage of the construction of the Zhijing National Tourism Resort to improve the waste water network in Zhuantang area and to connect it to the city's main pipeline networks.
5. *Financial compensation*: Financial compensation for the economic opportunities foregone due to strict environmental protection is one of the most important elements of the strategies adopted in the Zhuantang area. Transfers are made both from city government to Zhuantang as well as within the Zhuantang area. For example, the Zhijing National Tourism Resort and the local townships and villages can establish a benefit-sharing compensation system through tax transfer, i.e., the Zhijing Resort transfers a partial tax income to the Zhuantang District Government, or pays part for part of the infrastructure construction such as pipeline networks.

### 12.5.2 Experiences from the Study of the Zhuantang Area

Major issues for peri-urbanisation in the Zhuangtang area identified were:

- Combination of landscape and environmental protection with economic development
- Coordination of interests between the city and the local levels as well as between different governing bodies within the Zhuangtang area.

The combination of measures for environmental protection, promotion of tourism, cultural and creative industries with fiscal measures can be highlighted as an attempt to develop a comprehensive approach to the sustainable development of the area which takes advantage of its attractive cultural landscapes. Thus, the strategic approach recognises the specific needs and potentials of Zhuangtang. However, the success of these strategies is still uncertain. Although Zhuantang chose to promote environment-friendly economic activities, such as the cultural and creative economy, the scale of these industries is not yet large enough to make a marked contribution to local economy. The development of facilities and infrastructures for recreation by the nearby population of Hangzhou and for tourism is certainly still in its infancy compared to European cities. The compensation of economic gains foregone by environmental protection needs to be highlighted as an important mechanism to promote sustainable development in the Zhuantang area. Implementation of this mechanism is made possible due to the strong top-down government system. However, Zhuantang's attractive natural environment has now caught the attention of developers and speculators who wish to build up-market houses. This will greatly increase land prices and put green and open at risk. The city and local governments have not yet responded to these trends.

A range of problems and conflicts, in particular with ecological conservation, need to be solved in future urban development. The division of the management system between the Zhijiang National Tourism Resort and the local town government and the lack of corresponding coordination mechanisms limit to some extent a more coordinated development of Zhuantang as can be seen, for instance, in the conflict between the extension of rural homes and the protection of development zones and tourism zones.

## 12.6 Xixi Area: Restoring Lost Landscape Qualities While Coordinating Up-Market Residential Development

The Xixi area is located north-west of the West Lake District (Fig. 12.7). The study area comprises Xixi National Wetland Park and Jiangcun Township. Historically the Xixi area was renowned for its scenery and natural wetlands. However, the strong urbanisation process since the 1990s has destroyed a large part of the wetland area. In recent years the core section has been transformed into a national nature reserve serving ecological, conservation and recreation purposes. The former residents have been moved from the site and given new homes and compensation land on the edges of the newly established park. The Xixi area is therefore an interesting case where ecological restoration has been combined with coordinated planning of urban development.

### *12.6.1 Main Issues and Strategies for Sustainable Development of Land Use in the Xixi Area*

The delimitation of the study area was made according to administrative unit boundaries, in order to be able to use statistical data. The study area is a part of West Lake District. It covers 23 $km^2$ with 150,232 residents in 2006 (Hangzhou Statistical Yearbook 2006, Table 12.3).

The 'westward' process of urbanisation in Hangzhou is visible from the satellite images (Fig. 12.10): the urban land has expanded to the west year-by-year, while agricultural land has been decreasing and the wetland area has been shrinking. The original rural inhabitants have been forced to become urban citizens along with the demolition of their villages to be replaced by mass housing. Simultaneously, the inhabitants originally living in Xixi Wetland also constructed buildings and expanded their houses without permission. The eastern part of Xixi area had been totally urbanised by 2004.

In order to respond to these challenges, the municipal government of Hangzhou officially launched the protection and development of the Xixi National Wetland

**Table 12.3** Population structure in Xixi area (2006)

| | | | Part 1 | | Part 2 | |
|---|---|---|---|---|---|---|
| Xixi | Area (km$^2$) | Total population | Migrating population (%) | Household population (%) | Urban population (%) | Rural population (%) |
| Total | 23.06 | 150,232 | 35.5 | 64.5 | 92.7 | 7.3 |

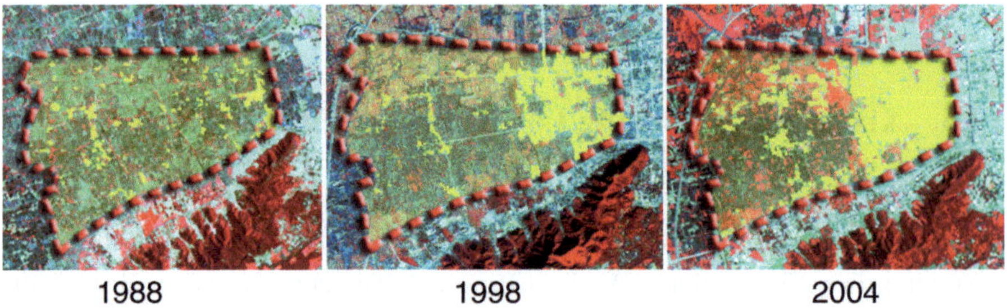


**Fig. 12.10** Development of built-up areas (*yellow*) in the Xixi area

Park in 2003, and the farmers who originally lived in the wetland were resettled outside the boundaries. Within this context, three strategies will be evaluated:

1. *Planning of Jiangcun New Town*: Residential development of the Jiangcun area was first driven by market forces from the early 1990s. Planning was largely absent. In addition, the market could not provide sufficient public facilities and infrastructure to support local development. To solve these problems, the Hangzhou government made a comprehensive plan for Jiangcun, and invested in the improvement of public facilities and infrastructures. The New Town of Jiangcun was established and will function as the main service zone for Zhejiang University and provide tourism services for the Xixi wetland. A large amount of land in the new town was planned for commercial, financial and mixed uses. To implement the plan, arterial streets were widened; the quality of pavements was improved along a number of roads; the Zijinggang Campus of Zhejiang University was established in Jiangcun, together with several schools and kindergartens. The development strategies also supported local business by building Western City Square, which is the largest commercial centre in the west of Hangzhou. These measures helped Jiangcun to become attractive for both living and investing.
2. *Supporting the livelihood of dispossessed farmers*: On completion of the Xixi Wetland project more than 2,500 rural households involving 13,000 farmers will have been moved. Great efforts have been taken to build resettlement communities for them. Dispossessed farmers have been prioritised for employment in the service, boating and property management sectors and they have also been employed in park construction and management. In Phase 1, 115 farmers from Jiangcun Township were employed, accounting for 39 % of the total of 295

employees. Moreover, West Lake District Labour Security Bureau has carried out a series of skills training programmes which are focused on the management of land acquired from farmers in terms of cultivation, gardening and green space maintenance.

3. *Construction of Xixi Wetland Park*: The Xixi area was threatened by the same problems encountered in the development of peri-urban areas. In response, the Hangzhou government began a large restoration project in 2004 to establish Xixi Wetland Park. The project is composed of three phases and its core area will cover 10.08 km$^2$ when completed. It aims to create a win-win situation between environmental protection and economic development. The wetland landscape with its associated ecosystem services will be restored. At the time of writing, the wetland park is almost completed and already now an attraction for recreation and tourism. Marketing strategies have been adopted to promote the Park. Various events, like dragon boat racing, World Wetland Day, Yue and Hang Opera have helped to reinforce the park's image as a scenic area.
   The project management received special attention. The management of the Xixi Wetland project mainly involves two levels of government, the Hangzhou city government and the West Lake district government. The former established a project coordination group and assigned a vice-mayor as the group leader. Meanwhile, Xihu district assigned a vice-mayor to lead a lower level management group. These two groups include more than 100 officials from all the major departments. Establishment of these two project coordination groups has helped to streamline decision-making structures and powers. As a result, the evaluation and approval process which normally take a very long time for big construction projects were expedited. The project management group was also able to deal effectively with farmers whose land was expropriated ('dispossessed farmers'). For example, in Phase 3 of the project 1,521 residents were relocated in just 17 days, including the process of contracting, evaluating compensation claims and relocating them. However, it should be noted that some rights, such as the right of these residents to appeal were sacrificed in exchange for this speed.

### 12.6.2 Experience from the Xixi Area Case Study

The current development of the New Town of Jiangcun is led by the government of Hangzhou and follows a well structured approach in accordance with planning for improvement of public infrastructures. From this perspective, the Xixi Wetland has experienced a shift from disorderly development to coordinated development, from 'property development first and public infrastructure construction second' to 'development in line with planning and construction of infrastructures at a simultaneous pace' and from 'purely residential development' to a 'new urban planning model characterized by urban clusters and suburban modular development' (Yuhang District Government 2008).

During rapid urban growth, it is difficult to protect natural resources effectively. The decline of the natural environment and historical cultural heritage poses a great challenge for achieving sustainability. In China, in particular, the protection of natural spaces is often neglected by local governments in their pursuit of economic benefits from the land. The Xixi wetland protection project, however, shows an example of where great importance has been attached to the protection and restoration of the natural environment, while the protection and utilisation of valuable historical and cultural resources are equally part of an innovative urban management concept.

The development of residential areas and new commercial services boosts local economic development, while the improvement of the surrounding environment will further upgrade the value of the area and increase the competitiveness of the urban core. This integration of economic, social and environment development created a multiple-win situation among the various interested parties. However, balancing economic development with protection of the environment and the rights of local people is still a major challenge for the overall area of which the wetland park is only a part. Therefore, the preliminary results of these strategies deserve further investigation and monitoring.

The long-term importance of the Xixi National Wetland Park project for the restoration of natural resources and quality of life in the western part of Hangzhou can hardly be overstated. Moreover, the Wetland Park creates a tourist attraction and it will further strengthen the character and the image of Hangzhou as a green city.

In Europe, it is almost unthinkable that an area of the size of Xixi wetland park would be restored in such a radical way by removing entire settlements, and particularly in such a short time, in order to remedy the mistakes of previous urban development. The implementation of this project had its price, though, in particular for the farmers who were forcibly resettled. Considerable efforts were made to compensate the farmers fairly but local people had no say in deciding on the future of the area. An investigation published by the Caijing Journal revealed that some farmers were reluctant to leave their lands (Yu 2007). Various ways were used by project managers to make those farmers leave, including putting some of them in jail. There are also some doubts on the fairness of the compensation rates used which is far below the value of the land when sold for commercial development.

European project partners wondered whether another approach would have been feasible by which local farmers would have be allowed to stay at least in parts of the park on their land in order to continue ecologically-adapted farming in a truly cultural landscape instead of converting the entire area into a park land with museum-like settlements. This may be a naive view given the realities of urban development in China in general, and the situation in the western part of Hangzhou in particular, but we believe it is important to continue and deepen this discussion in order to develop new concepts and models for peri-urbanisation.

## 12.7 Binjiang District: The Potentials and Challenges of Large Scale, Government Led Peri-Urban Development

Binjiang District is situated at the south bank of Qiantang River (Fig. 12.7). It is planned to become a civic sub-centre and the main location for science and high-tech industry for Hangzhou in the future. It is an example of large-scale peri-urban development led by the government, and therefore, represents peri-urban areas with a rapid and large-scale conversion of rural into urban structures and from low to highly productive commercial conditions.

### *12.7.1 Main Issues and Strategies for Sustainable Development of Land Use in the Binjiang District*

The District covers 72 km$^2$, with a population of 198,601 (Table 12.4). It has become the most vigorous economic growth zone of Hangzhou based on the software industry, IC design, animation and the communications industry. There are nearly 5,000 enterprises in the High-tech Industries Zone of Hangzhou (Binjiang District), 445 of which are foreign-invested. During the Tenth 5-year Plan period, the total production value in the entire district increased from 6.1 to 15.1 billion yuan, corresponding to an average annual growth of 19.8 %. In order to make way for development of residential areas a number of villages were demolished, resulting in a continuous decrease of agricultural population. The remote sensing images show the pace of urbanisation in 1998–2004 leading to a large scale conversion of farmland (Fig. 12.11).

In 1996, the High-tech Industry Zone was officially established, and the government made significant investments, largely into the urban infrastructure. Compared with the Xixi Wetland which is characterized by gradual sprawl, Binjiang District has been developed under the guidance of the government from the outset. Therefore, the road system was the first to take shape. From 1998 to June 2004, the urbanisation process in Binjiang District was very fast. The grid-shaped road system was completed and a large stretch of agricultural land was built up.

Binjiang District will be a major location for future residential developments, not least to relieve pressure from the crowded city core of Hangzhou. The Comprehensive Plan for Bingjing District defines one of its functions as accommodating the population emigrating from the city. Residential development will be combined with the development of public transport. Development will concentrate along railways and the Qiantang River.

The development pattern will include commercial housing, resettlement housing for dispossessed farmers and affordable housing for low-income families. The planned residential areas will cover 25 % of the built area. On completion, they are projected to accommodate 380,000 people with a per capita floor space of 34 m$^2$. A multimodal transport network is under development, including private

**Table 12.4** Population structure in Binjiang area (2006)

| | | | Part 1 | | Part 2 | |
|---|---|---|---|---|---|---|
| Binjiang | Area ($km^2$) | Total Population | Migrating population (%) | Household population (%) | Urban population (%) | Rural population n %) |
| Total | 72.02 | 198,601 | 36.2 | 63.8 | 86.4 | 13.6 |

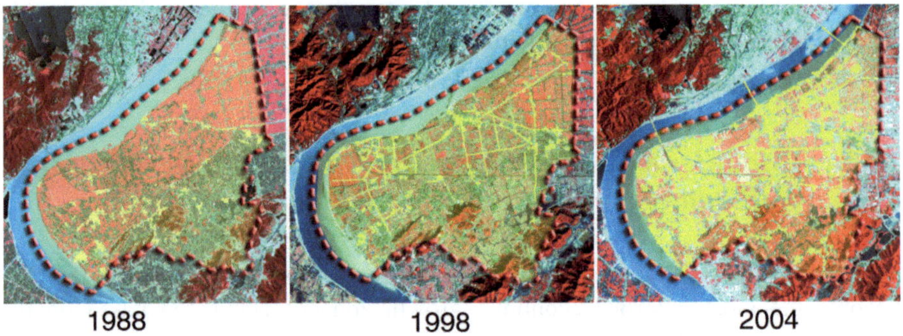


**Fig. 12.11** Development of the built-up areas (*yellow*) in Binjiang District

vehicles, public bus, and planned light rail. Commercial as well as educational facilities have also been established by the development of commercial streets, civic centres and state schools and colleges.

Major goals for the sustainable development of Binjiang District are:

- Development of public infrastructures and facilities on a large scale;
- Minimisation of environmental impacts;
- Attraction of desirable industries by incentives which are in line with the free-market economy;
- Integration of dispossessed farmers into the new town;
- Establishment of a streamlined government to manage the district efficiently.

Main strategies for development of Binjiang District that will be evaluated are:

1. *Establishment of Binjiang High-tech district*: The Comprehensive Plan for Binjiang District of Hangzhou, released in 2003, positioned Binjiang District as a high-tech centre with a specific focus on research and education, the communication equipment industry, the software industry and the semi-conductor and integrated circuit industry.
   The plan designated two industrial parks. They were both based on pre-existing industrial parks. The new parks allowed the original industries to stay, such as processing, manufacturing and machinery, while strictly forbidding polluting industries. In addition, high-tech industries, such as software development were promoted. The district government invested in public infrastructures and in facilities to improve public services. The plan also mapped out three high-tech research and design bases for the development of software, electronic and

information devices, biochemical medicine and new materials as well as the White Horse Lake (see below), where software development is the main industry.

2. *Supporting dispossessed farmers*: Binjiang District has acquired large areas of farmland in order to meet the demands for commercial and industrial uses. So in order to secure the livelihood of the dispossessed farmers, the government made great efforts to build new houses and employed special policies to ensure that the former farmers can afford housing, such as selling houses at a much lower price than the market. In addition, Binjiang District aimed to integrate former farmers into the urban employment system, including training to qualify them for jobs or helping them to become self-employed. The District also gives incentives and has preferential supporting policies to encourage enterprises in the District to employ the local dispossessed farmers and also provides social security for them.
3. *Establishment of the White Horse Lake Ecological and Creative Zone*: White Horse Lake area has various natural and cultural landscapes, such as water networks, wetlands, mountains, history-rich villages and bridges. In 2008 the White Horse Lake Eco-Creative Town Plan was completed. The plan aims to develop the White Horse Lake area as a multi-functional cultural and creative centre, combining recreation, living, retail business, research and development, design companies, etc. Furthermore, this area is expected to combine urbanity with the qualities of a rural environment. The ancient village and farmland landscape will be preserved, as well as the natural landscape.
4. *Preferential policies to support high-tech industry*: The development of Binjiang District benefits from exceptionally advantageous policies. The preferential policy that the state grants to high-tech industrial development, the special policy endowed by the municipal government of Hangzhou as well as the preferential policies that Binjiang District grants to foreign-invested enterprises all create a good environment for investment in high-tech and high-added-value industries. For example, the tax rebate policy for the software industry allowed taxpayers engaged in the software industry a tax rebate directly for the amount of VAT over 3 % from June 24, 2000 to the end of 2010. Foreign high-tech enterprises benefit from more preferential policies, such as low project application and review fees, lower rent and higher tax rebates.
5. *Streamlined government*: Binjiang District has all the functions of an independent governing body. Although Binjiang is just a district, the Hangzhou government granted it city-level administrative power as well as an independent budget system. The aims are to establish a 'small government with a full range of services' and 'simplification, centralization and efficiency of administrative structure', during which the original 37 departments of the district government have been reduced to 25.

### *12.7.2 Experience from the Study of Binjiang District*

In history, the model of new towns has been adopted elsewhere to expand urban space during phases of rapid development, i.e. moving industries and population outwards. Binjiang District is a civic city developed during the reorganisation of administrative regions. The main objectives for development of the District are adjustment of the industrial structure of the city and also a spatial strategy for supporting modern industrial development in a megacity and for alleviating the problems of overcrowding in the city core. The development is a typical example of peri-urban development entirely led by the government of Hangzhou.

The fact that Binjiang District has been developed as a new town by top-down planning has certain advantages such as the possibility to protect the natural environment and cultural heritage better by strong planning. From the beginning, environmental protection has been an important goal for development. However, some problems need to be noted. In particular, Binjiang suffers from a serious problem of conserving land for future use. Land market speculators expect that land prices will rise in the near future, so they bought land without intending to use it. They just want to sell the land or start to build commercial housing when the land price is higher. Thus, a large quantity of sold land has been left vacant. In addition, the preferential policies gave local industries fewer limitations on using land efficiently. Therefore, the investment intensity is low on many of the industrial sites, which has worsened the problem of low efficiency of land use.

There are also some uncertainties. Currently, Binjiang District's residential development is not balanced with its industrialisation process. A large number of new residential developments are still vacant—in some areas as high as 75 %. Therefore, the District needs to pay special attention to avoid it becoming a 'ghost town', where people work in the day time and leave at night. Although more residential development is underway, more time is needed to attract people to move to Binjiang. Importantly, the high price of commercial housing keeps most middle class and low income people away. The District needs to provide more affordable housing for people who really need it rather than up-market housing built by speculators.

Furthermore, seen through the eyes of the European project partners, the scale of development—from district level to design at street level—in Binjiang appeared to be rather intimidating and not fitting to human scales and needs. In Binjiang rather large scale mono-structures with high-rise apartment blocks, oversized streets and large distances between the high-rise buildings have been created. It is difficult to see how town layout can be developed into a well functioning city with urban qualities such as can be encountered in the inner city or in small traditional towns with their small scale mix of different functions and vigorous street life. European partners recognised as another critical issue the way open space and farmland is integrated into Binjiang.

Quantitatively, there is a high level of provision of open space, particularly when compared with the densely built inner city of Hangzhou. However, less concern seems to have been given to the quality and functionality of open spaces. A large proportion of the green spaces seem to have been solely designed for visual attraction, such as the wide belts of highly manicured plantings along the road network. Their use value can be doubted, however. So-called ecological corridors along canals mostly consist of intensively managed lawns and pruned shrubs and trees where little concern seemed to have been given to provision of wildlife habitats or to water quality.

A particularly critical and still unresolved issue in Binjiang as well as all other peri-urban areas, which will be taken up again in the overall conclusions of this chapter, is the way how farming and cultural landscapes are dealt with in the development of peri-urban areas. The official policy discourse now emphasises the need to protect farmland strictly but in reality it is still primarily considered as a resource for urban construction. The loss of farmland is certainly unavoidable given the extremely strong urbanisation pressures in China and particularly in the Yangtze delta. However, it is worrying how little it's potential role as a carrier of the future green structure of the metropolis with multiple economic, social and ecological and functions is recognised.

Binjiang still comprises large tracts of farmland interspersed with already built areas but concepts are missing for their future role. During field visits we also saw some areas where farmland—in fact very small, allotment like plots of land—is still managed by farmers but this is only considered to be a temporary use until urban development takes place. The potential value of these areas as urban allotments, which may help to integrate famers into the new urban fabric, were not seriously considered. Certainly, it may be debated whether it is appropriate to assess such developments as Binjiang by European standards of urban design and urban ecology. However, from this perspective it seems to be crucial to raise the level of debate. The success of developments such as Binjiang in solving some of the pressing needs of urbanisation in Hangzhou by providing floor space for living and high tech industry and forming a polycentric structure to relieve pressure from the inner city need to be recognised. However, more thought needs to be given to the question of how to create a liveable city at human scale which is able to integrate newcomers and the rural population alike.

## 12.8 Discussion and Lessons Learned

Before discussing the potentials and limitations of the planning strategies observed in Hangzhou, it must be noted that the situation is very different from that in Europe, even though some of the problems observed in Hangzhou were not. However, Hangzhou is undergoing very rapid urbanisation, during which the city core, the peri-urban and the rural areas are in fast evolution and transition. The economic, social and land use structures in the peri-urban areas are in a state of dramatic change. The core city area increased by 7 times or 668% and of the peri-urban areas by 4.5 times or 458% within only 16 years. It would be difficult to address the challenges resulting from this process successfully under any political and planning system.

Hangzhou was selected because it is known in China for its advanced state of planning. The three cases studied in Hangzhou showed that, by adopting appropriate strategies, peri-urban regions can fulfil more functions and help to establish a better urban spatial layout than being merely transitional zones.

However, the overall analysis of Hangzhou city region and the three selected case studies revealed that serious challenges for sustainable development in the peri-urban areas still persist, such as urban sprawl, low land use efficiency, the heavy pressures on the environment and natural and cultural landscapes as well as the need to preserve farmland and to integrate farmers into the newly urbanised areas. These issues relate to the conflicts among the various forms of control of land use rights, the fiscal system which encourages local governments to sell land in order to increase local revenues, the government-driven development in a project-orientated mode and the need to enhance the cooperation between different stakeholders in an equal manner. It is a challenge to address these successfully by effective urban development strategies and spatial planning under the existing political and planning frameworks. Three particular areas where improvement is needed are discussed in the following sections.

### *12.8.1 Deficiency of Development Strategies Under the Existing Framework*

Strong overall planning at city-regional level is considered to be of prime importance for development of a land use system that supports sustainable development under the current pressures of rapid urbanisation. Administrative reform whereby townships became districts of Hangzhou city proper laid the foundation for coordinated development of the spatial structure of the city. The shift from a mono-centric to a polycentric development pattern and the construction of a metro system were observed as responses to the challenges of rapid urban expansion in Hangzhou.

However, implementation of overall planning faces serious challenges. Difficulties arise from the coexistence of various forms of land use rights and the limited validity of legal planning instruments. In addition, government-driven development is characterised by a project-orientated approach for some specific goals with a particular emphasis on economic growth. This approach originates from the experience of development in the urban core area in which economic growth has priority. The transfer of this approach to peri-urban areas inevitably leads to conflicts due to the multiple interests and functions that meet and compete with each other in these areas. Moreover, the complicated and unbalanced distribution of authority results in the degeneration of the top-down planning system into fragmented approaches.

Realising the degradation of nature and farmland during the process of rapid urbanisation, Chinese cities began to think about being "green". Hangzhou takes the concept of the "Green City" seriously. Six green belts were designated and efforts have been made to protect and restore sensitive ecosystems, such as the outstanding cultural landscape of the West Lake Area and the Xixi wetland. Furthermore, the city government tried to make a transition to economic patterns with low environmental pressure such as the White Horse Lake Area within Binjiang. The development of tourism is another major economic theme as the Xixi and Zhuangtang cases have shown.

During the development process, the interests of farmers were to some extent considered by adopting various strategies such as establishing resettlement communities, training their labour skills or giving them employment priority in new developments. However, while these measures ease to some extent the consequences of urbanisation for local people, their multiple interests are still not seriously considered in the process or urban planning and its implementation. The life of dispossessed farmers has been greatly impacted as a consequence. Compensation measures alone, even if generous by Chinese standards, do not suffice to integrate the farmers into the urban system. In Zhuantang, on the other hand, the selection of green industries only provides very limited benefits to local citizens whereas it constrains their possibilities to improve their own quality of life through economic development.

Therefore, it may be concluded that the actual implementation of the "Green City" concept still reflects a limited understanding of the multiple social, ecological and economic requirements in the peri-urban areas. Economic growth still maintains a strong position as can be seen in the promotion of "hard" tourism and upper class real estate. The achievement of environmental goals is still mostly restricted to quantitative ones which can be measured by indices, e.g. for provision of green spaces and drinking water quality, whereas qualitative goals related to natural and cultural landscapes are far less important. The conservation of green spaces has a weaker legal foundation than the control of construction activity in the Chinese planning system. The conservation of green space, such as the green belt designated in the comprehensive plan, is not effectively implemented. As a result, green spaces are then gradually eroded by construction activities.

It is suggested that more active interventions are required to develop a well-functioning regional and urban structure. The development of transport infrastructure, especially urban mass-transport, which plays a pivotal role in urban expansion, is strongly needed. The speed of its construction is much slower than that of urban expansion. Per-capita car ownership in Hangzhou is still low by European standards but it is increasing at a dramatic pace. Moreover, Næss (2006) found that the current daily commuting distances are still moderate in Hangzhou when compared with European cities but this is likely to change as car usage increases.

### 12.8.2 Governance and Cooperation

In the hierarchical Chinese planning system, city governments hold strong powers to allow or restrict development opportunities in peri-urban areas. The main functions which serve the interest of the whole city-region area can be assigned to certain areas in the comprehensive plan followed by the allocation of financial resources and development rights for land.

Designated development areas obtain special support from the city government. Many other public sectors e.g. the state-owned enterprises, which are under the direct jurisdiction of city government, will also participate in the development process. By comparison, the other areas can obtain much fewer resources and encounter more difficulties in their own development. While it is desirable to control development strictly in some peri-urban areas for environment protection or conservation of cultural landscape, the problem is that they receive no or very little compensation for such restrictions.

These differences in development opportunities could be clearly observed in the three case studies. As a newly planned city centre, Binjiang District received a huge amount of investment from the government in order to improve its public infrastructure and facilities. In addition to the transfer of a series of authorities to the district government, the city government also granted high-tech industries preferential policies, such as tax rebates, so that more firms chose to locate in Binjiang. The City has provided Binjiang with a sufficient land supply and its strategic plans encourage Binjiang to expand eastward. Therefore, the District enjoys high fiscal revenues under the current system.

In the Xixi case, the development is mainly based on the protection project. The project team has formed a coalition between the city and district governments. It acted as a communication channel between the two governments and helped them to achieve common understanding. Therefore, the project management is highly efficient. The requisition of farmland for green space has been implemented at an extraordinary pace.

The situation in the Zhuantang area presents a more serious challenge. This area is located in one of the designated ecological belts. The city government put strict limitations on Zhuantang's ability to expand urban land uses and its economic development has been lagging behind in comparison to other peri-urban areas. Without the support from the city government, the three governing bodies in Zhuantang overlap in some areas but do not share the same interests. As a result, it is hard for them to reach consensus and to take action. Accordingly, local people have suffered from lower income and fewer job opportunities. The financial support for Zhuantang is limited to fiscal compensation, which is far from adequate to offset the loss of its development opportunities.

Further problems relate to the difficult relationship between the public sectors and private developers. Privately-driven development is often not systematically guided nor controlled by urban planning. The interests of private investors concentrate on some short-term projects. In particular, there is a lack of provision of infrastructure and public facilities. Moreover, because of the strict control of land resources by the city government, illegal projects are developed on collective land with the cooperation or acquiescence of the local government.

Public-private partnerships are also flawed by the restricted involvement of stakeholders and their unequal rights and powers. Cooperation is limited to a closed circle of public and private elites whereas local citizens are left out of the decision-making process entirely. Sometimes, the coalition between the developer and the government sacrifices the interests of local people. Although the collective land belongs to the local farmers, they are deprived of the right for urban development. Local citizens can only develop on a small scale in an illegal way to obtain short-term benefits.

### 12.8.3 Need for Creative and Locally Adapted Development Strategies

The three cases have shown great achievements of government-driven development under the hierarchical planning system. However, the project-orientated mode has resulted in a series of problems because of its oversimplified goals in the peri-urban area. It also greatly constrains the development of local creativity for finding context-specific solutions to the challenges of peri-urbanisation.

The shortcoming of local creativity is reflected, for instance, in the difficulty of transforming the dispossessed farmers' lifestyle from rural to urban. Although a series of meaningful measures have been implemented, such as establishing farmers' communities, providing training in re-employment skills, granting dispossessed farmers priority in obtaining employment as well as integrating those farmers into the urban welfare and insurance system, the governments are still firmly limiting the development rights of the local citizens.

Conversely, the potential of farming as a backbone of the future open space system still needs to be fully recognised. Generally, China now places much emphasis on protecting productive farmland but this is very difficult to achieve in rapidly expanding city regions. Recognising and promoting the role of farmland not only as a provider of food but also as a space for recreation and of ecosystem services within and adjacent to built areas is suggested as a way forward. This could be used to develop a multifunctional green infrastructure that is indispensable to the sustainability of the city and more resistant to the pressures of urbanisation. The remaining farmers may assume an important role as the managers of this green infrastructure. Experiences from Europe and also from other Asian countries such as Japan or South Korea may provide valuable inspiration for multifunctional farming in and near Chinese cities.

### *12.8.4 Mutual Benefit for the EU and China*

As an overall conclusion, the case studies of Hangzhou show the great challenges of peri-urbanisation in a rapidly growing city region, and it offers some examples of good practice for solving some of the associated problems. These examples can be of wider interest for other Chinese and Asian city regions, and they may even stimulate debate about planning in European city regions, for instance, regarding the speed and scale of the project-orientated mode, such as Xixi Wetland Park. The study also revealed some fundamental issues inherent in the Chinese government and planning system, which need to be addressed if peri-urban land use patterns should become more sustainable in China.

In particular, current patterns of development in the peri-urban areas reveal the crisis of the project-orientated mode for the implementation of long-term strategic objectives and in the cooperation between public and private stakeholders. Its hierarchical character has encountered greater and greater difficulties in the face of the ceaseless complication of socio-economic realities in the modernisation process. It is suggested that in particular a shift towards a more inclusive approach to planning and decision making is needed which equitably involves an increasing number of legitimate stakeholders. The fiscal system also requires reform to reduce incentives for urban sprawl.

From the previous sections it is concluded that there is a particular need to:

- Reform the tax and land ownership systems
- Reinforce laws and policies to control sprawl
- Strengthen strategic regional planning towards a poly-centric development pattern
- Develop a multifunctional green infrastructure, where urban forestry and urban agriculture play a prominent role
- Encourage engagement with stakeholders and encourage bottom-up initiatives
- Establish planning monitoring systems

Finally, it is recommended to adapt the scenario approach developed and applied to the European case studies so as to allow a look beyond the current situation and to explore the resilience of current strategies against alternative pathways of city regional development and assess potential impacts of drastic changes from radical scenarios such as oil shock. Further research would be desirable to elaborate on these issues in more depth.

## References

Ding C, Song Y (2009) Property tax for sustainable urban development. In: Song Y, Ding C (eds) Smart urban growth for China. Lincoln Institute for Land Policy, Cambridge, MA, pp 57–73

Hangzhou Municipal Commission of Development and Reform (2009) Report on the study of Zhuantang's development. Online from: http://www.hzdpc.gov.cn/hzdpc/0401/12568.htm. Jan 2010. Accessed 3 Jun 2010

Hangzhou Statistical Yearbook (2002–2006) Hangzhou construction yearbook, 2002–2006; Hangzhou national economic and social development statistical bulletin, 2002–2006
Hangzhou Statistical Services (2010) Hangzhou 2009 statistics. Online from: http://www.hzstats.gov.cn/web/#. July 2010. Accessed 3 Jun 2010
Leman E (2005) Metropolitan regions: new challenges for an urbanizing China. In: World Bank/IPEA urban research symposium, 2005. Online from: http://www.worldbank.org/urban/symposium2005/papers/leman.pdf. Nov 2008
Næss P (2006) Residential location and travel in Hangzhou metropolitan area. Norwegian Institute for Urban and Regional Research, Oslo, NIBR Report 2007:1, 357pp
Webster DR, Cai J, Muller L, Luo B (2003) Emerging third stage peri-urbanisation: functional specialization in the Hangzhou peri-urban region. Asia-Pacific research center, Stanford University, Monograph. Online from: http://aparc.stanford.edu/publications/emerging_third_stage_periurbanisation_functional_specialization_in_the_hangzhou_periurban_region/. Jan 2009
Yu D (2007) An investigation on land acquisition of Xixi Wetland Park. Caijing Journal, 2007 No. 1. Online from: http://www.chinavalue.net/Media/Article.aspx?ArticleId=1894&PageId=1. Feb 2010
Yuhang District Government (2008) The interpretation of Xixi wetland protection project plan. Yuhang District Government. Online from: http://www.yuhang.gov.cn/class/class_43/articles/94608.html. Accessed 3 Jun 2010

# Part III
# Synthesis

# Chapter 13
# Governance and Sustainability of Peri-Urban Areas: A Comparative Analysis of the PLUREL Case Studies

**Carmen B.E.M. Aalbers and Katarina Eckerberg**

## 13.1 Introduction

This chapter examines the analyses and strategies presented for each of the case studies in Chaps. 4, 6, 7, 8, 9, 10 and 11 and presents a comparison in order to draw out the main findings which may provide possible solutions for policy makers working in other regions. As explained in Chap. 1, there are a range of particular policy challenges in the urban fringe.[1] The policies of several adjacent municipalities or local authorities may be combined and managed by an indirectly elected regional authority, while different national and EU-level sectoral and regional policies frequently interfere with each other in such areas. Problems of implementation deficits with sector-based strategies are commonplace following institutional fragmentation, calling for increased co-ordination and better integration mechanisms (Tatenhove et al. 2000).

This chapter compares the findings on governance from the case studies, looking for examples of good practice and the policy lessons that can be drawn from them. More specifically, the chapter:

- Examines the ways in which specific regional and local strategies may be able to integrate concerns for sustainability,
- Analyses the role of the state in regulating various private interests and the land market in order to guide the development of these areas, and
- Draws conclusions based on these findings that may be used to improve governance elsewhere.

[1] We use the terms urban fringe and peri-urban area interchangeably.

C.B.E.M. Aalbers (✉)
Governance Team, Wageningen University and Research/Alterra, Wageningen, Netherlands
e-mail: Carmen.Aalbers@wur.nl

K. Eckerberg
Department of Political Science, Umeå University, Umeå, Sweden

K. Nilsson et al. (eds.), *Peri-urban futures: Scenarios and models for land use change in Europe*, DOI 10.1007/978-3-642-30529-0_13,

In the case studies five different types of strategy were examined.
Strategies for:

1. Comprehensive land use planning at the regional level,
2. Strengthening agriculture in the urban fringe,
3. Protecting biodiversity areas at risk,
4. Reducing housing and business building pressure, and
5. Integrating tourism and recreation in development.

## 13.2 Analytical Context

### *13.2.1 The Role of the State in Multilevel Governance*

The analysis of the governance patterns in peri-urban areas is based upon an understanding of this being a multi-level and complex policy situation. Multi-level governance as a concept was initially described a "system of continuous negotiation among nested governments at several territorial levels—supranational, national, regional and local" (Marks 1993, p. 392) but the term has come to include also the diffusion of decision making powers and processes to numerous informal and overlapping policy networks (Kohler-Koch and Eising 1999). At the heart of the definition of governance is the 'erosion of traditional bases of political power', i.e. the changing institutional position of the nation state (Pierre 2000). This erosion is based on three simultaneous processes: (1) the loss by national governments of many of the possibilities for control over international actors as well as over business corporations as a result of the deregulation of financial markets; (2) an increasing role for non-government actors in policy networks independent of states; and (3) a strengthening of the position of local and regional level actors. As a result, local governments, civic organizations and business networks are increasingly able to introduce their own policies and to influence policy-making processes at different system levels. This means, according to Pierre (2000), that a simultaneous movement of political power is occurring upwards to trans-national levels of government and downwards to local communities, but not in a coordinated manner.

Even if the state still has a considerable amount of control, the two movements of power identified by Pierre (2000) mean that local governments are gaining power, not only within the nation-state but also in an international setting. In contrast, this also suggests that units other than national governments can and will influence policy processes at the local level, through sub-governmental, trans-national networks and international organizations. This, for example, gives the European Union another route for changing the political behaviour of local governments, something that has become especially visible in the case of sustainability policies. Multilevel governance therefore potentially allows decision makers to adjust the territorial scale of governance to reflect heterogeneity, since for example, ecological conditions may vary from one area to another (Hooghe and Marks 2003).

Nevertheless, the state remains a key player in initiating and co-coordinating sustainable development planning processes and the absence of state leadership has been shown to have a detrimental effect on the implementation of sustainable development strategies and plans (Baker and Eckerberg 2008). The state frequently contributes to capacity-building through direct financing, institutional support and the provision of expertise to sub-national authorities, initiating and co-ordinating policy networks and retaining substantial power over the nature and functioning of network forms of governance (Baker and Eckerberg in press). Hence, rather than seeing a shrinking role of government, we are witnessing a shifting role as private and third sector engagement (such as voluntary organizations, charities, social enterprises and cooperatives) and interactions in policy making increase (Kooiman 1993). However, as noted by Baker and Eckerberg (2008) among others, it should be emphasized that governance and government are two intertwined yet distinct elements of the process of governing. To be effective, the active engagement of local civil society presupposes the commitment, leadership and guidance of local and national governments and international authorities (see also Evans et al. 2004). This can, in turn, generate the local resources, support and energy needed to deliver outcomes.

### *13.2.2 Analysing Multilevel Governance Through the Lens of Policy Arrangements*

In order to distinguish patterns of governance and their relationships in addressing sustainability concerns that allow for comparisons across the case studies, the concept of 'policy arrangements' (Tatenhove et al. 2000) will be used. Tatenhove et al. (2000) define environmental policy arrangements by distinguishing four policy dimensions: rules of the game, resources, policy coalitions, and policy discourses. The first three represent the organization of environmental policy and builds upon Giddens' structuration theory which conceptualises organizations as social systems comprising sets of agents that are nested in *structures* of rules and resources (Giddens 1984). In this context three main dimensions of organization can be identified: agents (coalitions), rules and resources. The second aspect of the policy arrangement concept, *substance*, operates through so-called 'policy discourse' (see also Dryzek 1997; Hajer 1997).

In the analysis of peri-urban policies these variables will be used to distinguish different patterns of policy arrangements which, in turn, relate to different types of multi-level governance. A short explanation of the variables based on Tatenhove et al. will be given in the following, together with some initial thoughts on how they apply to, and affect the analysis of the outcomes and process of urban fringe policies.

1. *Rules of the game*. Rules of the game consist of the legislation and allocation of tasks to the institutional structures. These exist alongside the informal planning culture. These rules of the game will also be explored and described further in Chap. 14. In Chap. 14 different types of territorial governmental systems will be distinguished, being combinations of unitary and federal states, and strong or weak local governments. At the level of the regions the rules of the game can also be examined in relation to the different types of strategies, for example to what extent multilevel government or multi-actor governance dominates over the different issue areas. The rules determine the procedures by which the allocation of tasks and division of competencies are used in the formation of strategies. They further delimit which actors are meant to be involved, how agendas are set and how decisions are supposed to be made.
2. *Financial and land resources*. This policy dimension reflects the allocation of power over land and the mobilization of financial and land resources. These are also central for explaining how agents maintain and transform their social or physical environment. The ability to achieve particular outcomes also depends on the autonomy and relationships between actors, and interrelations between state, market and civil society, which in turn may be classified as state-centred, corporatist, or liberal arrangements, where state agencies are dependent on companies and non-government organizations to create sufficient support and legitimacy for particular policies. For urban fringe policy, it is relevant to investigate what the broadening of policy coalitions (see below) means for the allocation of financial and land resources, and who takes over power if the state withdraws.
3. *Policy coalition development*. A policy arrangement can be characterized by certain coalitions of players (agents) who share the interpretation of a policy, share resources, identify similar policy goals and mobilize together to reach those goals. Some policy coalitions may support while others may challenge a particular strategy. Studying the development of policy coalitions and their achievements tells us something about how and why some actors' strategic conduct succeeds while that of others does not. It also provides evidence of the strength of public versus private governance networks. Empirical observations on the ways in which policy coalitions link local, regional, national and supra-national levels, and across policy sectors, may also inform us about the nature of multilevel governance structures and about the extent and in which contexts certain coalitions may be successful in achieving their aims. Questions of democratic legitimacy and accountability may also be raised in this context.
4. *Policy discourses*. A policy discourse refers to a shared way of apprehending the ways in which the problems are defined, reproduced and transformed through particular sets of practices. It provides assumptions, judgments and contentions for interpretative schemes by which meaning is given to environmental (or other) problems, and which form the basis for the design of policies. The mobilizing capability of different policy discourses may be investigated, while there may be variations in the predominant discourses across countries and among different policy coalitions. Such discourses may also provide statements and positions on the role of the state and the nature of interactions among actors

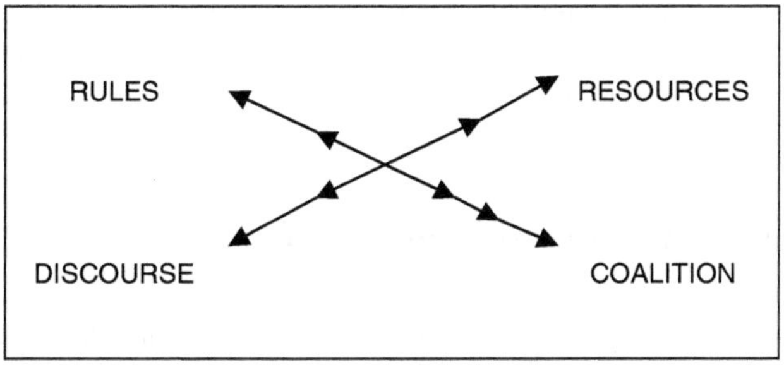


**Fig. 13.1** Template for illustrating the relative importance of rules of the game, financial and land resources, policy coalitions, and policy discourses in the governance of strategies. Length of the arrows indicates the strength of the dimensions

within policy coalitions. Furthermore, the nature of policy discourses may reveal patterns of policy 'branding'—i.e. how successful policies are motivated and conceived among actors.

The comparison of assessments from the seven case study areas will be made according to these four dimensions. To facilitate the comparison, the relative intensity of the four factors can be illustrated for each strategy under investigation by drawing arrows of different length (Fig. 13.1).

Since all types of issue areas were not found in all case studies, the comparison can only be based on the information available to the project. The impacts of the different strategies were assessed by both practitioners and researchers in each case study region. Since they contain inherent uncertainties in time and space no final conclusions on the outcomes can be made. Instead, indicative findings based upon observations from the case study research, which may then be related to the characteristics of policy arrangements for each strategy, will be presented.

## 13.3 Policies and Strategies for Peri-Urban Areas

Before comparing the strategies devised for the different purposes of reducing development pressure, strengthening agriculture, protecting biodiversity and integrating tourism and recreation in the urban fringe, an overview of the strategies presented in each of the seven case studies is required. When comparing them, the respective national and regional contexts provide some variation in the overall policy framework conditions, which will feature in the final conclusion. First, the main issues of each strategy will be outlined followed by a brief summary of the storyline for each case. A selection of nine strategies drawn from all seven case studies has been made, which will also serve to illustrate the more and the less successful practices. The examples begin with those strategies that build mainly on the support of only a few dimensions of the policy arrangements framework, then moving on to those with more comprehensive aspects.

## 13.4 Greater Manchester, United Kingdom

The Greater Manchester peri-urban areas contain considerable landscape diversity, and there is great social and economic disparity among the population in relation to the green environment, both as a residential location or as a place accessible for recreation. While physical expansion of the urban area is limited by spatial planning policy, there is continuing land use change away from agriculture to other uses.

There are in-built paradoxes in many developments, in that some of the most unique habitats are found on former industrial or infrastructure sites, while expanding leisure activities in green areas may have negative environmental impacts. In many places, however, local citizens groups (several hundreds of which exist) have become active in order to protect or improve the environment (Ravetz 2008, 2009).

### *13.4.1 Green Belt Policy Strategy*

Built development is generally prohibited within the designated Green Belt area except for certain 'strategic deletions'. There is an obligation to regenerate brown field sites within the urban envelope instead (Ravetz 2008, 2009). The containment policy of the greenbelt can be seen as a negative or defensive approach which often conflicts with other possible goals such as agricultural productivity and biodiversity (Ravetz 2008) and the policy is under growing political pressure for modification. Green belt policy is linked to local, regional and national planning levels (Ravetz 2009). There are no coalitions involved in the Green Belt policy as such, except for the strategic developments of economic interest, though of course these are not part of the strategy but threats to it, as further described below. Various other development partnerships also exist (Ravetz 2009).

The discourse on the Green Belt suggests it is poor at responding to needs because containment of pressures is the main goal (Ravetz 2008). There is little focus on the positive features of social, economic or environmental initiatives within the Green Belt. The statement of principles includes: to check urban sprawl, to prevent the growing together of towns; to preserve the setting of historic towns; to safeguard countryside against encroachment and to assist urban regeneration (Ravetz 2009). There is a general focus on housing development, employment development and maintenance of existing urban patterns and centres—that go against the goals of the Green Belt. This can mean that opportunities for more sustainable settlement models are automatically ruled out (Ravetz 2009) (Fig. 13.2).

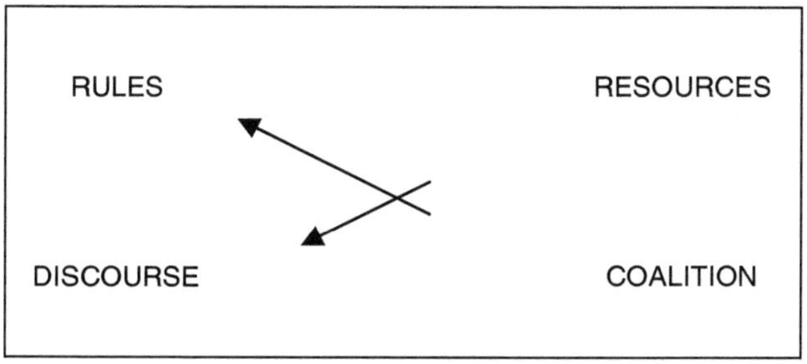


**Fig. 13.2** Policy dimensions of the Green Belt strategy

### *13.4.2 Performance*

The strategy arguably restricts farm diversification, leads to poor landscape management and is negative for the rural economy in general (see Chap. 7). Social and economic impacts have led to the leapfrogging of urban development over the Green Belt, where suburbanization by more affluent and mobile communities beyond the Belt, in the Pennines and the Cheshire plain, leads to increased commuter flows to and from the urban core through the Green Belt. The largest houses are found in or adjacent to the Green Belt. The containment also creates a risk of shifting urban pressures, such as the development of large shopping centres, to historic towns. The regime also helps to drive property value increases in the urban fringe. The Green Belt does not therefore strengthen the unique qualities of the area, it merely contains growth. The lack of involvement of a broader coalition that integrates social interests with those of the environment limits the achievements of Green Belt objectives. However, river valley initiatives and the identification of Special Landscape Areas aiming at maintaining or improving landscape quality, amenity and aesthetic values within the Green Belt have been successful. They seem important complementary actions to the containment policy of the Green Belt.

## 13.5 The Hague Region, The Netherlands

The Hague region is one of the most urbanized regions in the Netherlands, home to about one million people. The original, quite strong protection of green areas in the Hague Region, supported by the state, is undergoing rapid evolution due to increasing urban sprawl and rising land prices. The open agricultural areas are most susceptible to the development of various infrastructure projects, new housing and recreational facilities. In terms of governance patterns, parallel bottom-up and top-down developments can be seen. The possibility of financing those initiatives constitutes the major constraint to their further development. There is also a growing fatigue among many stakeholders who feel that too few concrete results emerge from extensive public consultation processes. The strategy for green open spaces is also heavily dependent on investment capital from outside the region.

Two strategies were developed by the Hague Region to spur the commitment of urban dwellers and politicians to the protection of green open spaces including cultural landscapes, which will be analysed below (see Chap. 6).

## 13.5.1 Strategy to Strengthen Citizens' Commitment to Green Open Spaces

'The Hague School Outdoors' project links the current landscape back to paintings of the same landscapes of the early twentieth century or before. The basic idea connects Green Rings for walking and cycling via historical waterworks and landscapes and other nature- and water-related structures, while presenting the landscape scenes at key locations en route. This project was initiated spontaneously as a kind of by-product of the PLUREL project. A coalition with some eight other parties (including an INTERREG project, The Hague city, private parties and cultural heritage funds) paid for its implementation, while the region only paid for one out of the ten large billboards used to present the landscape paintings. The discourse behind the projects is strengthening the identity of The Hague as 'the place to be'. The urban dweller is urged to fight for the rural areas, for 'The city of The Hague in a green region'. The Green Rings, an output of this strategy, should be an appealing concept, clearly identified with the regional authority, promoting its green open spaces and enabling cycle-borne urban dwellers to enjoy the green region. City dwellers and expatriate residents are also known to be interested in culture as well as nature, which is how the link of landscape painting to countryside comes to be so strong. Further information is provided via the website of the city of The Hague in order to promote a mental connection to the green open spaces of the region in the mind of residents and investors alike (Fig. 13.3).

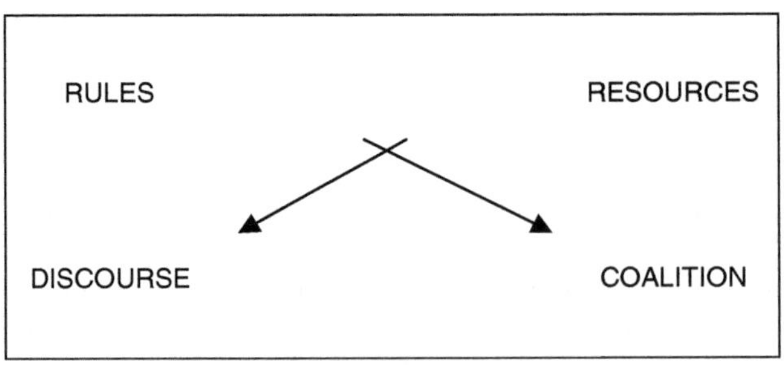


**Fig. 13.3** Policy dimensions of the strategy to develop citizens' commitment to green open spaces

### 13.5.2 Performance

The strategy helps to raise awareness among citizens, scholars and various authorities. A survey revealed that the use of the paintings in this way was appreciated by residents and that 46 % of residents (strongly) agreed that the paintings help in the discussion of the value of open landscapes. This appreciation was felt more by older than by younger residents, however. Which group, time or function are included in the identity concept of the Hague as the place to be, and which ignored, are decisive issues for the successful inclusion of sufficient different interests to make a successful coalition. For example, tourist offices are interested in the cultural dimension of the approach (the paintings) but not necessarily in the landscape, while nature groups may value the landscape but not the cultural or art aspects. Forming a coalition was needed because the region must negotiate most of the funds they can spend on green open spaces.

### 13.5.3 Strategy to Increase Local Politicians' Support for Green Open Space

This strategy links green open space with issues of major regional economic importance, namely the internationally competitive position of The Hague and its wider region. A network for the promotion of green open spaces was developed to stimulate financial investments to this end—investments that were not yet available. The strategy targets citizens as both users and advocates of the urban fringe (Aalbers et al. 2009). Green open areas that have the support of the urban population are being protected or restored. Links with other parties, especially the ones with different but compatible motives, have been made for fund raising, leading to coalitions between the international institutions and business, Delft municipality (with its international Technical University) and other municipalities in the region. They share the discourse of the attractiveness of The Hague region expatriates such as to international civil servants and knowledge workers as being an asset (Aalbers et al. 2009). Green peri-urban areas are also seen as recreational areas which should be adapted to the needs of the city dweller (Fig. 13.4).

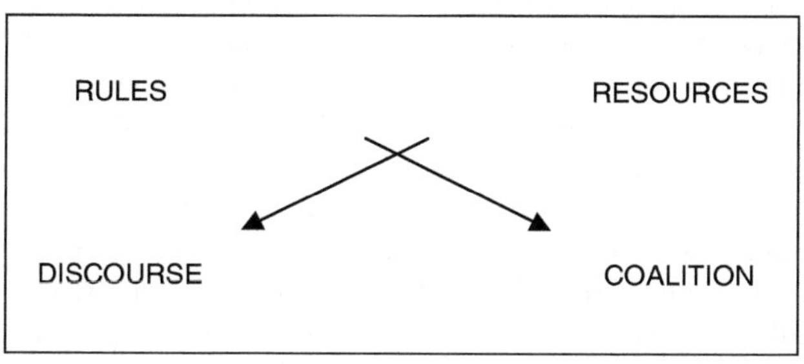


**Fig. 13.4** Policy dimensions of the strategy to increase local politicians' support for green open space

### *13.5.4 Performance*

The use of the attractiveness of the landscape to expatriate residents as an economic motive for governors in the regional board seems to work. In regional policy documents the discourse is promoted and funds are invested to improve the quality of green open space in the region. The 'green wishes' of expatriates are not very different from those of other groups (Veer et al. 2010) and expatriates find neighbourhoods with peaceful greenery, quiet green streets to live in and natural areas in their residential environment very important (Veer et al. in Aalbers et al. 2009).

## 13.6 Hangzhou, China

The major challenge in the Hangzhou region is how to balance environment and economic development interests in a context of extremely rapid urbanisation (see Chap. 12). While government authorities are acting in line with Communist Party-dominated and strong hierarchical planning traditions, private investors play an increasing role in land use development. In the process of the development of peri-urban areas, major problems include how to satisfy farmers who lost their land due to expansion of housing areas and how to protect nature. Farmers bought out by development projects often suffer from social exclusion. In general, there is a growing social and economic gap among the population in spite of the rhetoric about social harmony and that everybody should gain from current economic development.

### *13.6.1 Strategy: XiXi Wetland*

The strategy combines landscape and ecosystem restoration, tourism and recreation with up-market housing. The West Lake district administration covers the Xixi area. This is a district-level of government under the supervision of the city-level government which can overrule the West Lake administration's decisions. In 2005 the Xixi area was designated for wetland protection as the first national wetland park in China. Farmers were given urban resident status so that their rural collective landownership status could be taken from them (Yang et al. 2010). The collective land was then divided into smaller parts so that permission for development only needed to be obtained at the city level instead of the state level (Yang et al. 2010). Detailed regulatory planning was conducted by Hangzhou municipality, including planning of the supporting infrastructure (Yang et al. 2008). The expropriated farmers were reimbursed with new homes and compensation areas. The compensation rate for this is high in Hangzhou, since land prices are also high.

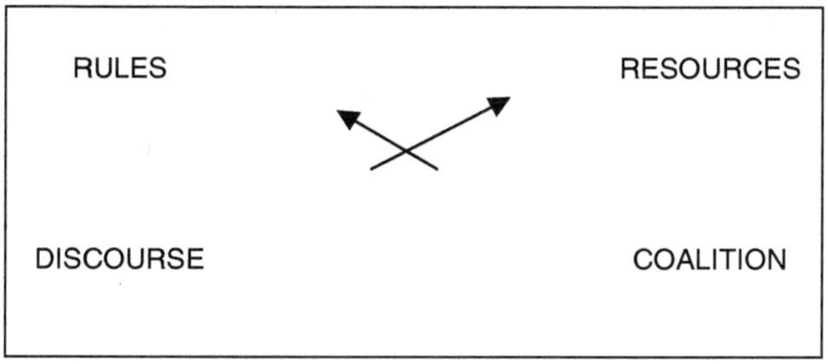


**Fig. 13.5** Policy dimensions of the strategy for Xixi Wetland

There are no coalitions involved in this strategy. The Municipality and real estate developers cooperate and there is also cooperation with tourist business, universities, research institutes and invited experts. But these are not policy coalitions. The Municipality is the sole authority; the others are listened to and worked with as long as they are in harmony with the municipality point of view.

The Xixi-wetland strategy has adopted a win-win principle of allowing for commercial development during the protection process as well as ensuring environment protection during development. It uses the identity of the landscape as heritage in the vicinity of residential areas in order to improve the image of the area and its attractiveness to potential up-market residents. Various events are used to reinforce the park's image as a scenic area but only with the limited consideration of educating the public about wetland protection. The political meaning of a discourse does not apply to the Chinese case. The government decides and does not engage in governance approaches with discourses to create coalitions for joint strategies (Fig. 13.5).

## 13.6.2 Performance

As a result of past developments the important wetland ecosystem was reduced from 60 to 10 km$^2$. (Yang et al. 2010) but with the new status as a protected wetland area this trend has been halted and historic ponds have been preserved. The Hangzhou government has great power to control development by top-down urban planning. The integrated approach combines the protection of local culture with microclimate improvement for the urban area, tourism and recreation. The Xixi area now represents the one high quality of life area in China attractive to up-market residents, bringing development opportunities and benefits to local people. There are still serious deficiencies in provision of public transport, however.

The brand effect of recreational tourism and strategic decisions such as the location of Zhejiang University have created unlimited opportunities for private business and for the positioning of Jiangcun as a tourism service base and upmarket residential location (Yang et al. 2010). There is also a release of tourist pressure from the West Lake area, yet the voice of the general public, especially of farmers, is not encouraged.

The urbanization of farmland has not only led to a loss of environment quality. Farmers, though reimbursed for the expropriation of their land, have no support from the city government for alternative job training and therefore they lack competitive skills in the labour market. They are only to a limited extent employed in garden development and management (Yang et al. 2010). The farmers are the vulnerable group in this process.

## 13.7 Warsaw, Poland

The transformation of the local and regional economy, accompanied by metropolitization processes, such as development of interconnections with regional and international networks and a heavy urban sprawl, is typical of transition countries in Central and Eastern Europe. A considerable part of the peri-urban development occurs in unincorporated (i.e. formally rural) territory and shows typical characteristics of a sprawling land consumption (see Chap. 6).

Existing strategies in the Warsaw region do not provide a sufficiently firm policy arrangement basis for sustainable development either in the metropolitan area or in the bordering areas. Pure economic development is favoured by politicians in most cases. Lack of coordination, along with limited financial resources, prevents the implementation of current plans. The legacies of the previous centralized planning system in the form of a dislike of central state cooperation and new dominance of local self-government tend to counteract any multi-level government coalition development. Also, there is no planning culture to build upon for increased societal involvement and strong vested economic interests dominate policy practice. The result is perceived by the stakeholders interviewed in Warsaw as messy and chaotic, while at the same time something that cannot be stopped or guided.

Social differentiation is one of the results of the multi-dimensional processes of centralisation and decentralisation in the region, allowed for by the flexible and development-orientated spatial policies. Different societal groups are empowered by the strategy of the Mazovia Regional Development. In contrast, the Warsaw Metropolitan Area Spatial Development Plan was not viewed by the assessment team to be enabling at all for any societal groups or grass roots initiatives. Since there is no legal, financial or cultural commitment to support bottom-up processes, one may conclude that engagement by citizens, business and NGOs is not systematically endorsed by the authorities in any of the strategies.

### *13.7.1 Strategy for Mazovia Region Development*

This strategy has not yet been implemented, so it is shown as such in the diagram (Fig. 13.6). The strategy aims to strengthen cooperation between municipalities (Grochowski 2009) and to harmonize development planning. The region aims to

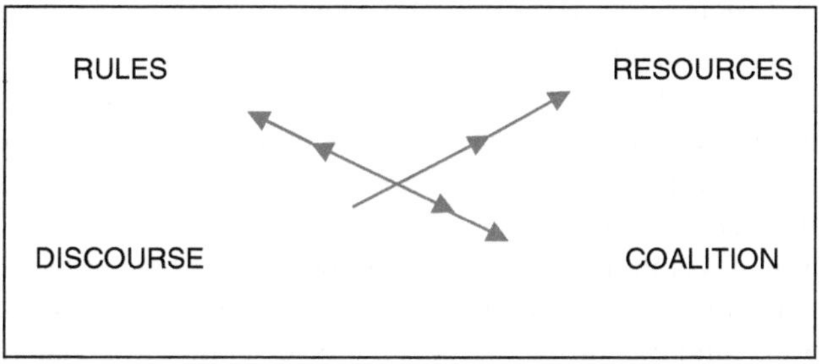


**Fig. 13.6** Policy dimensions of the strategy for Mazovian region development

create a cohesive system of legal regulations to protect environmentally valuable areas. At present, rules to settle controversies or to subordinate lower level plans to those of higher level are still missing and planning legislation is not precise, displaying gaps and inconsistencies. The region wants to make investments serve regional public purposes. Former post-industrial and military land are to be revitalized. From the case study report it is not clear where funds for all actions arising from the objectives will come from and where they will go. It aims at awareness raising and at cooperation within the Green Lungs framework, adopting a containment policy.

Opinions and remarks were collected from representatives of different authorities at all administrative levels but coalition development is still in its infancy. Cohesive or unifying concepts for social, economic and spatial development are absent. A good economic situation is considered a necessary precondition for sustainable development, to improve the quality of life for its inhabitants, reduce poverty, and to contribute to the development of rural areas as well as of Warsaw metropolitan functions, including technical and social infrastructure, and the environment. This fits with the public opinion or value of development being synonymous to economic development. The region aims to develop tourism and recreation but this is yet to be established as a discourse (Fig. 13.6).

## *13.7.2 Present Developments/Anticipated Performance*

The presently existing thematic discrepancies between the different sectoral plans are problematic and the organizational framework is fragmented between economic plans and concepts of spatial development (Grochowski 2009). Legal regulations are inadequate and coordination is almost absent. These contribute to a complex and dysfunctional land use pattern.

There is no redistribution of wealth from within the so-called Warsaw growth pole (Grochowski and Pieniazek 2008) because territorial cohesion and other mechanisms are missing that could have enabled Warsaw's metropolitan functions to give development incentives to surrounding municipalities and other parts of the region. The relationship between international competitiveness and territorial cohesion within the Mazovian region is strained.

Agricultural land is being converted into plots for other land uses, though the CAP has halted the previous decrease of agricultural land. The land is prone to changes due to congestion, pollution, spatial chaos, a dysfunctional system and a result also from lack of rigid spatial policies.

The impact on developments envisaged in the policy documents depends on the disposal of means of influence to enforce the developments. So far these are still weak (Grochowski and Pieniazek 2008). A public discourse dimension is missing in the Mazovian strategy-to-be, as well as the dimension of a coalition of private actors or the public. It seems likely that these gaps will reduce the potential of the strategy, in terms of missed opportunities for joined forces and public support. Also, the public desire for economic development entails a risk of the environmental side of development being overlooked.

## 13.8 Koper, Slovenia

Koper has undergone major shifts in land use in recent years. According to the assessment made by stakeholders, the predominance of economic interests to develop the port area further and to provide for economic development remains largely unchallenged, even though some growing awareness on the value of protecting the natural coastline and its remaining natural ecosystems can be discerned. In the future, the population projection in Koper predicts most land pressure by the expansion of housing in the peri-urban area mostly for the second houses for people from other parts of Slovenia (see Chap. 10).

There is growing bipolarity of economic development: new work is concentrated in the coastal belt, while in the hinterland there is a reduction of agricultural activities, leading to an impoverished rural economic basis and a demographic risk of out-migration. There is also a high percentage of commuters (Perpar 2009).

Some integration between strategies in terms of land use objectives can be discerned, but it is weak in terms of policy dimensions: no coalitions or discourses have been used as means of creating (public) support or to influence decision making, except for the Rural Development Plan. Concerning the guiding of these developments in the urban fringe in more sustainable directions, technical/expert solutions do not always play the main role, but political and economic power and interests are as important. Since the changes in spatial planning which took place in 1991, the spatial plan is now being developed by professional spatial planners as a technical/expert exercise instead of between sectors. This has reduced integrated planning via the spatial plan and reduced the inclusion of social aspects into planning.

Complementarity does exist between the three example strategies in terms of land use objectives but their integration is not supported by influential policy means: no coalitions or discourses are used either as means to create (public) support or to influence decision making. This can limit the potential impact of the strategies.

The green structure development in the future urban region of Koper is expected to contribute to a greater availability of green open space for recreation by peri-urban residents. Polycentric development is expected to combine social and environmental development with the predominantly economically-based discourse of those in power. The lobbying and political influence of investors seems a risk to land use efficiency. However, the Natura 2000 status is expected to protect important parts of the rural areas for nature.

### *13.8.1 Strategy for Land Use Efficiency and Protection of the Best Agricultural Land*

This strategy combines spatial planning with a new procedure for the classification of the best agricultural land. The University of Ljubljana, the Ministry of Agriculture and the Municipality of Koper cooperated to develop the new legislation and procedure. The improved procedure is expected to protect the best agricultural land and will be embedded in the land use plan. Improved land classification gives a better insight into the different qualities of agricultural land. Hence, better knowledge is also a major source of influence used by the coalition to influence decision making in favour of the preservation of the best agricultural land. The new procedure will allow inter-sectoral communication, which was impossible in the previous situation when information was strictly controlled by the Ministry of Agriculture. The data have now become publically available. Agriculture, environmental protection and forestry will harmonize their spatial requirements in advance—at an expert level of decision-making. This will help raise support for the new procedure. The previous procedure provided serious influences of the investors on spatial planning. A new financial system will be established by the Ministries of Agriculture, Forestry and Food in order to support the new procedure. Land use efficiency is the main objective and this could be promoted in the form of a discourse but this is not happening (Fig. 13.7).

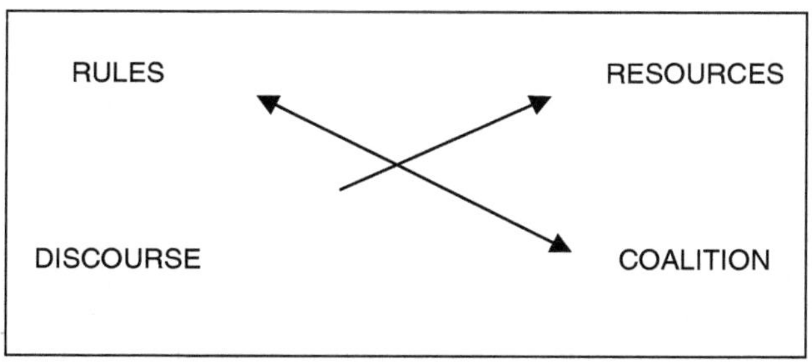


**Fig. 13.7** Policy dimensions of the strategy for land use efficiency and protection of the best agricultural land

### *13.8.2 Performance*

The national legislation underlying the strategy is expected to provide improved land use efficiency, in particular to preserve prime agricultural land. A strengthening effect on agriculture and empowerment of farmers is also expected to result. The strategy is incorporated into municipality spatial plans which the Ministry of Agriculture has the power and responsibility to approve or reject. The new approach is expected to be more effective and resilient because it is embedded in spatial planning and inter-sector communication. However, the idea is not yet known well enough among the general public to garner their support.

## 13.9 Leipzig-Halle, Germany

The Leipzig-Halle case study clearly shows the discrepancy between rhetorical policy statements in regional development plans advocating the protection of green areas in the urban fringe, and practical hindrances in terms of financial constraints as well as conflicting views on where development of housing and industry should best take place (see Chap. 8). The current taxation policy restricts the possibilities for peri-urban areas to support extensive protection of green areas, and any protection that happens is at the mercy of the availability of government subsidies.

Here, two strategies for the protection of green areas in the Leipzig-Halle case will be investigated—one which was initiated by the federal government and one which is more locally derived.

### *13.9.1 Strategy of the Green Corridors According to the Regional Plan for Western Saxony*

This strategy involves linking open spaces and unsealed surfaces in the environs of the city with urban green spaces. The corridors are also included in the Saxony (state level) spatial plans. The statutes of the region of West Saxony oblige municipalities to respect green corridors in local planning. The plan provides the framework for further plans at lower levels. Authorization of the plan is carried out firstly by the Regional Planning Association's general assembly and then approved by senior levels of the planning department of the Ministry of the Interior of Saxony (Bauer 2010). The Association consists of a coalition of mayors, district administrators and other authorities. Its elected planning board prepares the plans for the general assembly of the Association to approve (Sinn et al. 2008).

Financial support is lacking, however and this could lead to conflicts among municipalities, as the Green Corridors compete with funding for local economic development but do not come with a budget of their own. Regional development

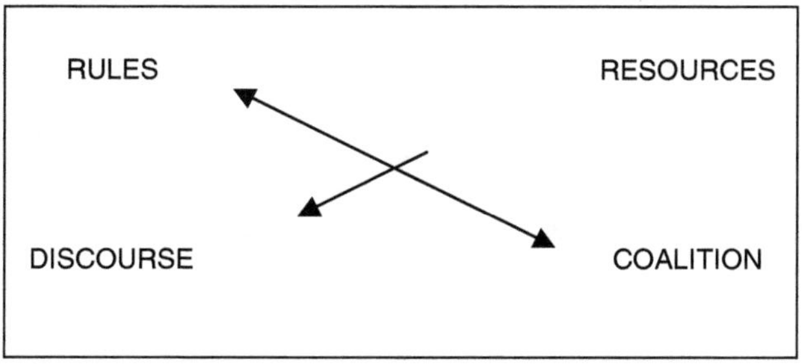


**Fig. 13.8** Policy dimensions of the strategy of the Green Corridors

funding is only available indirectly through the Green Ring that finances walking and cycling paths and environmental education.

Public involvement occurs only at the time of the presentation of the draft plan (twice), before it becomes statutory as part of the Regional Plan of Western Saxony. The combination of the plan with the highly valued floodplain forests contributes to the branding of the Green Corridors and awareness raising. The Green Corridors concept is easily understandable and convincing (Fig. 13.8).

### *13.9.2 Performance*

Investors seem to be capable of exerting pressure against Green Corridors (Bauer 2010). For example, building permits have been granted in the Northern Leipzig Green Corridor. Nevertheless the strategy contributes towards preserving the valuable landscape identity and natural habitats and agricultural land uses. The costs and benefits of the strategy are, however, unbalanced between the city and its environs: While everyone profits from attractive surroundings, it is only the development options of peri-urban communities which are limited. The financial support of Green Corridors through regional development funding is not sufficient to render other industrial/commercial investors unattractive to municipalities. The context of urban shrinkage favours investments in urban development because local authorities compete for residents and business with other municipalities.

### *13.9.3 Strategy for Inter-Municipal Cooperation in Parthe Floodplain Protection*

This strategy was developed by an inter-municipality coalition that integrates the preservation of the River Parthe Floodplain into the preparatory land use plans of three local authorities. It is enacted by the three relevant municipal councils. The Parthe floodplain agreements are binding for the official members. Expert staff,

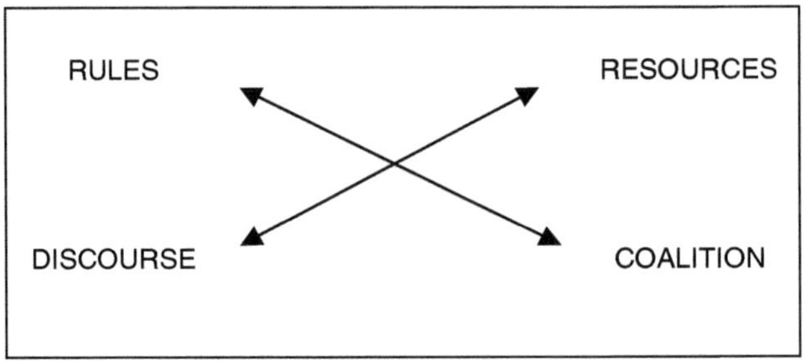


**Fig. 13.9** Policy dimensions of the strategy for inter-municipal cooperation in Parthe Floodplain protection

from each of the three municipalities, integrate protection goals into the preparatory land use plans. Landowners such as farmers are also involved in the coalition. Their resources are combined—project wise—with funding programmes. Financial resources are also drawn from the Green Ring budget by the city of Leipzig. Cooperation also occurs with the local tourist association, local inns, German Nature NGO, and other nearby municipalities (Bauer et al. 2010; Sinn et al. 2008).

'Parthe Floodplain protection' and 'Zwischengrün' are the main discourses. These are combined with art exhibitions along the river to make connections with parks and green spaces. For instance, there is a permanent exhibition on the floodplain's flora and fauna, guided excursions for schools, guided walks on ornithology and botany, as well as tips and recommendations for people interested in individual excursions.

There is a clear spatial entity, the Parthe floodplain, which is harmonised with the Green Ring Leipzig. The strategy's aims are tangible such as constructing cycling paths. Participating local authorities are constantly reminded of the importance of natural and landscape values for quality of life and as soft location factor for investors (Fig. 13.9).

## *13.9.4 Performance*

The Parthe floodplain cooperation depends on collaborative funding by its member municipalities, and achieves its goals by creating trust and continued support over a long period of time. Its success relies on convincing its members of the benefits of floodplain protection and on obtaining their political and financial support for such measures (Bauer et al. 2010). The strategy creates synergies in terms of sharing qualified staff and providing services in the field of landscape conservation and management, project management and funding applications, involving NGOs and business. The recreational value of the area is enhanced. However, commercial development in the floodplains is not completely successfully prevented. The combination of the different policy dimensions makes the strategy successful for the time being, while its resilience over the long term is not clear and depends on the effectiveness of the strategy in the eyes of the participants. Negotiations on the division of costs between local authorities are important for its legitimacy. The strategy's aims are tangible and therefore quite easy to control.

## 13.10 Case Study: Montpellier, France

The Montpellier region has recently experienced administrative changes through the decentralization of power for land use planning to the municipality level together with simplification and enhancement of inter-municipal cooperation which provide for a new territorial governance framework (see Chap. 9). In particular, the SCoT (Schéma de Cohérence Territoriale) planning approach has created space for a number of key local stakeholders to become increasingly involved in planning decisions in the peri-urban areas of this fast-growing French urban region. The tools provided by the SCoT are very influential in achieving cohesion among the lower municipal levels, while leaving room for spatial planning choices within certain limits. However, farmers' organizations as a whole are unable to defend their strongly held positions regarding peri-urban issues because of the considerable divergence of interests among different farming enterprises and interests. It is also evident that social segregation occurs in peri-urban areas, due to housing prices increases in the more attractive areas keeping less well-off people out. Leapfrogging of developments also occurs into the next agglomeration to the north of Montpellier, outside the jurisdiction of the Agglomeration de Montpellier itself. The influence of the policies of the cooperating municipalities of the Agglomeration does not extend to the whole urban region which results in a shift of pressure to areas still within commuting distance of Montpellier but outside the control of the Agglomeration (Buyck et al. 2009).

### *13.10.1 Scheme for Territorial Cohesion (SCoT)*

The Scheme for Territorial Cohesion is a strategy for peri-urban areas (Buyck et al. 2009) with an objective of cooperation between three levels of local authority: the Département, the Agglomeration and the municipalities of which it is composed. The SCoT was developed by the Communauté d'Agglomération, in cooperation with the Département, where both state officials and the local government are represented in the Conseil Général. The latter can intervene, in cooperation with Montpellier Agglomeration, to protect farmland in peri-urban areas. The obligatory SCoT spatial planning framework is applied at the inter-communal level and calls for coherence between urban plans and those of surrounding areas. In drawing up the SCoT (2002–2005) state officials were involved as advisors. The Agglomeration provides advice and also lends professional planning staff to the municipalities in order to

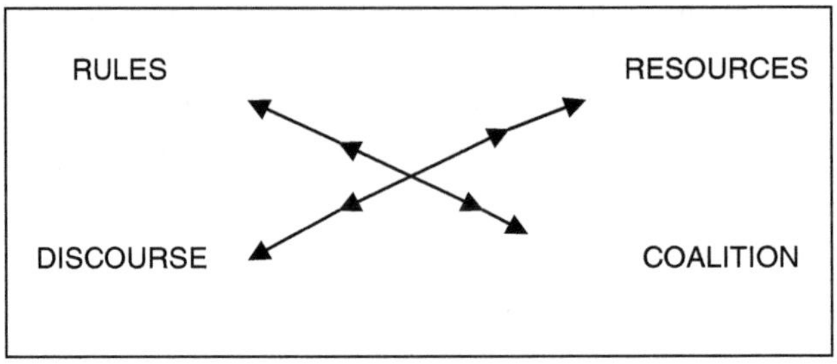


**Fig. 13.10** Policy dimensions of the scheme for territorial cohesion

develop local plans, contributing to a relatively high level of competence in spatial planning (Buyck et al. 2009). Politicians helped to avoid conflicts during the negotiations to set up the SCoT and in developing acceptable and shared solutions. A public enquiry was held and the Communautory Council approved the SCoT plan in 2006. The plan sets the direction for a decade. In total, however, only 31 out of 93 communes in the entire Montpellier region are included in the SCoT, even though the direct functional area of Montpellier is composed by over 40 communes and had about 500,000 inhabitants in 2008.

The SCoT is, by the application of a number of planning rules, also central to the municipal housing (PLU) and urban mobility plans (PDU), site developments (ZAC) and housing standards (PLH). Subsidies are obtained from the national government to finance the development and work of the Communauté d'Agglomération. The Agglomeration pays 50 % of the costs of urban studies to those communes that prepare their urbanism plan in coherence with the SCoT. Non-organized groups such as farmers with differing interests and illegal dwellers are often difficult to involve in consultations. Participatory methods mainly take the form of public utility enquiry.

The dominating discourse in Montpellier promotes the idea of territorial cohesion and of sustainable development. Demographic growth is valued as driver of economic growth. The 'shared city' is promoted, suggesting a concern for social relationships. Also the 'thrifty city', preserving resources, and identity and heritage are mentioned. The use of collective transport is also considered to be a lever of urban development. The 'landscape as a driving principle' sums up the spatial concept of green open space as a quality for sustainable urban development and is expected to help the acceptance of the SCoT by the wider public (Fig. 13.10).

## *13.10.2 Performance*

According to the assessment criteria formulated by case study practitioners and researchers the SCoT performs very well. The SCoT planning approach has led to negotiations at an inter-communal scale, contributing to the achievement of a balance between urban and rural interests, where the peri-urban has gained relative power. Inter-communal cooperation is built upon indirect representation, through the 'Social and Economic Council' representing the citizens. The shared learning process led to a new awareness and joint vision among the local authorities and citizens and to thinking at a higher level of integration (Buyck et al. 2009). Local politicians, both in their speech and actions, adhere to the SCoT. The strategy contributes to a sustainable allocation of more dispersed land use. The rule of 15 km as the maximum diameter of an urban region stimulated the Agglomeration to spread development in a more sustainable manner.

The promotion of green values as a vector for the region and a more controlled spatial layout are expected to strengthen the position of agriculture in the urban fringe and to limit the increase in land prices. The spatial details and building density rules guide municipal urban developments.

The development of planning tools such as the SCoT, that coincide exactly with the jurisdictional area of the regional territorial authority, is seen as an advantage by the local government. However, the fact that part of the functional urban region is not contained in the SCoT seems to be a problem, since urban development tends to leapfrog over the Agglomeration's jurisdictional boundaries. Regional coordination between the different Communautés now lies with the Département and even higher authority. The agglomeration is the most powerful local government body in the region and it has become more powerful than Montpellier city.

## 13.11 Comparison of Findings

In the following section a number of conclusions are drawn, based on the analytical framework and using the dimensions of the 'policy arrangement' model described earlier to explain the influence of the strategies summarised above on development and sustainability. Before comparing the cases in terms of each of these four dimensions, some general characteristics of the developments and strategies of the seven case study areas will be discussed in 13.12. The section following this comparison starts with some general reflections on the comparative research and draws lessons for policy makers.

### *13.11.1 The Urban Fringe as an Area of Conflicting Development*

The urban fringe is used not only for recreation by the urban population, but also for water storage, car dump sites, landfills, motocross sites, "horseyculture" and a range of other non-agricultural uses. At times it is seen as a location of great contrast, where both very wealthy and very deprived communities live. A distinctive feature is that the urban fringe is not considered to have its own intrinsic values but is seen merely as a problem caused by or a solution to urban problems and as a cheap source of land, thus only the location factor tends to count.

The case studies present a picture of congested urban cores versus rural decline with a gradient in between of increasing peri-urban character. Residents often leave the less-salubrious compact city centre areas in search of better living environments, investors make money from development or land banking. Those who either cannot afford to leave or who prefer an urban way of life stay behind, sometimes in emptying cores. Pushing the concept of the compact city too far can lead to social exclusion, with some groups deprived of access to or of sense of ownership of green open space. The primacy of power of regional or national capitals and their strength in attracting inward investments leads to seemingly ever-expanding metropolitan cores, while populations in other parts of the region or in other, remoter, regions, shrink. This leads to an unsustainable distribution of land use and represents an unsustainable form of growth. Suitable infrastructure is necessary to distribute development over wider areas and to include rural populations among the beneficiaries of economic development, providing them with access to higher standards of living, employment and income.

### *13.11.2 Comprehensive as Well as Sectoral Strategies Compared*

When comparing the different strategies adopted in the seven regions—to protect agriculture, to promote recreation, to conserve biodiversity, to reduce building pressure and to carry out comprehensive land use planning—it was found that in practice, it was difficult to analyze them individually. Instead, they were found to be integrated to a certain extent. The Koper strategy to protect prime quality land for agriculture appears to be the most sectoral of them all, while the Koper regional development plan and the SCoT from Montpellier are the most integrated in terms of area coverage. The other strategies lie in between, with varying degrees of integration.[2]

Hence, the different strategies should be seen as complementary at different levels and with varying degrees of precision in their aims. While some are or have been directly established with a clear focus on the purpose of protecting green belts and ecosystem services, the more comprehensive planning strategies are less rigid and leave more room for further interpretation in the way they balance competing interests. Comprehensiveness in spatial coverage as well as in terms of issues coverage (across sectors) is a matter that has particular implications for governance. In particular, such comprehensiveness calls for coordination and for accountability mechanisms, since it concerns achieving balances between different interests, between places, over time and among different social groups as well as with the

[2] This does not mean that the other regions did not adopt a comprehensive strategy but they were not always studied into detail by the research teams.

general public interest. It also raises concerns about providing appropriate tools for managing and solving conflicts between different interests both through traditional democratic mechanisms and through the inclusion of a wider range of stakeholders in the decision making processes.

### 13.11.3 Rules of the Game

In most of the European regions which were studied, the problems and opportunities of peri-urban areas fall into a gap between urban and rural policy regimes. As such, peri-urban areas are not easy to define, with no single agreed boundaries, and they vary greatly both within countries depending on the size and character of the urban pattern and across Europe due to diversity in urban cultural heritage. For example, neither of the areas of jurisdiction of the Association of Montpellier Agglomeration or of The Hague Region covers the whole functional rural–urban region. In the Montpellier case this leads to leapfrogging of urban sprawl to areas just outside the jurisdiction area. In The Hague Region there is hardly any peri-urban green open space available within the jurisdiction area; six enclaves remain. The nearby Midden-Delfland area, important for its agricultural produce and recreational potential, falls partly under the Rotterdam administrative region. European funds focus attention towards either rural or urban areas, but not towards a combined function of areas where rural–urban interaction is needed to balance developments in a sustainable manner.

In the former Eastern Bloc context—which applies to Leipzig-Halle, Koper and Warsaw region—land use planning from the Second World War until the early 1990s was traditionally centralised and the development pressure on the urban fringe was fairly low due to low land prices (in the almost complete absence of a land market) and public control. Spatial planning policy is now under transition towards the European Union standard (if such a thing can be described), with sustainable development as the guiding principle at least expressed in the public rhetoric but with few statutory coordination and control mechanisms available in practice, except for Leipzig-Halle region which adopted West-German planning standards following German unification. In parallel, agricultural enterprises have undergone similar changes due to land privatisation and more competitive agricultural markets. With rapidly changing economic conditions, lands previously protected for agriculture and nature conservation are now under greater pressure for commercial exploitation. However, there is little—if any—public control of current land use development or building permits. The result is a continuous erosion of the potential for ecosystem services provided by reducing green areas and diminution of local identity in previously rich cultural landscapes.

In the case studies in Western European countries—such as in Greater Manchester, The Hague Region and the Montpellier Agglomeration—globalised markets for agriculture and urbanisation of rural populations have also created new conditions for land use policy in the urban fringe. At the same time, new policy instruments have emerged to try to halt the loss of biodiversity and to provide recreational areas and access to green space for urban residents. The Green Corridor in Leipzig-Halle links inner-urban green spaces to peri-urban green spaces providing a network which traverses urban areas. Furthermore, urban perforation in Leipzig-Halle region as a result of population shrinkage is actively used to green the city, e.g. by creating neighbourhood parks on brownfield sites.

In all the case studies, and supported by enhanced economic conditions, preferences for living in the peri-urban or rural areas have increased pressure for development of housing with access to the rural landscape. This is a general phenomenon in Europe. If not accompanied by adequate transport infrastructure, it leads to severe congestion during commuting times, as noted in Warsaw. Even though Western European countries have mature spatial strategies and planning systems, they can still suffer from inadequate integration between parallel initiatives, as for instance in Greater Manchester.

The Chinese case study of Hangzhou represents a planning culture which is very different from the six European case studies. Here, the top-down government-controlled planning system is still in place, even though private investors seem to be gaining ground and diverting financial resources out of the region.

### *13.11.4 Financial and Land Resources*

All the case studies clearly point to the issue of steadily rising land prices in the urban fringe, with the exception of Leipzig-Halle, where shrinkage is taking place, and the accompanying pressure on local politicians and planners to allow for certain forms of economic development and the expansion of industry, commerce and housing at the expense of open space, such as green belts and nature reserves. The levy of local taxes also contributes to the tendency of peri-urban regional authorities or development agencies to attract new settlements in order to create a bigger tax base to pay for the provision of accompanying public services. It is possible to see the emergence of a complicated socio-economic mix of activity in peri-urban areas. There is also evidence of 'counter-urbanization' through the dispersal of population and economic activity out to a wider radius beyond the city, along with 're-urbanization' through the return to a more dense city lifestyle while maintaining a high material standard of living. Lower income groups are often left behind. Meanwhile, the effects of this 'displaced' urbanisation may ignore limited environmental capacities and weaken social cohesion in the peri-urban area.

The possibilities of financial compensation for farmers and others who help to provide ecosystem services in peri-urban areas can be important, particularly in

Western Europe. The EU schemes for environment and rural development as well the European Regional Development Fund (ERDF) are also relevant, but neither of them seem able to facilitate integration at the level of the rural–urban region but target instead either the rural or the urban setting and ignore the peri-urban areas which must be counted as one of the other if it is to share in the financial pot. In China, the Hangzhou government's payment to those municipal areas that may lose development opportunities as a result of providing environmental protection such as preserving water and green areas for urban dwellers is also a case in point. Financial support to green development was provided in the Leipzig area, through the Saxony "Regio" funding mechanism (FR-Regio), Regional development concepts with a focus on environmental actions have also been established. These provide access to funding for the small municipalities that would otherwise have more difficulty in engaging in environmental protection and development initiatives. However, the strong management of the applicant parties is an important condition for the final success of the Regional Entwicklung Konzept (Bauer 2010). In the Montpellier Agglomeration, financial support to agricultural initiatives and cooperatives is a way of compensating rural populations that miss out on urban development income by investing in environmental quality in the region, similar to the approach of the Hangzhou government.

However, the issue of local tax income is a problem in situations of population shrinkage as it may create sharp competition between 'green' developments and urban development, since shrinkage, leading to lower house prices, generally attracts lower-income groups with lower tax-paying potential. In Montpellier, the reduction of tax revenues from industrial or commercial activities has significantly reduced one of the main fiscal resources for local government. Emphasis on the "rewards" of inter-municipal cooperation, such as task-sharing and improved prospects for regional tourism, such as the Green Corridor in Leipzig and the Funding rules of Saxony seem to offer a way out of the vicious circle competition between municipalities in a shrinking situation.

The lease of state- or municipal-owned land to businesses, including farmers, might be a solution to consider as a means of guiding land management. In the European case study regions land ownership often obstructs a sustainable dispersion of urban and green functions (e.g. The Hague Region need for land banking to support Green Blue Services by farmers). In Hangzhou the state leasing of land to business embodies a degree of flexibility in planning, more state control and a continuous source of income to the municipality. In the Netherlands, the Green Blue Services concept which as yet has limited financial resources (1 million Euros) and means of regulation is expected to be more successful in preserving agriculture in the urban fringe when accompanied by land banking. Land banking by a cooperative combined with lease arrangements might be helpful for older farmers who depend on the value of their land for their retirement pension and want to preserve the rural identity, such as was found in the Montpellier region.

### 13.11.5 Policy Coalition Development

In many regions there is considerable lobbying from commercial interests for flexible spatial planning procedures that allow for the expansion of housing and business into peri-urban areas. Politicians are thus faced with having to find a balance between environmental protection and socio-cultural interests on the one hand and powerful land buyers and developers on the other.

Environmental groups generally advocate the increased protection of valuable natural areas, and sometimes form coalitions with other socio-cultural interests concerned with rural development. Even though they are supported by state authorities and legislation, those interests frequently have less access to the municipal political elites. The general pattern is that a lack of participation by the general public remains a major challenge to policy development in peri-urban areas, especially since many constituencies are only loosely tied to these geographical areas and social and economic cleavage lines provide obstacles to open policy procedures. The Montpellier Agglomeration, however, seems more successful in achieving a sustainable distribution of land use in the region through using professional planners instead of politics while mainly relying on informing the public for obtaining support.

Authorities or other actors with the greatest access to the levers of power have the strongest influence on planning decisions in a given region. It can be seen that most regional policies are weak in social and environmental performance because the most powerful actors have a biased discourse/vision, aimed mainly at economic development. The visions of professional planning staff in the interest of sustainable development or in developing a culture of sustainable development are either weak or absent, or else they do not succeed because the influence on their local council is inadequate, as seen for example in the Koper case study. The Montpellier Agglomeration is a positive exception. Here, there appears to be a political willingness in both speech and action to develop the agglomeration towards a sustainable dispersion of land use and to high quality urban development. The notion that the regional attractiveness is at stake has become widespread among the locally elected representatives and the coalition *Communauté d'Agglomération de Montpellier* is powerful. Also, the *communauté* possesses legal, statutory, financial and cultural competences to guide land use development. The region is a large one with a relatively low population density and hence can more easily realize such policies than can, for instance, The Hague Region.

The strength of the developers in land acquisition for housing and industrial or commercial development stands out as a difficult—as well as economically disrupting—element in the creation of powerful policy coalitions in support of wider public interests. For example, the onset of the post-socialist transformation in the Leipzig-Halle region beginning in 1989 eventually led to a huge oversupply of housing, office space and developed land in general (Nuissl and Rink 2005). Speculation in the agricultural economy in the urban fringe is also commonplace in several of the case studies and causes major problems for the safeguarding of public access to green areas. It sometimes also leads to legal challenges and appeals by developers against Green Belt planning decisions, as reported from the Greater Manchester and Warsaw regions.

### 13.11.6 Policy Discourses

Two parallel strands of public discourse can be discerned that concern the peri-urban areas: firstly on environmental protection and secondly on economic development and expansion of city infrastructure and urban values. The latter is embraced by local politicians and private developers. Economic development and job creation is what the public wants. There is often limited awareness of concepts of sustainable development among the electorate which leads to the election of few politicians with a broader vision of development than mere 'economic development'.

A lot of rhetoric is expended on nature and biodiversity protection and on providing access to cultural and natural landscapes for urban populations. For example, Green Belt policies have traditionally been strong and enjoyed extensive public support in both Manchester and The Hague. Similarly, in Montpellier, Warsaw and Leipzig new public discourses on preserving green areas are becoming more common, at least as reflected by some of the public servants in the case study research. However, discourses and coalitions that are not provided with sufficient financial means or land resources will not succeed. It can be concluded that in the case studies economic interests still prevail over sustainability and social justice. The case studies show that economic motives dominate spatial and land use planning decisions (see e.g. Perpar 2009), but this was not always the case. In Leipzig-Halle region there was also a notion of social planning that was central to traditional spatial planning under the socialist system and which was lost after the reunification of Germany. In Manchester the traditional Green Belt policy has prohibited general development for a long time but more recent strategic projects are often exempted. These are major economic enterprises leading to large scale employment and important infrastructure such as motorways and even business parks. In Warsaw Metropolitan Area residents in the urban fringe consider the first task of municipalities to be to the secure conditions for economic development. But the discourse of economic development is very one-sided. All other values seem to lose ground in the face of economic considerations. In most of the case study areas built development is uncritically considered to be the means to prosperity, whereas flows of people show that above a certain threshold level of prosperity other factors emerge that make people leave these same areas in search for a better quality of life. In particular, the social argument appears weak in comparison to the concept of economic growth or ecological motives aimed at protecting the environment and nature rather than considering it as part of the way of life and thus of society.

Perpar showed in Koper that ironically, the social achievements of the new integrated planning practice carried out by a few experts are relatively poor in comparison to the social achievements of the former socialist planning in which experts from different disciplines cooperated in the development of plans for an area.

However, the notion of 'landscape' seems to appeal to broader constituencies than does mere 'nature protection'. As noted by Buyck et al. (2009) landscape is integrated into the discourses of Montpellier as a link between disparate sections of the urban and peri-urban territory, putting centre and periphery at the same level. On the one hand, the discourse on 'landscapes' led to the reunion of the different towns from Montpellier agglomeration, while on the other hand there are fragmentary ideas on landscape, often associated with a particular actor's profession, where landscape becomes 'the area of projects', a 'great field of experiments'. In the Green Infrastructure Montpellier the landscape is seen as a vector in the development of the territorial cohesion scheme. Ravetz (2008) presents a similar notion through the term "spatial ecology", with green infrastructure seen as the main identifier and anchor of 'local place' and hence also promoting local identity.

## 13.12 General Reflections and Lessons for Policy Makers

The application of the fluid 'sustainability' concept to peri-urban land use and peri-urbanisation raises many questions. This is due to the diverse perceptions of the notion of sustainable development, which supposedly combines and balances economic, social and environmental goals, both locally and globally, and in both the short and longer term. The implication is that the peri-urban sustainability agenda is complex and inter-connected, with many layers which need careful analysis. In practice, the findings from this comparative study support those of several others, namely that the concept of sustainable development has been widespread on all levels of government at least at the declaratory level (Owens and Cowell 2002; Evans et al. 2004; Baker and Eckerberg 2008). However, the interpretations of what is sustainable vary greatly across countries and case study contexts, as well as in the allocated responsibilities for public planning which are far from politically neutral. In line with Owens and Cowell (2002), we find that the weaker conceptions of sustainability seem to dominate, those which underplay conflict but which in many instances may still end up promoting incremental greening and which increasingly accept that environmental losses should be compensated for. Indeed, the quest for opening up spaces for dialogue about the claims of environmental integrity, social justice and a dignified quality of life and on the substantial moral and political task of adjudicating between claims that cannot always be happily reconciled (Owens and Cowell 2002) is triggered by the notion of 'sustainability' and 'sustainable development' and as such it is an important contribution to our analysis.

As expected, different governance patterns can be discerned when comparing the Eastern Europe (and Chinese) situation with that of Western Europe. Public–private partnerships and voluntary action is commonplace in the west (as seen in Great Britain, France, The Netherlands and Germany), but rarely, if at all, occurs in China, Slovenia or Poland where the state dominates the policy-making of peri-urban areas. Having said that, however, it is also noted that part of the planning process is piecemeal and un-coordinated, which is at the cost of balancing between sectoral interests and the general public interest of retaining green open space in rural–urban regions.

In our case studies it can be seen that the role of government is changing; previously strict policies and state intervention to protect green areas are loosening up in some of the case study countries (Slovenia, Poland, The Netherlands) and leaving room for increased discretionary judgments when faced with multiple development interests.

The analysis of the case studies suggest that considerable regional governmental power is needed to steer development and to rebalance it away from economic and more towards environmental and societal needs. Weaknesses in the performance of well-intended strategies tend to be explained either by a lack of legal force or control over the land, together with the lack of financial means, inadequate cultures of spatial planning or a lack of cooperation at a regional level. The Montpellier region seems to have overcome most of these difficulties. Similarly, in the Parthe floodplain in the Leipzig-Halle area the three municipalities successfully combine legal jurisdiction through local land use plans and the involvement of all essential partners, including land owners, in the development of a discourse that enjoys the support of environmental NGOs and the general public. Such cooperation is absent in the Warsaw Metropolitan Area and the Mazovian region but seems badly needed there in order to resolve the current inertia in public planning for sustainable development.

The role for new governance actors—including environmental groups, neighbourhood groups, and local business and farmers—is stressed in many of the case studies. These groups can, while at times competing with the general public interest, contribute to innovative thinking and development of new options and solutions. Environmental groups were active in several of the case studies, contributing to promoting public awareness and applying political pressure to the local governments to prioritise the preservation of green areas and biodiversity. Farmers' organisations helped to promote the diversification of the rural economy through, for instance, Rural Forums in Manchester and the Farming for Nature project in The Hague region.

Several aspects connected to democracy and the accountability of decision-making concerning peri-urban areas may be raised here. First, there is the issue of how the constituency is defined. In most of our studied areas, public elections of regional politicians are absent. Second, functional rural–urban regions in Montpellier, The Hague, Leipzig-Halle and Greater Manchester are larger than the areas covered by the regional authorities. Thirdly, most peri-urban and rural parts of the rural–urban regions are politically weak through having fewer residents or voters. So the question may be posed as to whether the present democratic system allows the interlinkages between urban and rural areas to be taken into account. In the Hague Region, for example, the local authority of Midden Delfland is a green area without any urban settlement and thus has less voting power. Therefore, former small municipalities have deliberately merged to form one entity, the Midden Delfland municipality, in order to obtain a stronger position vis-a-vis the adjacent towns.

Another reflection concerns the issue of the short horizon of policy making, with frequent changes of local government according to election cycles, sometimes paired with a limited scale of budget and financial powers which largely constraints the scope of action. Both Warsaw and Koper suffer considerably from such risks. In such situations, it may be quite understandable that local politicians are subject to great pressure from developers who offer immediate economic benefits that are more easily grasped than those of long term investments in 'green' sustainability values. Here, it can be noted that the term of office for local politicians in Montpellier is 6 years, or 5 in the UK, compared with 4 years in the other European countries. A longer term in office might provide politicians with a longer time to achieve positive results from otherwise unpopular but necessary policies.

A more sustainable form of development in urban fringe areas that serve the residents of both urban, urban fringe and rural areas will require first of all that peri-urban areas are recognized as entities in their own right, rather than merely that of a spill-over or transitional zone. Inevitably, different interests will clash and need to be resolved and this will require strong government intervention in order to restrict the ongoing conversion of, in particular, green open spaces into privately secluded, or otherwise non-accessible land. Partnerships between public and private stakeholders and NGOs can provide opportunities for new alliances, investments and innovative thinking, but require statutory agreements in order to avoid ad hoc building permits that further social disparities as well as environmental deterioration of peri-urban areas, and profit specific interests at the cost of interests of non-organized groups.

## References

Aalbers C, van der Jagt P, Westerink J (2009) Assessment of strategies, case study The Hague Region, the Netherlands. PLUREL, working paper, unpublished

Baker S, Eckerberg K (2008) In pursuit of sustainable development: new governance practices at the sub-national level in Europe. Routledge, London

Baker S, Eckerberg K (in press) The role of the state in the governance of sustainable development: sub-national practices in European states. In: Duit A (ed) Mapping the politics of ecology: comparative perspectives on environmental politics and policy. MIT Press, Cambridge, MA

Bauer A (2010) Assessment report on local and regional government and planning strategies. PLUREL, working paper, unpublished

Bauer A (2010) Assessment report on local and regional government and planning strategies. Helmholtz Centre for Environmental Research. UFZ. Germany. 2010 Unpublished reviewed final draft

Buyck J, Chery J-P, Jarrige F (2009) Assessment report of strategies, Montpellier case study. D.3.3.9 PLUREL, unpublished report

Dryzek J.S. (1997) The Politics of the Earth. Environmental Discourses, Oxford: Oxford University Press

Evans B, Joas M, Sundback S, Theobald K (2004) Governing sustainable cities. Earthscan, London

Giddens A (1984) The Constitution of Society. Outline of the Theory of Structuration, Cambridge: Polity Press

Grochowski M (2009) Assessment report on local and regional governance and spatial planning strategies. Case study Warsaw Metropolitan Area. Institute of Geography, Polish Academy of Sciences Warsaw Poland. PLUREL, working paper, unpublished

Grochowski M, Pieniazek M (2008) Analysis of regional spatial planning and decision making strategies and their impact on land use in the urban fringe. Case study: Warsaw Metropolitan Area, Poland. PLUREL, unpublished report

Hajer M.A. (1997) The Politics of Environmental Discourse: Ecological Modernization and the Policy Process, Oxford: Oxford University Press

Hooghe L, Marks G (2003) Unraveling the central state, but how? Types of multi-level governance. Am Polit Sci Rev 97(2):233–243

Kohler-Koch B, Eising R (eds) (1999) The transformation of governance in the European Union. Routledge, London

Kooiman J (ed) (1993) Modern governance: government-society interaction. Sage, London

Marks G (1993) Structural policy and multilevel governance in the EC. In: Cafruny A, Rosenthal G (eds) The state of the European community. Lynne Riemer, Boulder, CO, pp 391–411

Nuissl H, Rink D (2005) The 'production' of urban sprawl in eastern Germany as a phenomenon of post-socialist transformation. Cities 22(2):123–134

Owens S, Cowell R (2002) Land and limits: interpreting sustainability in the planning process. Routledge, London

Perpar A (2009) Assessment report, assessment of regional strategies Koper case study, Slovenia, first draft-for internal use only. University of Ljubljana, Biotechnical Faculty. PLUREL, working paper, unpublished

Pierre J (2000) Introduction: understanding governance. In: Pierre J (ed) Debating governance authority, steering and democracy. Oxford University Press, Oxford

Ravetz J (2008) The city region in time and space. Analysis of regional spatial planning and decision-making strategies, and their impact on land use in the urban fringe. PLUREL, unpublished report

Ravetz J (2009) Strategic action for the Manchester rural urban region. Assessment report on peri-urban governance and planning strategies. PLUREL, working paper, unpublished

Sinn A, Haase D, Walde A (2008) Analysis of regional spatial planning and decision making strategies and their impact on land use in the urban fringe. D 3.3.6 Leipzig. PLUREL, www.plurel.net. Accessed 18 September 2012

Van Tatenhove J, Arts B, Leroy B (eds) (2000) Political modernisation and the environment: the renewal of environmental policy arrangements. Kluwer, Dordrecht

Veer P.M., Luttik J. (2010) To the Great Outdoors! Groen en recreatiewensen van internationale kenniswerkers in Nederland. Research report. Alterra rapport 1940. Wageningen

Yang J, He J, Pauleit S, Spiekermann M, Burkhardt I (2008) Analysis and assessment of regional and local government and planning strategies. Hangzhou Case Study. PLUREL, working paper, unpublished

Yang J, He J, Pauleit S, Spiekermann M, Burkhardt I, Yang F (2010) Assessment of regional and local government and planning strategies. Hangzhou Case Study. PLUREL, working paper, unpublished

# Chapter 14
# Sustainable Land Use in Peri-Urban Areas: Government, Planning and Financial Instruments

**Iván Tosics**

## 14.1 Introduction

In the previous chapter the comparison of the PLUREL case study regions focused on a selection of the strategies adopted in each, including Hangzhou, to try to limit sprawl and to safeguard ecosystem services and quality of life as well as to secure a role for agriculture in the peri-urban. This chapter switches focus to examine the kinds of instruments available to local or regional governments in order to see what potential there is in different regions to control land use change. The chapter looks at the broad picture of Europe and in this case ignores the Chinese case study of Hangzhou because of its completely different governmental model. Within the broad overview the case study regions are used to illustrate different points. However, it should be noted that in the framework of different systems found across Europe, the case studies do not represent all possible types (see Chap. 3).

On the basis that the assumptions used in the PLUREL modelling are correct, the scenarios for land use change in the next 20 or more years suggest that in Europe, the continent with the slowest population growth and, in several countries, demographic growth fuelled solely by in-migration, urban development is likely to lead to continuing sprawl. Furthermore, sprawl might continue not only in growing (in population terms) but also in shrinking urban regions. The further spatial dispersal of the stagnating (and from 2040 decreasing even with in-migration) European population and of businesses is likely to lead to serious consequences, questioning the basis of the economic, environmental and social sustainability of urban development.

Urban sprawl, by its very nature as an outward expansion of urbanising pressures, concentrates on the peri-urban areas which represent the largest land reserves for all types of development initiated by nearby urban centres. From a

I. Tosics (✉)
Metropolitan Research Institute, Lonyay 34, 1093 Budapest, Hungary
e-mail: tosics@mri.hu

K. Nilsson et al. (eds.), *Peri-urban futures: Scenarios and models for land use change in Europe*, DOI 10.1007/978-3-642-30529-0_14,

sustainability point of view the crucial question is how can the future development of Europe be de-coupled from the factors leading to urban sprawl?

The first step for any new development in peri-urban areas is land use change. From the perspective of market actors, land owners and developers, this is the 'weakest link' in the whole development process, as they have to get approval from public authorities for their planned land-use changes as a first step. This is precisely the point where the sustainability of future development can be influenced by public actors in the name of protecting public interests.

This chapter focuses on the possibilities of exerting public control over market-driven land-use change processes. Public control is a major problem in peri-urban areas, which are—as demonstrated in Chap. 2—areas 'in between' urban and rural and, in most cases, with no clear affiliation to any public institution or policy which aims first and foremost for sustainable development. In the absence of strongly represented public interests in peri-urban areas, the profit motive of land owners and developers, combined with opportunity-led planning systems, most often leads to significant land-use change, increasing the risk of unsustainable new developments.

Following a brief overview of some of the main theories in this area, the potentials of public control over land-use changes are discussed in three main ways. The first approach assesses the formal, broad framework of the territorial government system and the spatial planning policies which together constitute the institutionalized factors on which any land-use change decision has to be based. The second approach examines the financial systems, sectoral policies and legal regulations which influence the interests and behaviour of private and public actors. The third focus evaluates the options for influencing formal processes by informal means (governance building, strategy development). The chapter concludes by summarizing the lessons learnt from the perspective of local municipalities, who play key roles in the control of land-use change.

The chapter is based on a literature review and the author's own investigations into European formal structures, financial and regulatory policies regarding land use change and spatial development. This country-level knowledge is then confronted with and tested using the information acquired from the case study regions.

## 14.2 Theoretical Background: Diverging Individual and Public Interests

The URBS-PANDENS (2005) project suggested that when analysing the process of urban sprawl it is useful to distinguish between three types of actors:

- Mobile actors in the Rural–urban Region (RUR): households (low-middle-high income), industry/businesses, offices, retail/leisure centres, etc.;
- Non-mobile private actors in the RUR: key actors in the development process (landowners, developers);

– Non-mobile public actors in the RUR: local and supra-local governments (carrying out planning, regulating, financially influencing, investing/taxing).

In the market economies of Europe spatial processes are determined by the location choices of the "mobile actors". If the free-market logic of land development prevails, and in the absence of strong instruments such as the UK Green Belt system discussed in the previous chapter, the usual outcome is urban sprawl, as the myriad location decisions of individual private actors are usually based on short-term, self-interested considerations where the costs of public infrastructure and other external effects (e.g. increasing travel times and car use) are not taken into account.

The sustainable development of the city and its functional urban area can be considered as a public value. Market actors (households, developers, businesses, etc.) strive to increase their individual, private benefits which, however, can only be satisfied at the expense of public interests. In game theory this situation is described as the 'Tragedy of the Commons' (Hardin 1968). The problem lies in the fact that the gains (returns) and the costs are neither accrued to nor born by the same actors. Moreover, the gains are often abstract and lie in the future (and in classical economics are discounted back to the present) while the costs are concrete and fall due in the present.

How can this situation be conceptualized for the case of peri-urban development? For many people the move from, say, a small flat in the city centre out to a single family house in a newly developed housing estate in a peri-urban area is considered as giving a significant increase in their standard of living and quality of life. However, with each additional move the environmental quality tends to decrease, as the density of the developing areas increases. Similarly, for the individual resident the use of car is considered to be much more convenient than public transport. However, with urban sprawl comes growing traffic, while increasing car use causes air pollution and road congestion (which contributes to economic losses borne by people other than the new residents).

Thus the sprawl of the urban environment can be characterized by a strong contradiction between private interests and public values.

Hardin's essay on the tragedy of the commons also addresses potential management solutions to the problem. These include privatization, regulation and the principle that the polluter pays. The privatization and public regulation options correspond to the historical analogy, meaning, in the case of the "enclosure" of common lands in Britain and elsewhere, the change from the unregulated access to a wide range of values by all to systems in which commons are "enclosed" and subject to differing methods of regulated use with access prohibited or controlled by the few.

Hardin argues against the reliance on conscience as a means of policing commons, suggesting that this favours selfish individuals over those with greater foresight. Thus the conscience of private actors cannot be relied upon, necessitating the intervention by public actors on behalf of the public interest who can provide a solution to the dilemma, halting the tragedy of the commons. Based on this analogy, the competition between the private and public interests in the development of peri-urban areas cannot be solved by 'self-regulation', trusting to the conscience of the private actors. Only the intervention by public actors, in the name of the public interest can bring any solution or stabilization of the situation.

This chapter concentrates on the local and supra-local government structures of the rural–urban region. Their influence is crucial, even (or especially) in those countries where most of the land is in private ownership. As Larsson (2006) suggests, ownership rights do not mean development rights, which need approval via formal plans: "Most Western countries nowadays consider it a power of the local authority to decide if, where, when and how a development may take place."

Thus local and supra-local governments hold key positions to influence land development in the RUR. They can determine the conditions for new developments and through this influence the suburbanisation and peri-urbanisation processes. There are, of course, substantial differences between countries as to how the decision-making power is allocated across the different levels of governments in the RUR area.

This chapter aims to evaluate which factors determine the local governments' interests and what tools they have to counter urban sprawl, through influencing the decisions of mobile actors. Besides this, the supra-local levels will be investigated, to see how these can help to create more cooperation between local governments with conflicting interests.

## 14.3 The Institutional Framework: Government Structures and Spatial Planning Systems

In the next section a general analysis of European government and spatial planning systems will be presented, followed by some specific aspects which emerged from the study of the case study regions.

### *14.3.1 The Strength of Different Systems of Government and Spatial Planning Policy in EU Countries*

Throughout the EU-27, land-use change decisions are based on legal instruments and protocols established by national governments which also produce national spatial planning policies. There are, of course, significant differences between countries, in terms of how these formal systems and policies are shaped and how effective they are for decision making in specific cases at the local level.

In the course of the PLUREL project the formal systems of government and associated planning policymaking and implementation of all EU-27 countries were reviewed, by synthesising information from a series of national reports (see Tosics et al. 2010). The aim was to compare their potential strengths in exerting public control over land-use change.

The qualitative information collected from the national reports was converted into a comparable form in two steps. The first step was to evaluate the strength of the two factors (government structure and spatial planning mechanisms) for each country, in the following way:

- The government structure was explored in terms of which administrative level of government is or are responsible for land-use change (re-zoning) decisions, or at least which levels can exert some significant influence in the particular country. Then the average size of this main decision-making level was compared to the average size of local governments (based on population numbers per local government unit). The outcome was a classification of countries into different categories along a scale according to our hypothesis that the bigger the responsible administrative units are and the larger (the more integrated) the local governments, the greater the chance of sustainability considerations being reflected in land-use change decisions.
- The spatial planning policy aspect (what kind of regional/spatial policies influence land use change decisions) was assessed in terms of whether any supra-local policy had an influence on the land-use change decisions at the local government level. The outcome was a classification of countries into different categories along a scale of relative influence of supra-local and local levels.

In the second step the two aspects were integrated using a combined matrix. The following table summarises the potential combinations of the two scales, with some simplifications (these result from the assumption that if the supra-local levels are very weak, their size does not influence the strength of the system, while if they are very strong, the size of local authorities loses importance as their competence is very limited and they are strongly controlled) (Table 14.1).

Thus the indicator values in the table reflect the variations in the potential power of the public hand to resist the push of market actors towards more urban sprawl, ranging from a minimum (1) to a maximum (7) degree of power. Based on the results of the analysis fed back to the circumstances of each country as described in the reports, an overall EU-27 pattern can be presented (Table 14.2).

The results show a wide variety of government and planning systems in the EU-27 countries in terms of the potential control over urban sprawl through the intervention of national and local government and the respective planning systems. The North-west European countries (e.g. Denmark, the UK and the Netherlands) show higher levels of potential control mainly because of their consolidated[1] local government systems, while Southern European countries showing a higher potential (such as Cyprus, Greece or Portugal) have more fragmented local government

[1] Local governments are large and integrated with high level of professionalism as opposed to other countries where local governments are small—closer to the people but less able to work professionally.

**Table 14.1** Strength of public control over land-use changes according to the combined government and planning dimensions

| Spatial planning system: control from supra-local levels | Supra-local level of government deciding or influencing land-use changes (average size in millions) | Local level (average size in thousands of population) | Indicator of the strength of public control |
|---|---|---|---|
| Strong supra-local spatial policies | Large (>1m) | Any | 7 |
| | Medium-sized (0.5–1m) | Any | 6 |
| | Small (<0.5m) | Any | 5 |
| Medium level of control | Large (>1m) | Large (>30) | 6 |
| | | Medium-sized (10–30) | 5 |
| | | Small (<10) | 4 |
| | Medium-sized (0.5–1m) | Large (>30) | 5 |
| | | Medium-sized (10–30) | 4 |
| | | Small (<10) | 3 |
| | Small (<0.5m) | Large (>30) | 4 |
| | | Medium-sized (10–30) | 3 |
| | | Small (<10) | 2 |
| Weak supra-local level of control | Any | Large (>30) | 3 |
| | | Medium-sized (10–30) | 2 |
| | | Small (<10) | 1 |

**Table 14.2** Results of the country-level evaluation

| Value | Countries |
|---|---|
| 7 | |
| 6 | Denmark, Netherlands, Portugal, United Kingdom |
| 5 | Belgium, Cyprus, France, Germany, Greece, Ireland, Lithuania |
| 4 | Italy, Spain, Sweden |
| 3 | Austria, Bulgaria, Finland |
| 2 | Estonia, Latvia, Luxembourg, Malta, Poland, Slovenia |
| 1 | Czech Republic, Hungary, Romania, Slovakia |

systems, but stronger planning control at supra-local levels. Most new EU member states show a weak control potential, with the notable exemption of Lithuania (where the tradition of strong planning is based on the presence of the former Western Soviet planning institutions) and to a certain level Bulgaria (with a consolidated local government system).

It must be emphasized that this picture does not depict the real strength of public control over land-use change, since in practice these powers can be made more effective in different ways or conversely the potential strong effects are not realised for various reasons. Because of this, these values should be seen more as a potential resulting from the combination of governmental and planning systems. A weak (low level) potential for control is hard to overcome even if the willingness is present, while a high potential may or may not be completely applied, depending on the intentions of the public bodies in power.

### *14.3.2 PLUREL Case Study Examples on Government and Spatial Planning*

In this section the six European case study regions of PLUREL will be assessed according to the scheme presented in the last section. These do not, however, cover all the different types of government/planning systems arising from the above analysis.

From Table 14.3 it can be seen that three types of European government systems are not represented in PLUREL:

- Classic unitary countries (where the role of local authorities is weak),
- Centralized unitary countries with strong but non-integrated local authorities (strong power given to the local level but fragmented local governments),
- Regionalized unitary countries (where regulations can be very different region by region).

Consequently, the results from the PLUREL case study analyses cannot be generalized for the whole of Europe. The case study material, however, is very good for illustrating problems and demonstrating innovative solutions.

In the following sections the two dimensions of the matrix will be presented separately, highlighting selectively some good practices as well as problems from the perspective of public control over land-use changes in RUR areas.

## 14.4 Governmental Systems

In Poland—contrary to most of the other post-socialist countries—a strong, elected middle tier of the formal governmental structure had developed by the end of the 1990s. The Warsaw case study describes the general picture of decentralized, fragmented local government, which is a common feature in most post-socialist countries after the transition to a free market economy. The unique aspect of this case study is the story of the Warsaw Metropolitan Area. To develop strong cooperation between the settlements in the RUR of a capital city is not an easy task anywhere in Europe. In the Warsaw case, efforts to develop this remain for the moment unsuccessful, since neither the city nor the administrative region really supports the idea. Besides this, the strong political power given to the municipality level makes any bottom-up cooperation efforts illusory in the face of the much stronger interest in competition between different local municipalities.

The case of Manchester is interesting from the perspective of the quickly changing situation. Successive UK central governments have changed the conditions for territorial development frequently and sometimes quite radically, resulting in very different opportunities for cooperation between municipalities in functional urban areas. Recent developments by the latest government following the 2010 general

**Table 14.3** Draft typology of territorial governmental systems in the EU27 plus Norway and Switzerland

| Government structure | 1. Classic unitary countries | 2. Centralized unitary countries with strong, but non-integrated local authority level | 3. Centralized unitary countries with strong, integrated local authority level | 4. Decentralized unitary countries with strong local and strong regional level | 5. Regionalized unitary countries | 6. Federal states |
|---|---|---|---|---|---|---|
| EU-15 and EFTA countries | Greece<br>Ireland<br>Luxembourg | Portugal | Denmark<br>Finland<br>Netherlands<br>Sweden<br>Norway | France<br>United kingdom | Italy<br>Spain | Austria<br>Belgium<br>Germany<br>Switzerland |
| New member states | | Bulgaria<br>Czech Rep.<br>Hungary<br>Romania<br>Slovakia<br>Cyprus<br>Malta | Estonia<br>Latvia<br>Lithuania<br>Slovenia | Poland | | |

Based on Tosics-Dukes (2005), with alterations based on ESPON project 3.2 (2006).

election have abolished the regional level while offering some new opportunities for bottom-up cooperation in the RUR.

The multi-layered Dutch system allows for all types of cooperation and planning at the city-region level. The case study of The Hague Region describes the efforts to supplement the weak government and planning power of the city-region through innovative attempts to create more cooperation around topical agendas.

Montpellier is probably the case study with the most developed city-region level government structure, in the form of the creation of a top-down settlement association. The Montpellier Agglomeration has existed since 2001 as a compulsory association of 38 municipalities. Disputes over voting resulted in 7 municipalities opting out, thus the Montpellier agglomeration finally brought 31 municipalities together covering spatial planning, public transport, water management, housing policy, economic development and cultural policy. The case study describes in detail how the new administrative framework contributed to the change in the mentality of local leaders away from competition towards cooperation.

## 14.5 Planning Policies

The Manchester case study describes the very different planning powers and traditions at the different levels: the politically strong national green-belt policy has no similarly strong local counterpart, while the regional level has recently been abolished. Both politically strong and weak planning policies have their problems: while the fixed green-belts are sometimes ‘leapfrogged over’ by development (and cause in this way their own problems), the flexible bottom-up planning methods do not always achieve the necessary cooperation between municipalities.

In Germany, as a federal country, the sub-national Land (federal state) level is the most powerful in terms of the governmental and planning aspects of land-use changes. The federal states have developed their own internal structure and planning regulations—as a consequence of which the building legislations, funding programmes and spatial planning policies differ between neighbouring states. This may present problems for cooperation in RUR areas which cross the borders of federal states—as is the case in the Leipzig-Halle case study. In such cross-border situations even the relatively well-developed German bottom-up government cooperation and flexible planning practices may achieve only very limited results.

The Dutch planning system is also multi-layered, without being fragmented by politically strong regions. As well as national guidelines, funding systems and cluster policies, the provincial level provides the framework for the local planning system. Thus, the case of The Hague Region should be understood as finding a special role for the city-region within the planning responsibilities already shared between the state, province and municipality. The main role of the city-region level is to prepare the regional structure plan with the aim of coordinating different ideas (concertation). In addition, more power must be given to the administrative level of the province, which is closer to the size of the RUR.

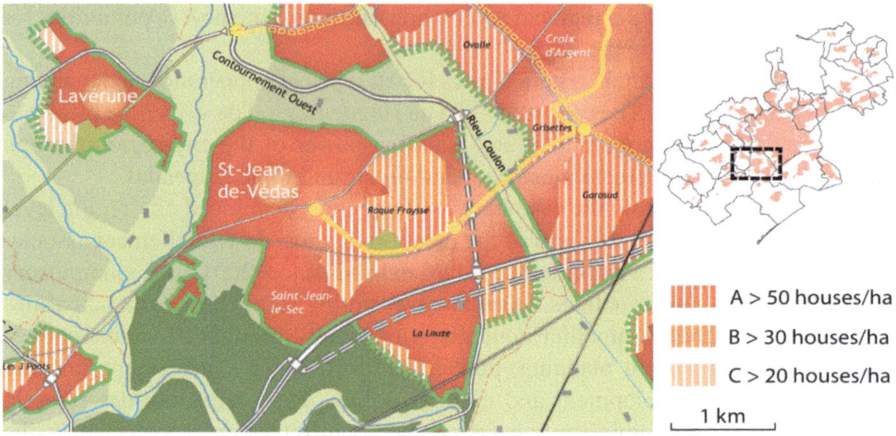


**Fig. 14.1** Montpellier SCOT territorial plan. One of the schemes of the Montpellier SCOT, showing the planned expansions of the built-up area, defining three categories of the minimum density according to the available level of public transport (RED: extension with more than 50 units/ha density; ORANGE: extension with more than 30 units/ha density; YELLOW: extension with more than 20 units/ha density) (Source: http://webarchive.nationalarchives.gov.uk/20110118095356/http:/www.cabe.org.uk/files/client-guides/lsud-examples-design.pdf)

As a compulsory settlement association, in 2005 the Montpellier agglomeration adopted the SCoT planning system (Montpellier Agglomération, 2006), for a period of 15 years, which is the most important tool to handle peri-urban issues. The SCoT defines growth areas and determines three categories of minimum levels of density, depending on local environmental aspects and public transport accessibility (Fig. 14.1). The public sector has pre-emptive land acquisition rights and is able to intervene to achieve higher densities where necessary. The SCoT is also innovative in protecting open space. Estimates for the effect of the SCoT suggest that over the 15 years of its life the amount of sprawl will be half that which would have occurred if no SCoT had existed.

The local plans of all the 31 municipalities of the Montpellier agglomeration have to be adjusted to fit the approved SCoT. In order to do so, some incentives and supporting tools are offered and special funding is available for urban studies before extensions are planned.

The SCoT planning system is an innovative approach to the difficult and controversial planning issues of metropolitan areas. This system could handle the development dilemmas of the whole RUR of Montpellier. Unfortunately the Montpellier agglomeration with its 31 settlements covers only a part of the RUR, which extends to 93 municipalities. In this larger area there are more inter-municipal associations with no coordination between their SCoTs. Another sign showing how difficult it is to achieve changes is that 4 years after the SCoT was introduced speculative strategies still exist and farmers still think that selling farmland for development will be their best decision.

## 14.6 Financial, Sectoral and Regulatory Factors Influencing the Processes in the Rural–Urban Regions

Within the formal framework established by governmental structures and the planning system, there are a range of further factors which influence the behaviour of public and private actors in a RUR. These factors were assessed on the basis of information gathered from the case study regions. The following factors will be discussed in detail:

- The local government financing system (from where and according to what parameters local governments receive their revenues).
- The taxation system (the existence of different types of taxes local governments are allowed to levy and the spatially relevant consequences of these).
- Sectoral (infrastructure, economic development, transport, housing, etc.) planning policy, regulation and subsidy systems.

The public sector has various tools to guide the land development process (through influencing the parameters of new developments which determine the interest of the mobile actors). Among these the following deserve the most attention:

- Regulations applied to new land developments in general, such as sectoral conditions attached to permits for larger-scale new land developments (in order to balance between jobs and homes, transport services, physical and social infrastructure requirements) and financial regulations, such as taxes on green field investments, subsidies for brown field redevelopment;
- Rules applied in the case of specific decisions on larger-scale land developments: possibilities for the public sector to capture some part of land value increases when allowing the re-zoning of a piece of land or at the moment when building permission has to be issued for the planned project.

From the overview of these factors it is possible to highlight the problems of peri-urban development. Such an overview also leads to ideas as to how legal and financial regulations could be adjusted in order to direct the development process in more sustainable directions.

### *14.6.1 Factors Influencing the Interests/Motivations of Local Governments to Attract Mobile Actors*

Local governments are embedded within the national regulations for public sector financing, taxation and sectoral policies. These regulations are usually not studied in detail when discussing the dynamic processes in the RUR context, although they strongly influence the motivation and the "behaviour" of local governments.

#### 14.6.1.1 Local Government Financing Systems

There are major differences among European countries regarding local government financing systems, determining from where and according to which parameters local governments receive their revenues. There are at two main questions here, from the spatial planning point of view. One is the method by which financial resources are transferred from supra-local (central, regional) levels towards the local government. The other is the level of independence enjoyed by local governments to raise their own revenues—this will be discussed in the next section dealing with taxation systems.

A taxonomy of the grant system made by Bahl-Linn (1992) takes into account three different dimensions. The first is the determination of the size of the total amount of money to be distributed in a given year, that is the divisible pool. The second is its allocation among local governments. Once the amount of the pool is determined, allocations among local governments are typically made in four ways: by returning shares to the jurisdiction from which the taxes were collected, that is, using a derivation principle; by formula (e.g. pro rata according to the number of residents), ad hoc or by reimbursing costs.

From the perspective of local government autonomy the third dimension is also very important, whether the transfers or grants are unconditional (or general purpose) or conditional. In the first case local governments have full control over the use of the money, while in the second the freedom to use the money is strictly defined *ex ante*. Block grants offer a third variant: the transfer can be freely used by local governments but only within a certain defined area of service provision.

In the higher-level financing of public services it is of crucial importance to know to what extent the real net costs (deducting fees collected for the service) of the services are covered by these transfers. If the centrally provided normative support only covers e.g. 2/3 of the real net costs of providing the particular service, the rest has to be secured from the budget of the local government entity which provides the service.

The grant allocation principles influence the behaviour of local governments to a large extent. If the share of general purpose transfers and/or grants calculated by a simple formula (e.g. number of residents) is high, this creates pressure to increase the amount of population in general. Regarding service-related conditional transfers, local governments tend to have more “positive interests” in those services which are financed by higher level transfers to a greater extent, while this interest is much lower in the case of the centrally less-financed services.

Some public services have spill-over effects (or externalities) on other jurisdictions. Examples are pollution control (water or air), inter-regional highways, education and social services, recreation facilities (which may be used by neighbouring areas), etc. The spatial consequences are clear: local governments are less ready (less eager) to run a public service where the larger share of the real net costs has to be financed from their own budget. If a public service gets little higher-level financing, the interest of the local government is not to run this service but convince the population to use the service provided by another municipality. In this case the population is using the service as

**Fig. 14.2** New garden city without any social infrastructure built by a developer. The picture shows one of the new'garden cities' around Warsaw where new housing development is not accompanied by the development of new social infrastructure e.g. schools or highways

"free rider" in the other municipality (not being resident there, thus not contributing by taxes, etc., to the general budget of this settlement).

As a consequence, families moving out to peri-urban settlements often face a lack of services—especially those services which are funded less by higher level financial transfers (Fig. 14.2). In such cases families are forced to use the services in the urban core from which they moved, contributing both to the increase of traffic and to the increase of financial difficulties in the city's budget.

To solve such problems metropolitan-wide programmes can be set up to deliver local public services which would otherwise have inter-jurisdictional spill-over effects. This is, however, only possible if a higher level law makes this compulsory (see e.g. the French inter-communal associations or the German transport associations) or the municipalities of the area voluntarily cooperate with each other. A financial agreement has to be reached, in the framework of which the municipalities not offering such services contribute to the costs of those municipalities which run them, proportional to the number of their residents (the free riders) who use the services.

#### 14.6.1.2 Taxation Systems

Local governments are allowed to levy different types of taxes and these can result in different spatial consequences based on the decisions of mobile actors, such as residents and businesses Kovács (2009).

Although most productive revenue sources are usually reserved to the central government, a certain portion of financial sources can be assigned directly or indirectly to the local level as part of fiscal decentralization. There are several arguments for levying taxes at lower levels of government:

- The provision and financing of local services at a local level might lead to an improvement in public attitudes towards taxation, because those who pay taxes directly receive the benefits from their use. This may make tax collection easier and more efficient.
- Higher local tax revenue might reduce the dependence on state budget support.

The share of public financial sources which are assigned, either directly or indirectly, to local governments, indicates the level of their financial independence. Needless to say, there are major differences in this regard across Europe. According to a British analysis (House of the Commons 2009):

> Danish municipalities raise 60% of their revenue from local taxes (mainly income tax), and a further 14% from charges for services.... Similarly, Swedish municipalities raise 69% of their revenue from local taxes (mainly income tax), and only 15% in the form of government grant. The local government representatives ... felt that the clear link between local tax payment and the delivery of local services led to a strong engagement in local democracy. In England the financial situation is reversed. Local government raises, in total, only 25% of its revenue locally—mainly through the council tax.

Another country where municipalities have little freedom to raise taxes themselves is The Netherlands, where the bulk of their budget comes from the state, the amounts being based on the number of inhabitants in each municipality.

A basic division in tax systems is between "benefit" and "non-benefit" taxation (Oates 1996). Benefit taxation refers to systems in which taxpayers are taxed according to the benefits they receive from the public purse. In the alternative "non-benefit" tax system there is no link between taxation and local government services (one form of this is "ability-to-pay" taxation).

There are some "rules of thumb" suggested by economists to enable sub-national governments to control their own revenues and to be able to act responsibly when making their own policy choices. Thus, a lower level of government should rely as much as possible on benefit taxation, raising revenues from mobile economic units, including individuals and mobile factors of production. It also essential that: "to the extent that local governments make use of non-benefit taxes, they should employ them on tax bases that are relatively immobile across local jurisdictions" (Oates 1996, p. 36).

Bird (1999) notes that property taxes, excise taxes, personal income taxes, sales taxes and taxes on business are the only economically acceptable categories which can be levied by local government. Property taxes and/or personal income taxes are the economically and socially most justified types for lower level of government. The property tax is a stable form of revenue, allowing for only limited tax exportation. It also acts as a rough form of benefit charge: the value of real estate such as housing is strongly influenced by the level of local services such as schools. Nonetheless, property tax is an unpopular form because it is a visible burden and citizens confront it directly. Property taxes levied on plots and buildings may also discourage investment in improvements. The other important and frequently used form is a local personal income tax, generally collected in the form of a surtax on the national income tax, which is much less visible to taxpayers and is neutral in terms of its effect on property values, for example.

It is a general opinion that local taxes on business are very weakly efficient and are often criticized for distorting business location decisions. Even so, such taxes are widespread and generally popular amongst politicians (as being easy to collect and might result in substantial revenues) and by the population (as they do not have to pay it) (Bird 1999).

Local taxation is based on property taxes in the United Kingdom, France, Spain and Poland. On the other hand the Scandinavian countries provide examples of personal income tax dominated local tax systems.

From the point of view of peri-urban development the most important question regarding local taxes is the topic of tax-competition within the RUR. According to Goodspeed (1998) tax competition occurs "... when the tax system of one government entity affects the tax system of a second governmental entity, usually through an effect on the second entity's tax revenues." If local governments within the RUR are entitled to levy a certain type of local tax, and can also determine by themselves the parameters of this tax (within the limits allowed by the tax law), the result is usually tax competition: some municipalities will lower their rates in order to become more attractive to mobile actors. Tax competition exists in some of the PLUREL case-study regions (e.g. Leipzig, Warsaw).

It is easy to assume that tax competition between the municipalities of a RUR is harmful because it distorts the location decisions of mobile actors. However, the economic literature suggests that this is not always the case. The analysis of Goodspeed shows that horizontal (i.e. between local governments) tax competition can result in an efficient allocation of resources if the taxes used are benefit taxes. In such cases "... firms and individuals will sort themselves among jurisdictions to obtain their most preferred tax-expenditure package in the same way that individuals shop in the private market. Benefit taxes reflect social marginal costs and therefore lead consumers and firms to choose jurisdictions efficiently. If taxes do not reflect benefits, however, externalities are created so that tax prices diverge from social marginal costs. This creates incentives for inefficient location decisions. ... Consequently, one cannot gauge the level and effects of tax competition simply by analyzing differences in tax rates" (Goodspeed 1998).

On the basis of this overview, the public actors in a RUR should evaluate the local taxation systems applied across their constituent local authorities. When exploring these taxes the major question to ask is: which of these create horizontal tax-competition within the region and how strong is this competition (to what extent does it influence the location decisions of mobile actors)? In order to assess the real effect of this competition on the functioning of the RUR a tax should be analysed in terms of to what extent it can be considered as benefit tax. Finally, the potential of higher level regulations (e.g. in order to limit such competition) should be explored where these are feasible (such as where there is a regional-level authority with power to do this).

#### 14.6.1.3 Sectoral Planning

As the following examples will show, the internal development of the RUR is not only affected by general financial and taxation issues but also by different sectoral policies, regulations and subsidy systems. From the broad range of possible sectoral issues, this section only deals with three: economic development and infrastructure, transport, and housing.

Economic Development and Infrastructure

Economic development and infrastructure is usually planned in parallel at different government levels or by different departments or agencies. While national plans and policies deal with macro-regional aspects (such as international competitiveness and regional inequalities), regional and sub-regional ones are more concrete and territory specific.

There are wide differences in Europe regarding the strengths of the supra-local tier of administration: while in some countries elected regional governments with strong planning power exist, in others only some weak planning and symbolic administrative roles are allocated to non-elected regional bodies. In more centralized countries public control can be executed through functional (sectoral) bodies. In the most decentralized countries upper-level public control is weak: neither administrative nor strong functional control exists over local municipalities.

From the perspective of development within the RUR it is of crucial importance as to which level economic and infrastructure development decisions are taken. If such decisions are taken at a regional or other administrative level which covers the RUR, there is a good chance that the sustainability aspects of development will be taken into account—new economic investments will accord with other infrastructure conditions for development, also taking environmental and social aspects into account. The risk of unsustainable development is much greater when these decisions are decentralized to the local government level.

If the strength of supra-local administrative levels to steer economic development processes is weak (e.g. limited to drawing up non-compulsory structural plans or regional economic development concepts), then the real decisions to attract economic investors will be taken at the local level. Local governments can decide the level of local taxes, as discussed in the previous section, as well as direct or indirect subsidies (e.g. in the form of subsidies for infrastructure development) given to investors, in order to attract new developments to their territory. Agreements between local governments and investors might be kept confidential, claiming them to be a private business matters. In the absence of higher level regulations municipalities do not even have to take into account the interests of the neighbouring areas (whether the planned investment will cause any harm to other municipalities through increased traffic, pollution, etc.). In extremely decentralized systems, probably the only regulation to be taken into account is the EU competition legislation which forbids the giving of too many subsidies to investors which would distort the functioning of the free market.

In the national patterns of infrastructure development similar differences can be seen across Europe. Here also the biggest problem from the perspective of sustainable development is the strongly decentralized system with the almost total control by the local government over new infrastructure investments. In such cases not only is RUR-level coordination difficult but so is the possibility to achieve concerted actions between neighbouring municipalities.

### Transport

Transport is one of the main challenges in urban areas from sustainability (air pollution) and competitiveness (congestion) viewpoints. Urban traffic and congestion is expected to grow as a consequence of increased urbanisation and greater urban sprawl. Congestion reduces the logistic efficiency of deliveries, which suffer from the 'last kilometre' problem (virtually all transport tasks/routes start and end in dense urban areas).

Cities may not be able to accommodate any larger volumes of private cars because road space to drive and park may be limited. Many cities might have to find radical ways of dealing with congestion (European Commission 2009).

Transport policies, if developed in a coordinated way for the RUR or even larger areas, can contribute in many ways to the required direction of changes:

- In urban areas, which are the main nodes in the transport system of larger regions, the links between the different transport providers (local and suburban transport companies vs. long distance transport providers) of different size and ownership (public or private) have to be optimized. Coordination and multimodal types should enable better access to public transport and reduce the need to use the car to travel to the urban core. "Co-modality" should be achieved in such a way that the share of private car use decreases.
- Financial regulations, such as taxes, road charges and emission trading systems should under no circumstances create incentives for car use. On the contrary: such tools should ensure the internalization of the external costs of transport, while maintaining users' free choice of transport modes. Only real market prices, which fully reflect costs, can lead to economic efficiency.[2]
- In the case of new developments (commercial, industrial, office, housing, etc.) above a certain size, a compulsory element of the planning process should be the

[2] The German commuters' tax allowance (Entfernungspauschale) is an example of subsidizing commuters' expenses. People commuting from more than 20 km distance can reduce their taxable income, independently of the real expenses and of the transport mode (foot, bike, motorbike, public transport or car). From the ecological point of view the commuters' tax allowance is very critical as it contributes to the spatial detachment of residence and work, i.e. work in town, live in the countryside ("Premium for urban sprawl").

**Fig. 14.3** Road user charging system in Oslo. The 2007 picture shows one of the entry points into Oslo, where the road-user charge had to be paid (this system has been fully automatized in the meantime; there are no collection booths any more)

preparation of a mobility or transport audit, the results of which should be taken into account in the location decisions, within the framework of strict land use policies.

- On top of all these measures, in the most dense urban areas—besides the improvement of public transport services—the application of special mobility management tools, such as green zoning and urban road pricing[3] should be considered (Fig. 14.3). These measures should contribute to the behavioural change of transport users without decreasing their mobility options.
- Public authorities have an essential role in providing the planning, the funding and the regulatory framework for a complex system such as that of transport. They should also act as catalysts in city region-wide public transport associations.

RURs include territorial units with very different densities, ranging from high-density core urban areas through medium density residential, office and commercial parks to very low density peri-urban areas and rural hinterlands. These different elements usually have very different dominant transport modes, ranging from high-capacity fixed track rail-metro-tram systems through buses to individual cars. The big challenge for the RUR is to find the optimal way of connecting these transport modes, ensuring very different aims at the same time: avoiding congestion in dense urban areas, assuring mobility options even in low density (especially the more

[3] The best known European examples of urban road pricing are London, Stockholm and Oslo, while the system introduced in Milan has a strong environmental character. In Manchester the idea for a large-scale congestion charging system was rejected in a local referendum, although a 1.5 bn GBP government fund was offered for the expansion of public transport network.

rural) areas while controlling for the sustainability of transport and the reduction of urban sprawl over the whole area.

The organization of transport across the RUR area, based on a public transport association for the denser urban areas and extended by opportunities provided for changes between different transport modes (Park-and-Ride, etc.) can become one of the strong identity elements in the strengthening of the RUR cooperation.

### Housing

The mobility of households is naturally very closely linked to housing policy. Both market processes and public interventions (subsidies, building regulations and available financial resources) have to be assessed in terms of their spatial consequences as a result of the movement of households.

New residential locations significantly influence the environmental and social conditions of an area and they also influence the spatial distribution of wealth or poverty through the taxation aspect. Cooperation in land-use and housing policies between the municipalities of a RUR (based on jointly accepted regulations) might help to avoid urban sprawl and its unwanted negative externalities, such as traffic congestions, air pollution or social segregation, and might contribute to a more balanced, cohesive and sustainable functioning of the RUR. If housing is also included in multi-sectoral strategic planning, this might help to establish stronger links between the development of residential and workplaces in the RUR, which is necessary to avoid increases in traffic.

Thus, the housing sector can potentially play an important role in strengthening cohesive development, sustainability and the competitiveness of RURs. This is a strong argument in favour if controlling the housing sector by public policies generated at a higher territorial level, such as the functional urban area or the RUR. If agreements can be reached on housing policy cooperation (controlling access to and distribution of housing development possibilities, such as developable land and infrastructure) this might also ensure the environmentally and socially sound functioning of the whole RUR. Moreover, new housing development strategies agreed for larger areas, if well connected to economic development potentials, might contribute positively to the competitiveness of the RUR. The spatial distribution effects of new housing investments might easily create conflicts between housing and urban aspects. For lower income groups, in the absence of any public intervention, housing affordability requires the construction of new housing in outer and remote areas (with lower land values), which, at the same time increases transport demand and is therefore in clear contradiction to the basic principles of sustainable urban development.

To resolve this contradiction there are different possible forms of public intervention. One possibility is to ensure the affordability of new housing which is built according to the principles of sustainability. One method to achieve this is to prescribe by law or by conditions placed on the approval of building permission the minimum share of social (affordable) housing in all new developments.

**Fig. 14.4** Montpellier social housing in city centre. In this model (mock-up) new houses can be seen in Montpellier: one of the buildings is social housing, the other private, the third university dormitory—from outside no difference can be seen, which is which...

In France all new housing developments have to include at least 20 % social housing. Municipalities which do not comply with this rule have to pay a penalty to the regional government, which is then used to support social housing in other parts of the region (Fig. 14.4). Similar regulations are in force or under discussions in other countries (Germany, UK).

A second possibility is to ensure the sustainability of new housing built according to the principles of affordability. This can be achieved through creating links between new social housing and spatial planning decisions (e.g. to ensure that new social housing is close to public transport, as has been attempted in France), through public interventions into the land market, such as land banking and by selecting investors for the construction of new housing on land made available by the municipality by the criteria that in the new housing project both sustainability and affordability principles have to be applied.

European countries apply different models in the location of decision making power in housing issues. In the majority of countries housing matters are the exclusive right of local municipalities. In some countries, however, housing matters are transferred in whole or in part to a supra-local level. The UK and Netherlands are just two examples where decisions about the number of units to be built are taken "above" the municipalities.

A further way to influence the housing market is to issue compulsory guidelines about the spatial allocation of new housing construction. In the NW region of England, according to the NW regional spatial strategy 60 % of all housing has to be on brownfield land (in Greater Manchester this share rises to 80 %).

In the Netherlands prescriptions exist for the share of new housing units which have to be built within the existing borders of developable land of municipalities.

### 14.6.2 Tools for the Public Sector to Steer New Land Developments in the RUR

The public sector has a range of tools available in order to be able to steer the land development process. The point of departure is the demand from the side of the mobile actors (households, businesses, developers) for new development somewhere in the RUR. The municipalities (and/or higher level public actors) have some power to influence the parameters of new developments, which, in turn, have an effect on the decisions of the mobile actors.

The most direct intervention tool of the public sector is land acquisition. A municipality can directly intervene in the land market by buying up land for future development purposes (land banking). If the proportion of the municipality among the different types of owners of developable land is high, then it can affect land prices and can also impose conditions on developers who buy the land for their developments from the municipality. This was the case, for example, in Vienna: the city could initiate large social housing construction on land bought up in the 1920s. Later housing development was also influenced in a less direct way, by selling the land to developers under specific conditions. Several European countries, such as Netherlands, Sweden and Spain actively use land banking strategies (Larsson 2006, p. 57).

More indirect powers of municipalities to influence new large-scale land development consist of regulations applied to new land developments in general, and more detailed rules applied via specific decisions on larger new land developments.

#### 14.6.2.1 Regulations Applied on New Land Developments in General

Jorgensen et al. (2007) provide an overview of tools to manage urban growth. As well as examples from PLUREL, some of the examples quoted below come from the USA, where the strict planning tools used in most European countries (zoning regulations, clear limits of developable land, etc.) are not universally applied. Instead, there are many local attempts to control otherwise very un-regulated urban development and from these examples some interesting and innovative ideas can be found.

European practices are very much concentrating on "... restricting urban growth through land use regulations that hinder or forbid urbanisation in certain parts of the urban fringe or the peri-urban areas ... urban growth boundaries, various forms of zoning, green belt protection, or the like. These strategies are thus connected very strongly to protection of land from urban development rather than attracting urban growth to the more densely built up areas" (Jorgensen et al. 2007).

Examples of tools from the PLUREL case studies include the following. In Montpellier a Green Tax is paid by people building or buying new houses, in order to preserve green or blue areas. In the Greater Manchester area 47 % of land is protected in principle as green belt (no development possible), but exemptions exist (e.g. road developments). Recently a policy change has been suggested from green belt towards ‘eco-belt’ in which eco-friendly activities should be allowed—as UK planning approaches move from direct regulation and fixed zoning towards partnerships, spatial management and flexible frameworks. In Leipzig an inter-communal land balance pool has been established in the Green Ring of Leipzig, in order to make use of the legal obligation for compensation in the process of land use change. In The Hague region some of the profits gained by municipalities from developing a residential area have to be paid into a green fund, ensuring the future of the remaining farmland. The state also bought up land for buffer zones between settlements for landscape protection.

Many of the American attempts can be summarized under the heading of “smart growth techniques” which aim to achieve more sustainable development through a combination of economic incentives and disincentives, rather than strict planning regulations. “Economic incentives comprise several instruments: development impact fees that internalise infrastructure costs via a fee on new development corresponding to the infrastructure services cost, transfer of development rights (TDR) from some areas to others in a sort of quota organisation, and tax reductions in areas that meet certain requirements (density, nearness to public transport) and direct support for denser and socially mixed neighbourhoods as well as for infill and revitalisation” (Jorgensen et al. 2007).

Some special versions of smart growth techniques (applied in some especially innovative US cities) are the following:

- Balance between jobs and homes: larger office developments are required to pay a contribution to the city fund from which affordable housing is being built (Santa Monica, California)
- Physical and social infrastructure requirements: as a result of new private developments (office, retail, housing) the level of public services should not decrease below a certain threshold—if this would be the case, the private developer has to contribute to the development of the public service or has to postpone the development (Boulder, Colorado) until the public sector increases the level of service
- Transport services: as a consequence of large developments, transport possibilities usually deteriorate—in Palm Beach county, California a plan of anticipated future road congestion has to be prepared and if the outcome shows a significant deterioration, the private developer has a choice of contributing to road development or waiting until the public sector is able to do this.

A special version of these techniques is that of Transit Oriented Urban Development (TOD), addressing the relationship between transport infrastructure and urban development. “TOD implies that the link between urban development; functions; and public transportation is designed at the project level or through targeted policies” (Jorgensen et al. 2007).

Among the European versions of the TOD approach, the Dutch ABC-location principle[4] and the idea of VINEX locations[5] for residential development is well known. In Denmark the general principle of proximity to railway stations is applied to the location of intensive land uses. The Montpellier SCoT plan defines different urban densities with special regard to increased urban densification around new tram stops (as part of priority for urban renewal in a strip of 300 m either side of the tram lines).

An important lesson to be learnt from the American examples, that "...local growth management programs without a state-wide framework have negative consequences, such as regional traffic congestion, increased housing affordability problems, income segregation, etc. ... therefore growth management need to be practised at regional or state level" (Jorgensen et al. 2007). This observation is also valid for the more traditional planning tools, such as zoning: restrictive measures only have real effects if introduced over a sufficiently large area, otherwise the development power migrates just outside the restricted area, causing even more trouble.

#### 14.6.2.2 Rules Applied in the Case of Concrete Decisions on Land Development

Local governments have key role in the process of land development, deciding on re-zoning of land and issuing building permissions for planned projects. In both cases the public sector, usually the local government, aims to capture some part of the land value increase which is connected to public action. This is the idea behind value capture financing which aims not only to share the risks and costs of urban development between public and private sectors, but also the rewards. "VCF sees some of the costs associated with making urban development succeed internalised within the balance sheets of the developments themselves" (Huxley 2009, p. 7). Regarding land development this might mean that through a land-related tax,

[4] The ABC location policy (since 1988) is designed to help reduce the growth of car travel. The policy aims to match the mobility needs of businesses and amenities with the accessibility of different locations. The core element of the ABC location policy for companies is the classification of types of locations (good or bad accessibility by public transport and by car) and types of companies (number and mobility needs of workers, visitors' intensity, need for transport of goods). "The policy aims to locate each company on a location with an accessibility profile in accordance with its mobility characteristics" (Martens and Griethuysen 2000).

[5] VINEX locations refer to the Dutch planning policy, introduced in the early 1990s, aiming at locating new residential, work and recreational areas and public facilities in and as close as possible to large and medium-sized cities" (VROM 1993, pp. 6, 7, quoted in Roo 2003, p. 62, where also the critical analysis of this planning policy can be read).

increases in private land values generated by a new public investment are "captured" in all or in part.

The re-zoning of land leads to land value increases which can be very substantial. According to a brochure of the Town and Country Planning Association (TCPA 2007) agricultural land in England had at that time a market value of 8–10 thousand GBP/ha, while this value increased to at least 800,000 GBP/ha when the owner was given planning consent for new housing. At the time of issuing building permission to a specific project another financial issue also emerges: the share of the costs needed for infrastructure development. Given the fact that the building permission contributes to a further increase of land value, the local government might try to place a condition in the building permission on the developer that they contribute to the costs of the necessary infrastructure developments such as schools.

There are very different conditions for negotiations between local governments and private actors regarding value capture and infrastructure contributions across EU countries. Wolff's analysis (2007) shows that at least four actors are in competition with each other to obtain the largest possible share of the value increase created by re-zoning or building permission: the landowner, the developer, the owners of the new homes and the public hand. The latter has in principle several tools to obtain a share of this increase: "having the land owner pay for or take care of some measures, by following an integrated land development procedure, entering into an agreement, levying a tax or obligating the land owner to in kind contribution" (Wolff 2007, p. 4).

These options, however, are not easy to apply and depend on the legal system of the country concerned. The TCPA brochure (2007) describes the four attempts of the UK government to ensure a share for the public sector from land value increases. The first attempt was in 1947 when landowners were obliged to obtain permission from local authorities for any land development. This and the following two government attempts were unsuccessful, the measures introduced by a Labour government usually being withdrawn by the subsequent Conservative government. The last attempt was the 2004 Labour proposal to introduce a Planning Gain Supplement to impose a charge, levy or tax on the increase in the development value of land attributable to the grant of planning consent.

Austria is another country where the capture of the value increase of re-zoned land is difficult. According to a decision of the High Court in Austria it is forbidden to combine public and private law—thus zoning cannot be connected to taxation. Therefore the only possibility for the public actors is to be more active in the land market, as the public sector can only get back money if participating in the process when acting as a development company.

In the Netherlands, after long debate, the outcome is a new regulation which allows the local municipality to capture some part of the land value increase—not as a "profit tax", however, but in a regulated process to force the landowner to pay for quite broadly defined costs the public sector incurs in connection with the development. Because of the many possibilities left open by the law, the system allows more than simple cost recovery. Conversely, the process is made transparent, insofar as the public costs have to be reported and proved that they are used for the area and improve spatial quality.

In Hungary the legal possibilities for value increase taxation are absent. Recently a flexible tool has been introduced into the building law, that of "development agreement". This allows a local government to negotiate with the developer over the financial conditions for new developments, in the cases where a municipal decision is needed because of re-zoning of land or of giving building permission for the project. Even if such development agreements clarify the situation to some extent to, they do not compensate for the lack of value-increase taxation. Neither is the problem of a lack of territorial cooperation between municipalities solved. The development agreements are purposely very flexible, there being no prescribed content. As a consequence, in Hungary a competition situation emerges between the municipalities as to who offer better conditions for the developer. From this it logically follows that developers will determine the place and the sequencing of their developments according to the financial criteria the different municipalities insist upon. In this way the financial conditions and not the substance of the development determines the location decision. As a further problem municipalities usually do not have sufficient trained officers to negotiate good agreements with the developers—and also the danger of corruption is high.

The contractual relationship between the local municipality, the landowner and the developer is broadly applied in European countries. The content of the development agreements differ across countries and besides the economic and financial conditions, in many cases the timing is also specified—in some countries after a given number of years the land can be expropriated by the municipality at pre-development value if the developer does not perform its duties (Larsson 2006, p. 59).

The British system, being grounded in common law, is based on cases, not rules. Although the size of local governments is large (not fragmented, as in France or Hungary), they are not independent in their decisions, the influence of the central government being substantial, exercised through the development agencies. The procedure can be highlighted by the example of the Royal Woolwich Arsenal, which lies in one of the poorest parts of East London. The task of development of this former military area was given to the London Development Agency. After a general plan was passed by Greenwich Council, the main developer was selected by competitive bid. The developer had to get the building permissions for each block separately, until which the value of the plot had not been decided. Thirty percent of the new flats have to be affordable, the costs of social infrastructure and the preservation of heritage values also have to be covered by the developer. As a result, in the final evaluation, the price of the plot may even be zero, if the developer has many public tasks to perform. However, if a profit is achieved (above a given limit) this has to be shared with the local government.

### 14.6.3 PLUREL Case Study Examples

The PLUREL case study regions offer some interesting examples of financial policies and regulatory tools:

- The Hague region introduced an area fund from which payments are given to farmers for their "services". In this way the green-blue services are strengthened with the contributions of those members of the urban populations who enjoy these high quality areas.
- Montpellier agglomeration is a tax-sharing association of 31 municipalities. A new legal tool, the ZAD enables local authorities to prevent land speculation, as the future development areas of the SCoT are put under price control. The precise delimitation of development areas gives clear signals to land owners and developers.

The more detailed comparison of case study regions has shown that in order to ensure public control over land-use change, important aspects of the processes have to be well regulated:

- The value of land before and after development has to be known.
- Transparent procedures have be in place for issuing permissions, sharing benefits (value increases) and sharing costs (infrastructure development).

In most European countries the developers are gaining in importance and the public hand is losing opportunities to fight for sustainable development—opportunistic planning is gaining ground as job creation becomes more and more politically important.

## 14.7 Informal Structures: Governance Procedures, Actors and Their Interrelations

The analysis of the governance aspects in this section is not as detailed as that of the more 'hard' factors of peri-urban processes, discussed in the previous sections. The main reason for that is that the governance aspects were analysed thoroughly in Chap. 13, focusing on 'Strategies' as discourses around which coalitions can be built in PLUREL case study regions.

### 14.7.1 Governance Relations and Their Potential Role in Public Control Over Land Use Changes

The previous sections of this chapter have shown that the 'hard' elements of territorial systems, such as the government and planning framework and the financial, sectoral and regulatory factors show large variations across different EU countries. Now we turn to the more ‘soft’ elements and aim to explore the role that these (governance) elements can play from the perspective of public control over land-use processes: can they contribute to the strengthening of the control or just the opposite, do they weaken the potential control based on the hard elements? Governance practices involve vertical and/or horizontal links between government levels and non governmental units in both formal and informal ways, influencing the functioning of the RUR and influencing the strength of public sector control over market processes.

It is assumed that informal structures, governance and participation factors can modify the strength of public control over market processes in both directions: making it stronger or weaker. Strong governance relations and practices (cooperation, win-win strategies, etc.) can result in greater public control over the market as far as the institutional/planning framework in the given country and region will allow. We will illustrate this point using the example of the case study region of Leipzig-Halle (further examples could have been taken from The Hague or Manchester).

The alternative view is that the procedures of public participation, the functioning of pressure groups and lobbying might lead to bottom-up resistance against planned public policies or to pushing through some changes against the will of politicians and planning procedures. If these processes become very strong, they can lead to a substantial decrease in public control over market processes involved in land use changes. Some aspects of such developments will be illustrated by the Warsaw case study.

## *14.7.2 PLUREL Case Study Examples*

### 14.7.2.1 Leipzig-Halle

The special (though not at all uncommon) feature of the Leipzig-Halle case is that the functional area of Leipzig and Halle straddles the border of two federal states, Sachsen and Sachsen-Anhalt (Philipp 2009[6]). This fact made the German post-reunification development of public policies even more difficult. Private development without regional planning in the first 5 years after reunification proved to be disaster. Large and territorially uncontrolled subsidies given for green-field developments (retail, industry and housing in peri-urban areas) has led to the fast development of sprawl, some communities growing quickly from 500 to 5,000 people without the necessary infrastructure being put in place.

---

[6] Notes taken by the author from a presentation given by N. Andrea Philipp, ICLEI Europe, at the DG Regio, 2009: EC-DG Regio Urban-rural conference, January 2009.

Since the middle of the 1990s spatial planning has been strengthened and the Saxon Metropolitan Area (Dresden-Leipzig-Halle) has been established. The hierarchy of centres around Leipzig has been established. 'Green Ring Leipzig', a voluntary association, has been formed by 13 municipalities to enable joint coordination and alignment of regional development, the main aim being to preserve the cultural landscape and biodiversity while highlighting the values and attractiveness of the area.

New strategies have been designed against sprawl, based on the rehabilitation of the inner city while designating the urban fringe for recreation and concentrating development on existing land uses in peri-urban areas. The New Leipzig Lakeland idea has been developed—based on the creation of 90 new lakes on former opencast brown coalfields, 190 sq. km in area, to be ready by 2050 and connected to Leipzig city centre by waterways.

In the opinion of the local politicians, Greenland was a positive idea dating from 1996, which attracted cooperation by neighbouring municipalities. Lakeland became an even bigger idea, turning the whole area into something qualitatively different. The cooperation which started in landscape planning turned into cooperation regarding industrial development: small municipalities joined their efforts to offer place for large new industries (e.g. Porsche, TNT and DHL) around the airport.

In the Leipzig-Halle case strong regional planning was the initial step of key importance. This was extended by voluntary cooperation—first in landscape planning, later expanding to other, more difficult areas of development. The result is an increased readiness for shared land-use planning to respond demographic changes and applying a coordinated mix of instruments for urban-rural linkages.

The difficulties of strengthening multi-level governance across state borders are still substantial. Some results, however, can be achieved through development strategies which obtain funding and require voluntary cooperation.

#### 14.7.2.2 Warsaw

In Poland, similar to the other post-socialist countries, local municipalities obtained a large degree of independence in the early free market period of development. Though Poland was more successful than the other countries in developing a strong, directly elected middle tier of government (16 regions), by the end of the 1990s when this happened, local governments had already became very strong and it would have been politically risky to constrain their power. The planning code enacted in 2003 increased liberalism by disqualifying the previously existing hierarchical planning system and providing for the possibility of investor-driven development. As a result, urban sprawl has become very substantial in the metropolitan area of Warsaw; settlements compete with each other, to see who can attract more and bigger developments to their territory (Fig. 14.5).

These preconditions are not very advantageous for any governance ideas aimed at strengthening public control over land use change. The Warsaw case study describes the attempts towards inter-municipal cooperation, which is a relatively weak cooperation model but the only one with any real chance of success, as it

**Fig. 14.5** Warsaw suburban development. The picture shows a congested 2*1 lane road, just outside the city border of Warsaw, heading to new suburban housing, office and shopping development—which are connected to Warsaw only with this narrow, always congested road

would leave the policy domain of local governments intact. In 2002 the Metropolitan Warsaw Association was formed as a result of a grass-root initiative by local leaders from municipalities of the metropolitan area of Warsaw.

Though the city of Warsaw joined the association in 2006, it was never an active supporter of the idea. A nationwide political debate started at the end of the 2000s about the possibility to create metropolitan areas as compulsory or as voluntary settlement associations. However, this issue is politically risky for many politicians regardless of whether they are active on the local, regional or national political scene. Without the active support of Warsaw and the region, the Metropolitan Warsaw Association remained a largely unsuccessful attempt to create cooperation in the peri-urban area, which is still dominated by direct competition among individual municipalities.

### *14.7.3 Summary*

The analysis of governance processes in the case study regions has shown that they can be effective if they are based on a solid formal government and planning system—otherwise they have only very little potential to modify the strength of control over market processes in peri-urban areas. It is highly unlikely that informal policy agreements can be reached without strong backing of the formal government system and the financial and regulatory processes. Thus the correct setup of the formal institutional systems, policies and regulations is of prime importance for resisting urban sprawl and while governance processes can become useful extensions, they cannot substitute for these in the RURs.

Equally, governance procedures and policy arrangements between local governments are of crucial importance for cross-sectoral and cross-territorial integrated development, as these are the most important content-elements of the functioning of the formal government and planning system.

## 14.8 Conclusions

Sustainable urban development requires territorial government for large enough areas which also include the peri-urban zone and extend into the rural hinterland. Multi-functional, integrated development is needed in these territories, led by the public sector, also including all the other important actors in governance arrangements.

Public money is likely to be less available in the future, while the tasks and challenges will be bigger. Thus the public sector has to use all available means—regulatory power, planning tools, financial and indirect tools—to influence the behaviour of market actors to follow more sustainable directions. Municipalities who want to exert more control over land-use change processes have to:

- Fight for supra-local level compulsory settlement associations and/or compulsory comprehensive integrated planning in the functional rural–urban areas
- Aim to gain control over the financial, sectoral and regulatory tools, in order to decrease their sprawling effects
- Attempt to create informal cooperation, bottom-up partnerships in general or along specific strategies, with all the important actors involved in territorial development.

These options are not mutually exclusive: they can and should be combined with each other. Both the hard (strengthening the role of the province/region, planning control, tax equalization, etc.) and soft tools (e.g. funds for green-blue services) are needed to enable sustainable development, based on plans with regional leadership.

Though the analysis has proved the importance of all three types of factors listed above, results show that these are not of equal importance: first and foremost are the "hard factors" (government and planning systems, financial regulations, sectoral policies, land use regulations) which should be strengthened. In the lack of strong institutions, planning processes and regulations, informal governance attempts will not be able to solve all major problems: voluntary governance procedures are not enough on their own to achieve systemic results. In the case of strong formal institutional and planning frameworks, however, the role of governance relationships could be bigger so as to achieve changes towards more sustainable development and thus avoiding unnecessary urban sprawl.

The government and governance analysis explored two extreme types of countries from the perspective of public control over urban sprawl processes in peri-urban areas. There are a few countries (among the case study countries these include the Netherlands, the UK and France) where the potentially strong control assured by the formal government system and planning policies over RUR processes seems to be underpinned by financial, taxation and sectoral policy and regulatory tools which are used in practice. These countries have good opportunities to control peri-urban development and to avoid urban sprawl—especially if informal governance agreements can also be reached among the partners in this direction.

The other extreme is the group of countries (mainly the new EU member states) where the formal government institutions and planning policies are weak and do not allow for sufficient control over the RUR processes. In these countries neither the financial, taxation and sectoral policies and regulatory tools, nor governance processes assure such control, so that the public sector is weak in all aspects in terms of being able to control peri-urban development. The most likely consequence is extensive urban sprawl.

This PLUREL government and governance analysis has highlighted the importance of effective formal institutions, planning, financial and sectoral policies at the RUR level, especially targeted to the peri-urban areas. In the absence of such formal systems the more informal (governance) processes can achieve only very limited results at controlling urban sprawl. The PLUREL synthesis report *Peri-urbanisation in Europe: Towards a European Policy to sustain Urban-Rural Futures* (Piorr et al. 2011) presents some hypothetical models as to how the European level could contribute to the strengthening of the basic formal elements of public control over peri-urban processes.

The author wishes to thank Antal Gertheis and Konstantinos Lalenis for joint work on the common deliverables which served as background to this paper. Further thanks to Carmen Aalbers, Katarina Ekerberg, to the authors of the PLUREL case studies, to Dimitris Kalergis, Hanna Szemző and Dóra Illés.

## References

Bahl R, Linn J (1992) Urban public finance in developing countries. Oxford University Press, Oxford

Bird RM (1999) Rethinking subnational taxes: a new look at tax assignment. IMF working paper 99/165

de Roo G (2003) Environmental planning in the Netherlands: too good to be true. From command-and-control planning to shared governance. Ashgate, Farnham

ESPON project 3.2 (2006) Spatial scenarios and orientations in relation to the ESDP and cohesion policy. IGEAT—Institut de Gestion de l'Environnement et d'Aménagement du Territoire, Université Libre de Bruxelles (Belgium). http://www.espon.eu/main/Menu_Projects/Menu_ESPON2006Projects/Menu_CoordinatingCrossThematicProjects/scenarios.html. Accessed 14 Mar 2012

European Commission (2009) Focus groups' report on the future of transport, 20 Feb 2009. http://ec.europa.eu/transport/strategies/studies/doc/future_of_transport/2009_the_future_of_transport.pdf. Accessed 14 Mar 2012

Goodspeed TJ (1998) Tax competition, benefit taxes, and fiscal federalism. Natl Tax J 51(3):579–586

Hardin G (1968) The tragedy of the commons. Science 162(3859):1243–1248

House of the Commons (2009) The balance of power: central and local government. House of the commons. Communities and Local Government Committee, London

Huxley J (2009) Value capture finance. Making urban development pay its way. Urban Land Institute, London. http://www.uli.org/ResearchAndPublications/Reports/~/media/Documents/ResearchAndPublications/Urban%20Investment%20Network/Value%20Capture%20Report.ashx. Accessed 14 Mar 2012

Jorgensen G, Nielsen TS, Jensen MJ (2007) Urban growth management. PLUREL D2.4.1, Unpublished report

Kovács G (2009) Local taxes: theory and the practice (manuscript). http://www.nispa.sk/_portal/files/conferences/2009/papers/200901301309430.KovacsPaper.doc. Accessed 14 Mar 2012

Larsson G (2006) Spatial planning systems in western Europe. IOS Press, Lansdale, USA

Martens MJ, Griethuysen Sv (2000) The ABC location policy in the Netherlands. The right business at the right place. TNO Inro, Delft

Montpellier Agglomération (2006) Schéma de cohérence territoriale de l'Agglomération de Montpellier: Rapport de présentation. Projet d'aménagement et de développement durable. Document d'orientations générales. Plans de secteurs, p 211 + ann

Oates WE (1996) Taxation in a federal system: the tax-assignment problem. Public Econ Rev (Taiwan) I:35–60

Piorr A, Ravetz J, Tosics I (eds) (2011) Peri-urbanisation in Europe: towards a European policy to sustain urban-rural futures. Forest & Landscape University of Copenhagen/Academic Books Life Sciences, p 144

TCPA (2007) Planning and land value creation. Town and Country Planning Association. Great Britain. http://www.tcpa.org.uk/downloads/20070531_planningandlandvalue.pdf

Tosics I, Dukes T (2005) Urban development programmes in the context of public administration and urban policy. Tijdschrift voor Economische en Sociale Geografie 96(4):390–408

Tosics I, Szemző H, Illés D, Gertheis A, Lalenis K, Kalergis D (2010) National spatial planning policies and governance typology. Deliverable report 2.2.1 PLUREL. http://www.plurel.net/images/D221.pdf. Accessed 16 Mar 2012

Urbs Pandens (2005) Urban sprawl: European patterns, environmental degradation and sustainable development. Coordinated by Gerhard Petschel-Held, Potsdam Institut für Klimafolgenforschung, Potsdam. Unfortunately, the link to the project's website was disabled during the writing of this book and we were not able to substitute the information presented in the chapter with another reference

VROM (Ministerie van Volkshuisvesting, Ruimtelijke Ordening en Milieubeheer) (1993) Vinex Policy, Den Haag

Wolff H de (2007) The new Dutch Land Development Act as a tool for value capturing. ENHR conference. www.enhr2007rotterdam.nl/documents/W09_paper_Wolff.pdf. Accessed 14 Mar 2012

# Chapter 15
# The Future of the Rural Urban Region

**Kjell Nilsson and Thomas Sick Nielsen**

## 15.1 Introduction

In the Introduction to the book the spectre of urban sprawl was introduced. In subsequent chapters we have presented a detailed analysis of the dynamics and pressures facing the peri-urban zone of rural–urban regions and have examined a number of case studies. These revealed a range of different political, economic, spatial and environmental contexts where different combinations of problems have been addressed using a range of different strategies, some more successful than others. Chapters 13 and 14 attempted to see what lessons could be learned by comparing these case studies and whether there were general aspects which could be extracted from these by forming recommendations for policy makers at a city region level (as opposed to a national or pan-European level).

This final chapter looks at the future of the rural–urban region, and of the peri-urban part of it in particular, in order to try to gauge the likely trends. This appraisal is based on the results of the PLUREL project. The project as a whole started out with a set of scenarios and used these as possible futures against which to judge different aspects of the research undertaken by the different teams. It must be stressed that these scenarios are not predictions of what will happen nor even suggestions as to what might happen—they are just possible imagined futures based on a qualitative interpretation of some of the main drivers affecting the world—economic, demographic, environmental and technological aspects. Given that the PLUREL project was funded under the “Global Change and Ecosystems”

K. Nilsson (✉)
Center for Forest & Landscape, University of Copenhagen Rolighedsvej 23, 1958 Frederiksberg, Denmark
e-mail: kjni@life.ku.dk

T.S. Nielsen
Department of Transport, Technical University of Denmark, Bygningstorvet 116B, 2800 Kgs. Lyngby, Denmark

K. Nilsson et al. (eds.), *Peri-urban futures: Scenarios and models for land use change in Europe*, DOI 10.1007/978-3-642-30529-0_15,

theme of the Sixth European Framework Programme for Research and Technological Development it made sense at the time to align the scenarios with existing work, although only time will tell if these bear any resemblance to the real future as it unfolds or whether other, completely unforeseen circumstances will arise. As it was, the project started before and ended during the global financial crisis which changed the economic landscape dramatically in ways that caught almost everyone by surprise. This means that any work which tries to look ahead and see how the future might be must be taken with a very large pinch of salt. Nevertheless, for planning purposes, some assumptions must necessarily be made. One aspect which is known rather better is the demographic picture because the adults living in 2020 or 2030 have already been born.

How might the future look? In PLUREL four scenarios for future development based on the global scenarios of the Intergovernmental Panel on Climate Change (IPCC), known as the Special Report on Emission Scenarios (SRES) were defined (see Ravetz and Rounsevell 2008) and presented in Chapter 2:

A1 “Hypertech” describes a world of rapid economic growth, the rapid spread of new technologies and declining energy prices.
A2 “Extreme water” describes a ruralised world, economic development is primarily regionally-orientated, and per capita economic growth and technological change are fragmented and slow.
B1 “Peak oil” describes a future resilient society of environmental and social consciousness—a global approach to sustainable development, and a dramatic increase in energy prices.
B2 “Fragmentation” describes a divided world of slow economic growth and a fragmentation of society in terms of age, ethnicity and international distrust.

Scenario A1 suggests that small poly-centric towns would become even more popular and it may lead to increased peri-urbanisation of rural areas, while in B1, most people attempt to return to larger towns and cities since high transport costs will limit commuting distances. In A2 huge sums of money are spent on adaptation to climate change and people are attracted to live in small, self-supporting communities, and in B2 cities become more dispersed as younger immigrants dominate city centres while the older “natives” escape to the outskirts and enclaves outside the city. However, regardless which future scenario we choose, urban expansion will continue by 0.4–0.7 % per year, which is more than 10 times higher than the development of any other comparable land uses, e.g. agriculture, pasture land or forestry (Fig. 15.1).

The pattern of urban growth in Europe was assessed by applying a ‘regional urban growth model’ (RUG) allocating the urbanization pressures to every 1 km grid cell in Europe. According to the model and based on the assumptions contained within it, most of the development is likely to take place in peri-urban areas, i.e. discontinuous built development containing settlements of less than 20,000 inhabitants and with an average density of at least 40 persons per $km^2$ (Loibl and Köstl 2008). Such areas are currently growing four times faster than urban areas, and at a rate which would double their total area of 48,000 $km^2$ in 30–50 years. The highest share of peri-urban areas in Europe is along the

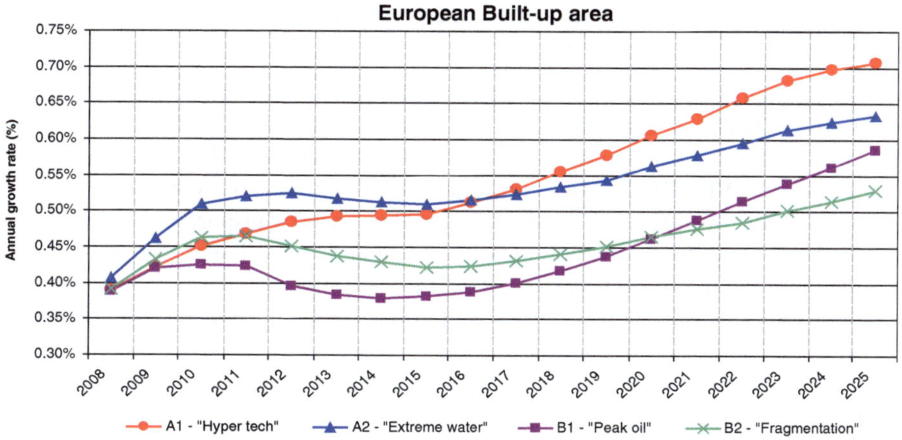


**Fig. 15.1** Development of built-up areas based on the four PLUREL scenarios. The figures are calculated from the NEMESIS econometric model developed by the Research Laboratory ERASME in Paris

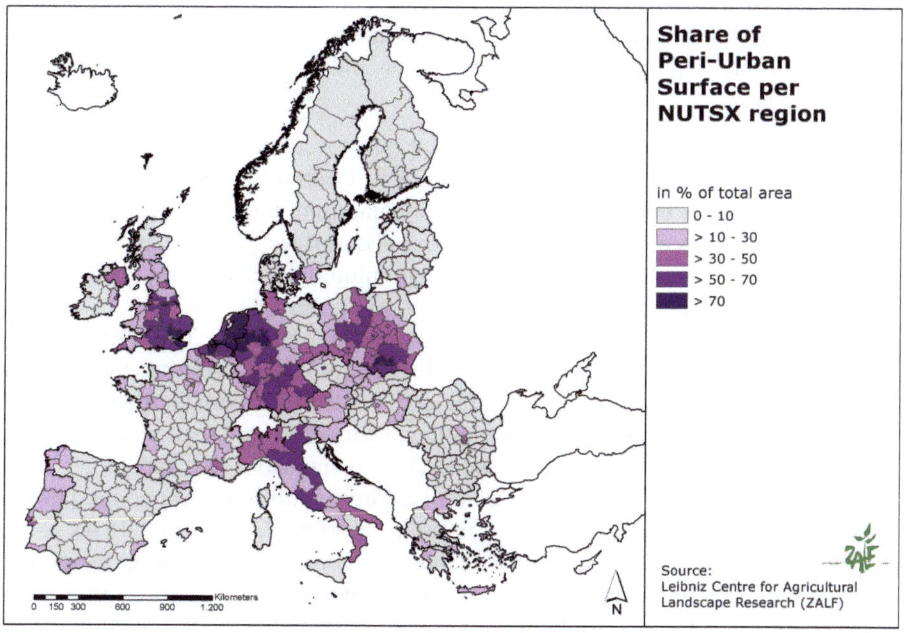


**Fig. 15.2** Peri-urbanisation in Europe. Share of artificial surface per region (Piorr et al., p. 28)

'pentagon' of London—Paris—Frankfurt—Munich—Milan, with the highest concentration in Greater London and the Benelux countries, but also in large parts of Poland and in the Copenhagen/south Swedish Øresund region (Fig. 15.2) (Piorr et al., p. 28).

## 15.2 The Consequences of Urban Sprawl

There are many impacts of such rapid expansion, should it come to pass. In the EC document "Towards a strategy for the urban environment" (CEC 2004) urban sprawl is recognised as the most urgent of urban planning and design issues. The EU project SCATTER (2004) divided the effects of urban sprawl into five groups, namely public and private capital and operating costs, transport and travel costs, land/natural habitat preservation, quality of life and social issues. Sieverts (2003) pointed to three main shortcomings of peri-urbanisation or *Zwischenstadt*, namely that it is inefficient in transport terms, short on aesthetic appeal, and fragmented in political and administrative terms.

Based on the results of the PLUREL project the most important negative consequences emerged as:

- Consumption of land, especially loss of highly productive agricultural land
- Destruction of habitats and fragmentation of landscape structure and ecosystems
- Less open space, longer distances to attractive recreational areas, and less healthy lifestyles
- Increase in the dependency on the private car, in traffic congestion, longer commuting times and distances, and air pollution
- Decay of inner city areas; social segregation and larger gaps between rich and poor areas.

Consumption of agricultural land is expected to continue throughout Europe (Fig. 15.3). In large parts of Scandinavia, the UK, Central Europe and the Mediterranean coastal areas but also in parts of Romania, more than 5 % of the currently used agricultural area will be turned into sealed surfaces (note that in northern Scandinavia and the Baltics the loss of agricultural land is mainly caused by afforestation, and not peri-urbanisation) (Piorr et al. 2011).

Amongst the areas with major agricultural importance, The Netherlands, Belgium and the Mediterranean coast of France look like suffering the greatest and most concentrated loss of agricultural land, while in northern Germany, Poland and Hungary the degree of land consumption occurs in a more scattered pattern. In addition, highly productive soils associated with intensive use and larger farm structure will probably be affected, as well as areas with lower economic performance and a higher share of part-time farming (Piorr et al. 2011, Fig. 15.4).

Effective Mesh Size is a measure of landscape continuity, indicating the probability of individual animals of various species meeting (and breeding) in a landscape fragmented by infrastructure and human settlements (and so continuing to survive as local populations). Landscape fragmentation is concentrated in central Western Europe, where only small patches of open landscapes remain and are likely to become more fragmented (Zasada et al. 2010). With improving quality of life, changing lifestyles and consumption patterns, urban growth is likely to continue, especially in the economic convergence regions of South and central Eastern Europe and in the Iberian Peninsula (Fig. 15.5).

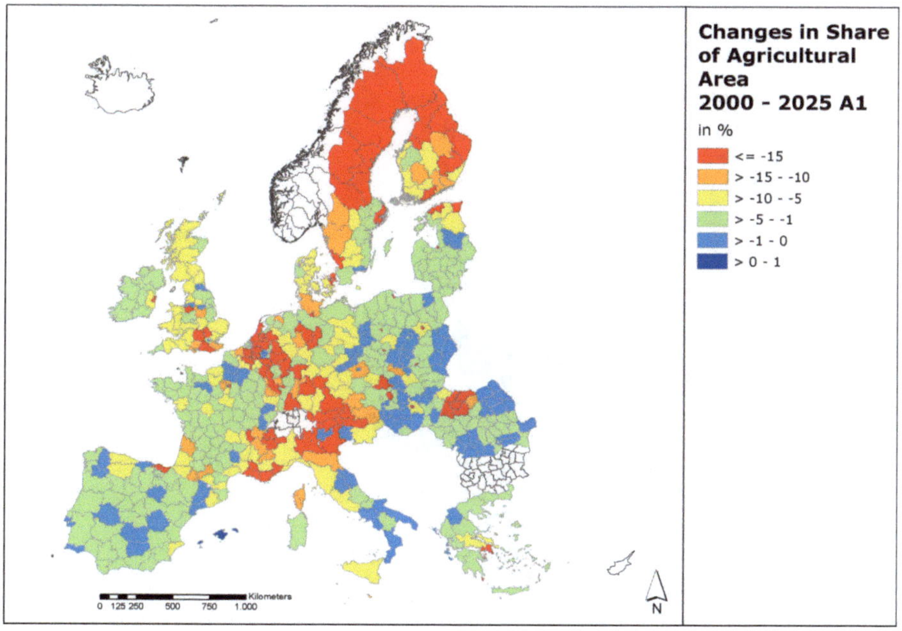


Fig. 15.3 Consumption of agricultural land in Europe 2000–2025

Fig. 15.4 Urban development into agricultural fields of high productivity soils. Jakriborg, Sweden (Photo: Kjell Nilsson) 2000–2025 (Source: Leibniz Centre for Agricultural Landscape Research ZALF)

A similar pattern can be seen when it comes to recreational capacity, as measured by a Green Background Index (Fig. 15.6) (Zasada et al. 2010). A reduction in capacity might have serious implications to people's health and well-being if it means that fresh air and exercise opportunities are less readily available.

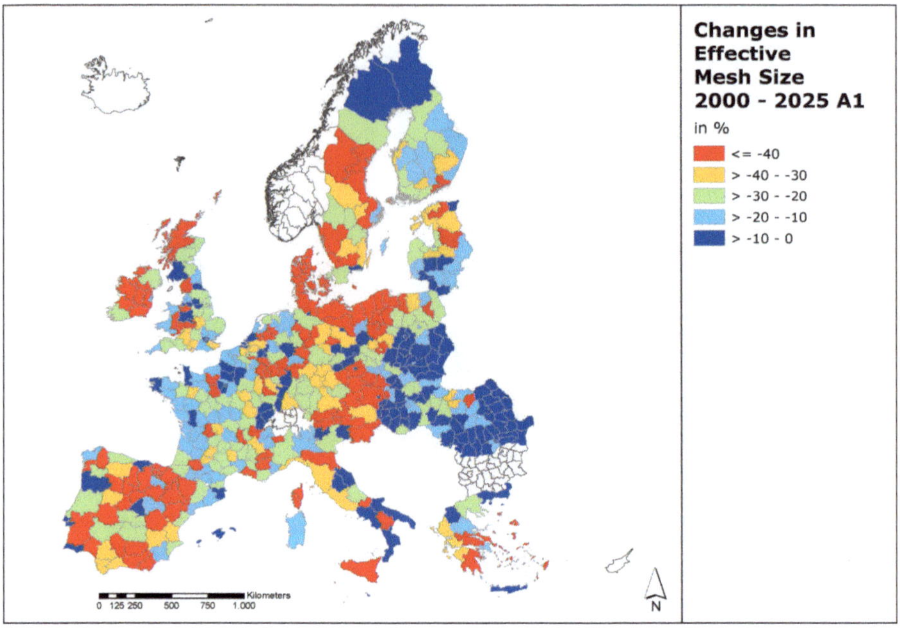


**Fig. 15.5** Landscape fragmentation in Europe 2000–2025 (Source: Leibniz Centre for Agricultural Landscape Research ZALF)

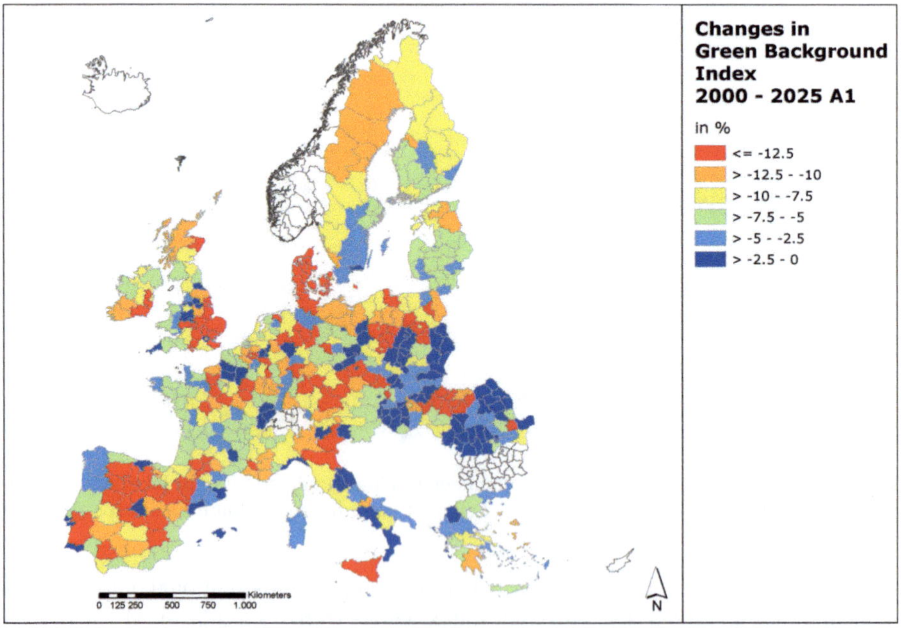


**Fig. 15.6** Recreational capacity in Europe 2000–2025 (Source: Leibniz Centre for Agricultural Landscape Research ZALF)

Society today is faced with increasing incidences of various forms of poor health related to modern lifestyles. Natural elements such as forests, parks, trees and gardens are known to provide opportunities to enhance public health (Nilsson et al. 2011). For example, activities taking place in natural outdoor environments are known to be good for mental and physical health. If more and more people are finding longer and longer distances between their homes and accessible spaces for outdoor recreation and exercise, they are likely to use it less since it is a well-known fact that the distance has a critical effect on the amount and frequency of people's use of green spaces.

The ultimate goal of the transport system is good accessibility to work, shops, cultural and recreation places. The highest levels of accessibility are usually reached in densely built areas where workplaces, shops, services and activity centres are in close proximity. Urban sprawl and low-density peri-urbanisation leads to longer commuting times and, if public transport is not so readily available, encourages car use, which tends to exclude poor and carless people, as well as promoting an unhealthy lifestyle (Frumkin et al. 2004; Helminen et al. 2009). One should also expect some higher particulate matter pollution, although $NO^X$ emissions from road transport are expected to decline despite an increase in traffic, due to the increased use of catalytic converters (Piorr et al. 2011). Despite lower emissions per vehicle, the impact of increased traffic congestion is likely to remain a serious health risk in the future as well as leading to higher economic costs due to time and fuel wastage (Levy et al. 2010).

Finally, at a certain point, an increasing spatial concentration of population and economic activity in metropolitan areas leads to an increase in their functioning costs above the level of total aggregated benefits. When this point is reached, many cities experience a counter-urbanisation, where people move from urban to peri-urban or rural areas in order to achieve better living conditions and quality of life.

In most European countries demographic growth is likely to occur solely as a result of immigration, mainly from outside Europe. These migrants, from low-income countries, tend to be attracted mainly to low status districts in the inner-metropolitan periphery leading to a growing social and spatial segregation. This process may cause a shift in the balance of attractiveness in favour of medium-sized and small towns in regions with lower population density at the cost of large metropolitan areas (Piorr et al. 2011). In the United States, cities like Detroit, Atlanta and Nashville have been named "doughnut-cities" because of expanding peripheral parts while the central parts turn into desolate and empty slums.

However, we have also identified some positive effects of peri-urbanisation, such as:

- Fulfilment of people's living preferences
- Under smart growth: attractive and competitive urban environments
- Proximity to consumers for local producers and greater potential for eco-friendly lifestyles
- More life and more economic development taking place in rural communities lying outside the main urban zones.

The main reason, especially for young families, for moving from urban to peri-urban areas is that they want fresh air, clean water, green surroundings and a safe environment for their children to grow up in. With the development of smart technologies for long-distance working and an improved public transport system, people tend to accept longer commuting distances between their home and the workplace. A related reason (in some countries more than others) is people's desire to live in their own house (as owner occupiers) and the simple fact that house prices generally decrease with growing distance from the urban conglomeration. The life in peri-urban and rural communities also gives their inhabitants opportunities for more environmentally conscious lifestyles, e.g. by growing one's own vegetables, buying meat from local farmers, composting organic wastes, etc. At the same time the newcomers bring new spirit, youth and liveliness to, in some cases, declining, depopulated and sleepy rural villages.

## 15.3 Strategies for Growth Management

How can sprawl be avoided while realising these opportunities? Development in peri-urban areas causes conflicts between urban, agriculture and nature values. It challenges and blurs the distinctions between urban and rural areas as applied in policies and regulations. The seven PLUREL case study regions provide examples of how these conflicts and the pressure towards expansion of peri-urban areas can be strategically managed in different development and regulatory contexts.

The assessment of growth patterns at the pan-European level was supplemented by more detailed modelling at the case study level, developed in cooperation with partners from each region. The modelling approach at the case study level involved an adaptation of the basic scenarios to provide the context of the individual case studies where important and likely aspects of the future, as well as planning responses within the scenario storylines, were elaborated through dialogue between researchers and local stakeholders and practitioners. The modelling was carried out using MOLAND, a cellular automata model which simulates urban development as a growth process (see Chap. 4), where the location of each grid cell, the suitability of the land, zoning regulations, as well as the status of neighbouring cells, determine the probability of land use change.

Adapted case study scenarios were chosen for modelling of land use change up to the year 2025 and an assessment of sustainability was made in parallel with the cruder set of results covering Europe.

The assessment of sustainability relies on a wide set of sustainability indicators as well as a detailed response function approach based e.g. on changes in land use along the rural–urban continuum. The consequences can only be fully exploited in cooperation with practitioners and stakeholders from the regions who can compare the signals coming from the sustainability indicators with the desired outcomes. From this they can subsequently revise strategies in order to aim for better desired outcomes in a future that is uncertain but may be approached through assessment of scenarios. The following sections present sustainability impact assessment results from the PLUREL project as well as the results of cooperative case studies based on detailed modelling at the regional level and analysis of strategies.

To summarize, the following strategies could be identified as important steps towards more sustainable urban–rural futures (Table 15.1 and see also Chap. 13):

1. Better coordination between transport, land use and open space planning

Copenhagen's master plan from 1947, the so-called Finger Plan, is a classic example of combined urban transport, land use and open space planning. Development has taken place along the transport corridors, "the fingers", leaving open space such as urban forest areas (e.g. the Vestskoven) in the wedges between them. An important part of this strategy is the near-to-station-principle, meaning that new workplaces should be located, if possible, close to public transport. Although, the fingers of the hand have become shorter and somewhat "webbed" over the years, the original vision behind the Finger Plan is still the principle for future development of Greater Copenhagen. The plan has an iconic value but remains a relatively detailed and binding plan for urban development zones, infrastructure, location of office space and green space protection (Jørgensen et al. 2010).

The Regional Structure Plan of The Hague Region and the Scheme of Territorial Coherence of Montpellier Agglomeration (SCoT) are examples of advanced modern strategies for better coordination of transport and land use planning (Fig. 15.7). The Regional Structure Plan of the Hague Region shows the main strategies that strengthen the relationship between Spatial Planning and Traffic/Transport, it coordinates sectoral ambitions and goals and is the basis for long-term, politically-supported agreements (Aalbers et al. 2010). In the Scheme of Territorial Coherence of Montpellier Agglomeration, a highly attractive region, there is a political willingness in both speech and action to develop the region into a sustainable dispersion of land use and high quality urban development (Aalbers and Eckerberg 2010).

2. Good governance and integrated policy approaches

"How can you govern a nation that has two hundred and forty-six different kinds of cheese?" complained the French president Charles de Gaulle (Mignon 1962). In most European countries, the formal powers and tasks of the region (in those countries where regions exist as formal political entities) with respect to spatial planning are limited, since the municipalities or their equivalent usually have planning autonomy. Therefore, a key issue is that the rural–urban region is frequently split up among many local government entities, while integrated territorial policy approaches are a prerequisite for better coordination.

Good governance, i.e. the sphere of public debate, partnerships, interaction and dialogue between citizens, organizations and local governments, is a precondition for achieving sustainable development (Evans et al. 2005). As examples of successful governance attempts in carrying out an overall strategy for protecting agriculture and green space in the urban fringe, the Regional Structure Plan of The Hague Region and the Scheme of Territorial Cohesion of Montpellier Agglomération have already been mentioned. A contrasting and unsuccessful case is the

**Table 15.1** Overview of strategies used by the case studies

| | Leipzig-Halle | Haaglanden | Montpellier | Manchester | Warsaw | Koper | Hangzhou |
|---|---|---|---|---|---|---|---|
| 1. Better coordination of land use, transport and open space | | Regional Structure Plan | Scheme de Coherence Territorial (SCoT) | | Warsaw Metropolitan Area Spatial Development Plan | | Binjiang development—combined housing with an effective transport network |
| 2. Good governance and integrated policy | Parthe Floodplain cooperation | Regional Structure Plan | Scheme de Coherence Territorial (SCoT) | Pennine Prospects | Mazowieckie Vovoidship Development | | |
| 3. Urban containment | Green Ring of Leipzig | Green Heart | Zoning | Green Belt policy | Zoning | | Zhuantang – combined urban development with protection |
| 4. Green compact cities | A variety of "Innenstadt" projects | 80 % "infill policy" – Green Metropolis by the Seaside | High density – High density | | | | |
| 5. Preservation of green infrastructure | Saxon Green Corridors | | | Green-Blue Infrastructure – Red Rose Community Forest | | Skocjanski zatok Nature Reserve | Xixi National Wetland Park |
| 6. Local production and short circuits | | | Longterm agricultural status – Agriparks | Incredibla Edible Todmorden | | Koper strategy for protection of agricultural land | |
| 7. Ecosystem services | | Green and blue services | | | | | |

**Fig. 15.7** Providing a good and efficient public transport system, as in the Montpellier Agglomeration, is a key component in a strategy for sustainable urban development (Photo: Kjell Nilsson)

situation of Warsaw, which is characterized by severe pressure on high-value nature and agricultural areas due to high growth rates and uncontrolled peri-urbanisation both in the area directly surrounding the city and in the Mazovian region as a whole. The attempts by the regional authorities to overcome this, as identified in existing strategies and plans, do not provide sufficient basis for sustainable development. The policies tend to be focused on specific problems, mainly of an economic nature, and the local dimension is superior to overall coordination and cooperation with neighbour communities (Groschowski and Pieniazek 2010). Not only planning and environmental, but also financial and sectoral policies contribute to urban sprawl in Warsaw.

Of course, coordination of the wills of 72 independent municipalities comprising the Warsaw Metropolitan Area is far more difficult than dealing with the nine of The Hague Region although The Hague Region itself is part of larger agglomerations such as the province of South Holland in the Randstad. Furthermore, Montpellier has succeeded in joining the ambitions of, for the time being, 31 municipalities in a strategy for more equally distributed development.

3. Urban containment by conservation and densification

Urban containment, or a clear spatial boundary between urban and rural land, is an important instrument of use in growth management for sustainable development. A clear physical delineation between the urban and the rural does not mean that urban and rural development issues should be treated separately. On the contrary, we need coordinated and integrated approaches involving all relevant planning authorities—this is especially important in countries with many small independent municipalities and a weak regional level—as well as including stakeholders and their organizations. Strong planning legislation as well as an open governance process are key factors in such an approach. Green Belts in the UK and the Green Heart in the Netherlands are policies that have proven to be successful in quantitative terms and for many years or decades, but we can also find negative consequences as observed in the Greater Manchester Area. A certain proportion of the development which would have taken place in the green belt is merely pushed further out, leapfrogging the belt itself, and leading to a greater impact on the countryside and longer travel distances, an impact that may become more acute as people are increasingly prepared to accept longer commuting distances and with the growth in tele-working (Ravetz 2008).

Urban densification is another common strategy for avoiding urban sprawl. An international survey of transport practices and urban structures showed that American cities are much more dispersed than European and Asian cities, and that their energy consumption for transport is significantly higher (Newman and Kenworthy 1999). The EC Thematic Strategy on the Urban Environment (CEC 2006) recommends more compact settlements, as well as better coordination between urban transport and land use planning. This has been taken seriously by The Hague Region—80 % of all urban development must take place within the existing urban fabric. However, quality of life in the city can suffer as a result of urban densification unless this is carried out properly with good urban and landscape design. For instance, densification at the cost of parks and other public open spaces is a risk. Therefore, in the Regional Structure Plan of The Hague Region, multiple functions and high quality public space, as well as integrated poly-centric development, e.g. limited sprawl in an organized way, are named as strategies to maintain quality of life in dense areas (Westerink et al. 2010).

A dense city with a lot of sealed surface and lack of green space is also more vulnerable to local climatic extremes, exacerbated by the Urban Heat Island effect. For example, increasing areas of sealed surfaces lead to bigger problems with flooding since there are fewer areas left over for the water to infiltrate into. Absence of shade or shelter also causes local discomfort to residents in many cities which are too dense and insufficiently green and increases the need for air conditioning which in turn adds heat to the local atmosphere.

The issue of urban density versus green space is an example of the potential conflict between mitigation and adaptation concerns (Carter 2008). If increasing density, in order to reduce energy by lowering travel demand and heating requirements, leads to the loss of green space, one consequence will be the loss of a vital adaptation resource. Results from a study in the city of Manchester suggest that an increase of green space of 10 % in residential areas may compensate for even the worst case/least likely temperature scenario by 2080 (Gill et al. 2007).

4. Development of a compact garden city with attractive inner-city areas

A critical discussion of the simplistic point of view promoted by European Commission policy on urban planning and sustainability is thus essential (Porter and De Roo 2007). To prevent residents from leaving compact cities with lower quality of life in search for better living environments in the urban fringe, new concepts have to be developed that makes the compact city more attractive. Is a

**Fig. 15.8** Decay of inner-city areas in Leipzig, Germany (Photo: Kjell Nilsson)

**Fig. 15.9** Urban renewal in the inner-city of Leipzig (Photo: Kjell Nilsson)

compact city with lots of green an oxymoron? The answer could be the "compact garden city". The concept of a green metropolis by the seaside in The Hague and a variety of urban renewal and social regeneration projects in Leipzig are examples of such ambitions. In Leipzig, due to outmigration and suburbanisation, there is a variety of vacant buildings and brownfield sites, so the urban core can be described as perforated rather than compact (Fig. 15.8). The overall strategy is to counter suburbanisation processes by enhancement of the city and improving quality of life in order to retain residents or to attract others to return (Sinn et al. 2008) (Fig. 15.9).

**Fig. 15.10** Tapiola (FI) is a good example of Scandinavian modernist building and planning tradition (Photo: Thomas Sick Nielsen)

The idea of the garden city originates from a utopian vision by the British writer Ebenezer Howard (1902) which initiated the garden city movement in Great Britain. A common misunderstanding is that Howard advocated low-density development, but according to Peter Hall (2002) his garden city would have had densities similar to inner-city London. Though the idea of the garden city comes from Britain, Beatley (2000) concludes in his book Green Urbanism that American cities can learn about sustainability practices from the Netherlands, Germany, and the Scandinavian countries. In particular, the Scandinavian planning and building tradition includes many good examples, e.g. Fredensborg Houses, an estate of low rise and dense housing north of Copenhagen built in 1962–1963, and Tapiola, a district of the Finnish town of Espoo, largely built in the 1950s and 1960s by the Finnish apartment foundation and designed as a garden city (Fig. 15.10). There are also newly developed, highly attractive residential areas in the inner-city with a garden profile, e.g. Pilestredet Park in Oslo, a former hospital site transformed into

**Fig. 15.11** Pilestrædet Park in Oslo (NO) is an example of a newly developed, highly attractive residential area with a garden profile (Photo: Kjell Nilsson)

an eco-friendly residential area in 2000 (Fig. 15.11), and the Western Harbour, Malmö, Sweden, built in 2001 (Figs. 15.12 and 15.13).

5. Preservation and development of blue and green infrastructure for biodiversity and health-promoting transport systems

Another way of integrating transport and open space planning is to develop the "green and blue" (vegetation and water) corridors for energy-saving means of transport such as walking and cycling, as well as for biodiversity and human health and well-being. The Green Belt of Leipzig, involving 13 municipalities in a spatial, environmental and recreational strategy, and the Red Rose Forest covering the six western districts of Greater Manchester are examples of this. The Red Rose Forest, one of a national set of community forests created in the early 1990s, forms, together with the adjacent Mersey Forest and the Pennine Edge Forest, an extended network of community woodland, often planted over mining spoil and landfill sites,

**Fig. 15.12** Western Harbour in Malmoe (SE) is another example of high quality ecological housing in Scandinavia (Photo: Kjell Nilsson)

**Fig. 15.13** Local infiltration of rainwater is, as well as green roofs, important elements in the ecological concept of the Western Harbour area in Malmoe (Photo: Kjell Nilsson)

where much of the previous contamination and dereliction has been greened and made safe and usable (Ravetz 2008).

Occasionally the peri-urban zone contains areas of high natural values, which thanks to their location give the urban population easy access to areas of more natural quality. An example is Skocjanski zatok Nature Reserve, the largest brackish wetland in Slovenia, in the urban fringe of Koper. The main attribute of the nature reserve is its rich flora and fauna, including a number of endangered species. The area was severely damaged in the 1980s, when the local authorities planned to infill the lagoon in order to develop the area as an industrial and commercial zone. In 1993, the NGO BirdLife Slovenia initiated a public campaign to protect the area,

**Fig. 15.14** Two Camargue horses in Skocjanski Nature Reserve (Photo: Kjell Nilsson)

**Fig. 15.15** Xixi National Wetland Park (Photo: Kjell Nilsson)

and after 5 years of persistent work the area was officially declared a nature reserve by the Slovene government. BirdLife Slovenia has obtained a license to manage the area, which is now open for the public to observe the birds and other forms of wildlife (Pintar et al. 2008, Fig. 15.14).

The Xixi area, situated northwest of the West Lake District in Hangzhou, China, used to be a plain with a large network of ponds and rivers. The area, known for its beauty and rich ecology, was dominated by farmland and fishing with only few dispersed settlements (Fig. 15.15). Despite its proximity to central Hangzhou, hardly any urbanisation took place before 1990, but in 1996 Jiangcun Village was incorporated in the West Lake District of Hangzhou and a large-scale urbanisation process started. In less than 10 years, the area had become one of the largest residential districts in Hangzhou, while the wetland area had decreased from 60 to 10 $km^2$. At the same time, the city authorities experienced a growing demand for new recreational areas due to the expansion of the city westwards. They reacted promptly, in November 2001, by approving the planning programme for "Xixi Wetland Cultural and Ecological Tourism Area" and in 2004 the Xixi Wetland Reserve Master Plan was approved (Jiangjun et al. 2008).

A fairly new initiative to preserve a green infrastructure is the establishment of National Urban Parks. The first NUP, the Royal National Park or the Eco Park in Stockholm, was protected in 1994, but it is in Finland that the idea has been most fruitful, and where the green infrastructure in towns like Hämeenlinna, Heinola, Pori and Hanko all have been awarded NUP status. An important criterion for the selection of a NUP is that the green areas should be connected to each other and to the surrounding landscape so that it is possible to walk or cycle through them from the countryside on one side of the city via the centre and out into the countryside on the other.

6. Promotion of local production and protection of agricultural land

In the old days—as in developing countries today—there were short circles in the flow of resources between towns and the surrounding countryside. In a more sustainable future urban–rural relationship we need a form of urbanism where recycling and circular resource flows are re-established in water and waste management, food and energy production, supply of raw materials, etc. The PLUREL case studies present examples of how this goal can be achieved through different ways of supporting local food production.

One example of an advanced sectoral strategy to protect the best quality agricultural land has been developed in the case study of Koper in Slovenia. Agriculture was one of the most important activities in the Koper region in the past, but today it has lost much of its economic importance. A major problem is the number of small and deteriorated parcels of land and an ageing farmer population. Another threat is urban sprawl on high quality agricultural land. However, analyses show that there are good opportunities for further development of agriculture in the region, mostly in connection with tourism and recreation, but also for local consumption. Therefore, spatial planning with a new approach to developing a classification procedure of the agricultural land is under development that includes soil characteristics, local meteorological conditions, possibilities for carrying out economically and environmentally sound production, and isolation from pollution sources (Pintar et al. 2010).

Montpellier Agglomeration also has an active policy for protecting agricultural land use in the urban fringe. As well as land use zoning, land price regulation and stimulation of shorter product chains from farmer to consumer, an "agri-park" is being developed in Nord Lez. The agri-park concept relies on multifunctional land use where recreational and other social functions are integrated with agricultural production as a cost-efficient way of protecting and maintaining open spaces in the urban fringe (Buyck et al. 2008).

The most ambitious initiative is to be found in the market town Todmorden, 20 km north of Manchester, with the aim of making the population self-sufficient in

**Fig. 15.16** The community of Todmorden has the ambition to make itself self-sufficient in food by 2018 (Photo: Kjell Nilsson)

food by 2018 (Fig. 15.16). The initiative, named "Incredible Edible Todmorden", has a true bottom-up perspective encouraging public and private bodies to help in finding land, removing legal constraints, and supporting local action.

## 15.4 Ecosystem Services

The surrounding landscape provides a lot of other goods beneficial for the urban community. We sometimes call these common goods "Ecosystem Services". Typical ecosystem services include:

| | |
|---|---|
| Biodiversity | Since 70 % of the European population live in urban areas, the experience of nature and wildlife close to where people live is important for the environmental consciousness of future generations. |
| Air quality | Forests and trees clean the air from particles and gaseous pollutants. |
| Water | Protection of groundwater resources and flood risk management. |
| Recreation | Since distance has proven to be the most important single factor for people's recreational use of the outdoor environment everybody should have access to attractive green spaces close to where they live. |

Ecosystem services can also represent considerable economic values. Economists have tried to put monetary values on them. Boman et al. (2010) provide an example of assessment of various values per hectare based on a literature survey. Estimates from different studies show a considerable variation, depending on the

**Fig. 15.17** Green Blue Services is an option for farmers in the Hague Region to gain extra income (Photo: Kjell Nilsson)

exact context in which they are valued. The examples below may give an idea about the values that ecosystem services represent:

| | |
|---|---|
| Biodiversity | €30–90/ha/annum for increased biodiversity in a national park, in Denmark (Jacobsen and Thorsen 2010) |
| Air quality | £900,000/annum for all woodland in UK (Powe and Willis 2004) |
| Ground water | €130/ha/annum (for clean water Hasler et al. 2007); |
| Recreation | €26/ha/annum (Zandersen et al. 2007) |

The possibilities for financial compensation to farmers for providing ecosystem services in the peri-urban areas are noted as important. In The Hague Region farmers are rewarded for their improvement of the landscape, so-called Green and Blue Services (Fig. 15.17). The system was made possible by a fund that was created by a number of municipalities from their revenues from housing projects. The Green Blue Service strategy provides an alternative to land purchase since the farmers gain extra income. The strategy is tailored to farmers' entrepreneurship. They negotiate the design, measures and the price. The positive aspects include the fact that while the strategy contributes to maintaining agriculture it also contributes to biodiversity, recreation and tourism. There is an interest from both farmers and the authorities to develop the model further, as well as to supply the green funds with money, but the strategy needs complementary measures, such as land banking to make a real difference (Aalbers and Eckerberg 2010). The freezing of land use status for 15 years combined with support to farming business as implemented in the territorial cohesion plan (SCoT) of Montpellier Agglomeration is another example of such measures.

## 15.5 Future Perspectives

Batty (2008) argues that cities mainly grew from the bottom up, where economy and transport behaviour are the key elements, while spatial planning and urban design plays a subordinate role. In the PLUREL case studies we have seen several successful bottom-up initiatives for the urban rural interface, e.g. the preservation of Skocjanski zatok Nature Reserve in Koper and the Incredible Edible movement in Todmorden. We have also experienced exemplary models for good governance as the Regional Structure Plan of The Hague Region and the Scheme of Territorial Coherence of Montpellier Agglomeration. But we can also see the need for strong government in the shape of legislation and an efficient spatial planning system. In this sense the regional perspective and the rural–urban region are the most strategic levels, but we can also see the need for initiatives from the EU at the pan-European level. Otherwise, we will see growing gaps between different parts—north/south, east/west, rich/poor—instead of a more sustainable and inclusive Europe.

The ability/power of the public hand to resist the push of market actors towards more urban sprawl are depending on two things: the strength of the planning instruments and level in the governmental system at which land use decisions are taken (see Chap. 14). The more decentralised the system and laissez-faire orientated the policy, the weaker is the steering potential and vice versa. The PLUREL case studies represent different spatial planning trends and traditions (Knieling and Othengrafen 2009), and different legal and administrative planning families (Newman and Thornley 1996). Manchester represents the British planning culture with a land use management approach, Montpellier the Napoleonic and regional economic planning approach, and Leipzig the Germanic planning culture with a comprehensive integrated approach. Warsaw and Koper represent the Eastern European tradition with its transformation processes, while The Hague belongs to the Napoleonic planning culture with a comprehensive integrated approach.

However, land use change also depends on other kinds of drivers such as economic and taxation systems, transport, housing and other sectoral policies. A comparison of the case study regions' potential ability to control peri-urban development shows significant differences (Tosics and Gesthesis 2010 and Chap. 14, Table 15.2). The Hague Region has the greatest potential to control urban development, followed by Manchester and Montpellier, while the Eastern European city regions such as Warsaw and Koper have changed from a centralised planning system (seen as very negative and belonging to the discredited Communist system) to a more liberal regime where the market has a freer hand. We have also identified the need for an integration of spatial planning and territorial cohesion policies with financial and taxation mechanisms, which often also create direct or indirect incentives for urban sprawl, and that more tools should be given to the public sector to use value capturing more efficiently.

A better balanced and more sustainable development requires more policy attention to the urban–rural interface at the regional level, but there is also need for a new agenda for EU policies and funds with linkage to spatial development.

**Table 15.2** Strength of the public sector to control urban development (Tosics and Gesthesis 2010)

| | Haaglanden | Manchester | Montpellier | Leipzig | Koper | Warsaw |
|---|---|---|---|---|---|---|
| Financial transfer system | 0.67 | 1.00 | 0.67 | 0.67 | 0.00 | 0.33 |
| The local taxation system | 0.67 | 0.67 | 0.67 | 0.33 | 1.00 | 0.67 |
| Local government financing systems | 0.67 | 0.83 | 0.67 | 0.50 | 0.50 | 0.50 |
| Economic development and infrastructure | 1.00 | 0.50 | 0.50 | 1.00 | 0.50 | 1.00 |
| Transport | 0.83 | 0.67 | 0.50 | 0.50 | 0.17 | 0.33 |
| Housing | 1.00 | 0.67 | 0.67 | 0.33 | 0.33 | 0.33 |
| Sectoral policies | 0.94 | 0.61 | 0.56 | 0.61 | 0.33 | 0.56 |
| Tools to steer development | 0.75 | 0.50 | 0.50 | 0.75 | 1.00 | 0.25 |
| Summary | 6.53 | 5.45 | 4.74 | 4.69 | 3.83 | 3.97 |

When it comes to policy-making, the time has come to challenge the historic distinction between urban and rural issues. Instead, we need a more holistic, territorially oriented perspective to shape future EU agricultural and structural policies. Integrated urban–rural development should be accepted as a general requirement and put as a condition for EU support from Structural and Cohesion Funds and the CAP (Common Agricultural Policy). The Green Paper on Territorial Cohesion (CEC 2008) is a step forward in this direction.

Research has an important role to play in supporting sustainable urban–rural relationships, which was the main point of PLUREL. We have also identified some knowledge gaps we were unable to fill. First of all, the effects of different decision-making mechanisms need more attention. We are still too uncertain as to how effective different kinds of steering mechanisms, e.g. planning legislation, taxation rules, housing policies etc., are when it comes to land use changes. Secondly, knowledge about the efficiency of different transport systems in relation to the form of urban agglomeration still is very elementary. Thirdly, the vision of the Green Compact City needs a better knowledge base when it comes to the optimal balance between urban densification and green infrastructure. Fourthly, forestry and agriculture will have a new and important role to play in the future urbanised society. Research should therefore generate knowledge that will improve the planning and management of forests as well as how to integrate local food production into urban settings better. Fifthly, ecosystem functions and services provided by the green infrastructure and their potential for adaptation to climate change is a topic that needs to be further developed. An important part of this is to develop good methods for calculating and exploring the monetary value of these goods and services.

Finally, urbanisation is a global phenomenon—the World Resources Institute (2001) estimated that urban areas in developing countries will account for nearly 90 % of projected world population increases between 1995 and 2030. Therefore, managing urban population change will be one of the most important challenges over the next few decades. In developed countries, the urban future will involve dealing with complex changes in the composition of urban populations and containing urban sprawl beyond the suburbs to retain the critical ecosystem services that will sustain population growth, although this is expected to be modest and to decline in places. In developing countries, where 80 % of the world's population resides, central issues will be how to cope with an unprecedented increase in the number of people living in urban areas, and with the growing concentration of these urbanites in large cities with millions of residents and declining availability of natural resources. Urban sprawl is a small problem in Europe compared to the size of the problem in developing countries. Therefore, we should do everything we can to support these countries with knowledge and experiences accumulated during our urbanization process in the twentieth century.

## References


Aalbers C, Eckerberg K (2010) Governance patterns and performance of regional strategies in peri-urban areas. Comparative analysis of seven case studies in Europe and China. PLUREL Deliverable 3.3.8

Aalbers C, van Dijk T, van der Jagt PD, Wetserink J, Rodenberg J, Velinova T (2010) Analysis of regional spatial planning and decision making strategies and their impact on land use in the urban fringe. The case study of the Hague Region, The Netherlands. PLUREL Deliverable 3.3.1

Batty M (2008) The size, scale and shape of cities. Science 319:769–771

Beatley T (2000) Green urbanism. Learning from European cities. Island Press, Washington, DC/Covelo, CA

Boman M, Jacobsen JB, Strange N, Norman J, Mattson L (2010) Forest amenity values and the rotation age decision: a Nordic perspective. Ecol Bull 53:7–20

Buyck J, Chery JP, Jarrige F (2008) Montpellier case study. Analysis of regional and decision making strategies and their impact on land use in the urban fringe. PLUREL Deliverable 3.3.2

Carter J (2008) Climate change: a rural–urban region perspective. PLUREL Newsletter No. 3, April 2008, Copenhagen, pp 4–5. Online from: http://www.plurel.net/images/PLUREL_Newsletter_No3_screen.pdf. Accessed 14 Mar 2012

CEC (2004) Communication from the Commission to the Council, the European Parliament, the European Economic and Social Committee and the Committee of the Regions towards a thematic strategy on the urban environment. Commission of the European Communities, Brussels

CEC (2006) Communication from the Commission to the Council and the European Parliament on thematic strategy on the urban environment. Commission of the European Communities, Brussels

CEC (2008) The green paper on territorial cohesion. Commission of the European Communities, Brussels

Evans B, Joas M, Sundback S, Theobald K (2005) Governing sustainable cities. Earthscan, London

Frumkin H, Frank L, Jackson R (2004) Urban sprawl and public health. Island Press, Washington

Gill S, Handley J, Ennos R, Pauleit S (2007) Adapting cities for climate change: the role of the green infrastructure. Built Environ 30(1):97–115

Groschowski M, Pieniazek M (2010) Analysis of regional spatial planning and decision making strategies and their impact on land use in the urban fringe. Case study: Warsaw Metropolitan Area, Poland. PLUREL Deliverable 3.3.4

Hall P (2002) Cities of tomorrow, 3rd edn. Blackwell, Malden/Oxford/Victoria
Hasler B, Lundhede TH, Martinsen L (2007) Protection versus purification—assessing the benefits of drinking water quality. Hydrol Res 38(4):373–386
Helminen V, Ristimäki M, Kontio P, Rita H, Vuori M (2009) Response function for commuting. PLUREL Deliverable D2.3.7. Unpublished report
Howard E (1902) Garden cities of tomorrow. Swan Sonnenschein, London
Jacobsen JB, Thorsen BJ (2010) Preferences for site and environmental functions when selecting forthcoming national parks. Ecol Econ 69:1532–1544
Jiangjun Y, Youjoun H, Pauleit S, Spiekermann M, Burkhardt I (2008) Analysis and assessment of regional and local government and planning strategies. Hangzhou case study. PLUREL Deliverable 3.3.7.m. Unpublished report
Jørgensen G, Nielsen TS, Grunfelder J, Jensen JM (2010) Urban growth management. Effectiveness of instruments and policies. PLUREL Deliverable 2.4.1. Unpublished report
Knieling J, Othengrafen F (2009) En route to a theoretical model for comparative research on planning cultures. In: Knieling O (ed) Planning cultures in Europe. Ashgate, Farnham
Levy JI, Buonocore JJ, von Stackelberg K (2010) Evaluation of the public health impacts of traffic congestion: a health risk assessment. Environ Health 2010(9):65
Loibl W, Köstl M (2008) Report on a methodology to delineate RUR sub-regions. PLUREL Deliverable D2.1.4. Unpublished report
Mignon E (1962) Les mots du general. A. Fayard, Paris
Newman P, Kenworthy J (1999) Sustainability and cities: overcoming automobile dependence. Island Press, Washington, DC
Newman P, Thornley A (1996) Urban planning in Europe: International competition, national systems, and planning projects. Routledge, London
Nilsson K, Sangster M, Gallis C, Hartig T, de Vries S, Seeland K, Schipperijn J (2011) Forests, trees and human health. Springer, New York/Dordrecht/Heidelberg/London
Pintar M, Udovic A, Istenic MC, Zupan M, Prus T, Pirnat J, Hladnik D, Glavan M (2008) Analysis of regional spatial planning and decision-making strategies and their impact on land use in the urban fringe—case study Koper. PLUREL Deliverable 3.3.5. Unpublished report
Pintar M, Kralj T, Instenic MC Perpar A (2010) Report on enhanced planning and decision making for urban fringes, including land use scenarios for the case study Koper. PLUREL Deliverable 3.4.7. Unpublished report
Piorr A, Ravetz J, Tosics I (eds) (2011) Peri-urbanisation in Europe: towards a European policy to sustain urban–rural futures. Academic Books Life Sciences, Copenhagen
Porter G, De Roo G (2007) The end has no merit. In: De Roo G, Porter G (eds) Fuzzy planning. The role of actors in a fuzzy governance environment. Ashgate, Farnham, pp 1–18
Powe NA, Willis KG (2004) Mortality and morbidity benefits of air pollution, SO2 and PM10 absorption attributable to woodland in Britain. J Environ Manage 70(2):119–128
Ravetz J (2008) The city-region in time and space. Analysis of regional spatial planning and decision-making strategies, and their impact on land use in the urban fringe. PLUREL Deliverable 3.3.6. Unpublished report
Ravetz J, Rounsevell M (2008) Scenarios—crystal balls for the urban fringe. PLUREL Newsletter No. 3, April 2008, Copenhagen, pp 1–2. http://www.plurel.net/images/PLUREL_Newsletter_-No3_screen.pdf. Accessed 14 Mar 2012
Sieverts T (2003) Cities without cities. An interpretation of the Zwischenstadt. Spon Press, London/New York
Sinn A, Haase D, Walde A (2008) Analysis of regional spatial planning and decision making strategies and their impact on land use in the urban fringe. PLUREL Deliverable 3.3.6
The SCATTER (2004) project—sprawling cities and transport: from evaluation to recommendations. www.casa.ucl.ac.uk/**scatter**/download/ETC_**scatter**_gayda.pdf Accessed 14 Mar 2012
Tosics I, Gesthesis (2010) Review of further instruments with indirect effect on regulation of peri-urban land uses. PLUREL Deliverable 2.2.2

Westerink J, Kiers M, Kok K, Aalber C (2010) Scenarios and enhanced strategies. Case study The Hague Region, The Netherlands. PLUREL Deliverable 3.4.1. Unpublished report
World Resources Institute (2001) Earthtrends: the environmental information portal. Washington, DC. http://earthtrends.wri.org/
Zandersen M, Termansen M, Jensen FS (2007) Evaluating approaches to predict recreation values of new forest sites. J Forest Econ 13:103–128
Zasada I, Berges R, Piorr A, Schläfke N, Werner A, Toussaint V, Müller K (2010) Modelling approach for response functions on agricultural production, ecological regulation and recreation function. PLUREL Deliverable D2.3.9–10. Unpublished report

# Author Biographies

**Carmen Aalbers** works as a senior researcher at Alterra/Wageningen University. After her thesis on land use planning in the arid tropics she pursued her career combining societal needs with applied land ecology, formulating projects and policy advices. During her professional career she worked as expert in participation, Natural Resource Management and employment promotion for the United Nations, for the Office of the International Labour Organization. Since 2001 she works at Alterra, leading national and international research on social dimensions and governance of green open space in urbanized areas. Carmen developed the framework and coordinated the research in the seven governance cases studies of PLUREL. She was a front runner in involving local stakeholders in the formulation of PLUREL research issues and products.

**Tjaša Babič** has a Master's Degree in Geographical sciences from the University of Ljubljana. Currently she is a Senior Consultant for Spatial Planning at the Municipality of Koper, where she is responsible for the process of preparing the new Spatial Plan for the municipality, which is the most fundamental document in terms of spatial development on the local level. She is also in charge for the preparation of a new strategy for spatial development of the Municipality and the Municipal Spatial Plan. Her contribution to the PLUrel project, as a stakeholder of the Municipality of Koper, included, but was not limited to, providing data for the researching teams, supervising the outcomes of the various activities and providing insight into the local specifics of spatial development.

**Georgi Bangiev** has a Bachelor's Degree in Geodesy and Geoinformation and is currently finishing his Master's Degree in Urban and Spatial Planning. He is the Head of the Department for the Environment and Spatial Planning of the Municipality of Koper and has a decisive role in decision-making about spatial development and policies that impact the development of the region. Throughout his career he has been in constant involvement with policy making and spatial planning, thus developing a deep insight in the processes that shape the spatial interactions and development on the local and broader level. As a stakeholder for the Koper case study region, he supervised all the activities within the project, regarding Koper and

K. Nilsson et al. (eds.), *Peri-urban futures: Scenarios and models for land use change in Europe*, DOI 10.1007/978-3-642-30529-0, other case studies, with the aim to set up a clear and understandable overview of spatial development issues and possibilities with the case study region and to gather examples of good practices from his peers.

**Annette Bauer** works for a bavarian regional planning authority since 2010. Previously she was part of the research team of the European PLUREL project at the Department for Computational Landscape Ecology, Helmholtz Centre for Environmental Research—UFZ in Leipzig. She focused on analysing landscape planning strategies in the Leipzig-Halle region for a comparison of international case studies within the PLUREL project.

**Annette Bauer** was awarded a diploma in Geography in 2002 by the Technische Universität München. After working as a consultant at the German Federal Office of Migration and Refugees between 2003 and 2006, she went on to complete her Masters degree in Planning Research and Theory at the University of Sheffield. Her main research interests are applied land use scenarios and regional governance.

**Simon Bell** is a chartered forester and chartered landscape architect, Professor of landscape architecture at the Estonian University of Life Sciences and Director of the OPENspace Research Centre at Edinburgh College of Art, University of Edinburgh. He specialises in larger scale landscape planning and design, including urban forests and green spaces as well as social aspects of the environment, people's perception of landscape change and the dynamics involved. He has written or contributed to several important books on landscape planning and design and his broad field of research interests are reflected in these. He was chair of Cost Action E33 on forest recreation and nature tourism and within the PLUREL project he was social science coordinator and lead scientist on quality of life and land use change.

**Regine Berges** graduated in geoecology at the University of Potsdam, Germany by specializing on environmental planning and resource management. As a researcher at the Leibniz Centre for Agricultural Landscape Research (ZALF) in Müncheberg, Germany, she has been working in European and national projects. In PLUREL, she mainly elaborated cause-effect-relationships between urban growth and its environmental, agricultural and recreational impacts for peri-urban areas in Europe. In her recent work, Regine studies the contribution of urban agriculture for a sustainable urban development in the INNSULA project.

**Marco Boeri** has a degree in Economics and statistics from the University of Pisa (Italy) and a Ph.D. in Environmental Economics from queen's University of Belfast. He has been involved in several projects as Project Manager Assistant and in two projects under the 6th EU Framework Program, EXIOPOL and PLUREL, during his Ph.D. Marco Boeri is currently researcher at Queen's University and Associate Researcher at "Fondazione ENI Enrico Mattei in Milan. He is interested in cost benefit analysis and in non-market valuation techniques applied to energy, Public Health, transport, land use, and biodiversity.

**Irene Burkhardt** is a landscape architect and urban planner in Munich, Germany. She is the owner of Irene Burkhardt Landschaftsarchitekten und Stadtplaner planning office.

For a decade she was research associate at the Chair for Landscape Ecology at Technische Universität München where she specialized on ecological model studies. She was also staff of the Department of Urban Planning and Building Regulations of the City of Munich where she focused on the ecological strategy for the redevelopment of the former airport and sustainable city development.

From 1998 to 2004 she chaired the Bavarian regional branch of the Association of German Landscape Architects. She is also a member of the Consultative Committee for Nature Protection in Bavaria as well as member of the managing board of the Bavarian section of the German Werkbund.

**Jennifer Buyck** is a french architect, junior professor of urban design at the Institute of Urbanism in Grenoble (France). She specializes in aesthetics and town planning of metropolitan areas. As shown in her Ph.D. thesis completed in 2010, the role of contemporary landscapes and nature is also one of her issues. She wrote several papers on this theme and finally published an important joint publication about potential relationships and the shifting boundaries between urban areas and agriculture: Agriculture métropolitaine / Métropole agricole (Paris, Maison des sciences de l'homme, 2012).

**Jean-Pierre Chery**, Ph.D. in Geography, Lecturer-and researcher in Spatial Analysis at AgroParisTech (French Institute of Technology for Life, Food and Environmental Sciences), and UMR TETIS, joint research laboratory of Geomatics for Agricultural, Environmental and Regional Planning Engineering Research. His main research fields are spatial analysis and spatial modelling in regional, environmental systems, spatial indicators and spatial scenarios. Within the PLUREL project, he contributed to and managed the French team involved in modules about land use relationships in Rural–urban Regions and about the governance and strategic planning scenarios, applied to the Montpellier case study.

**Davor Deranja** has a B.A. in Geography and currently works as a spatial planning consultant at the Municipality of Koper. His responsibilities include drafting and reviewing municipal spatial plans and planning spatial and environmental development on a local level. Apart from the role of stakeholder for the Koper case study, within the PLUREL project, he closely collaborated with the MOLAND team in preparing data and setting up the simulations of land use changes and with the Quality of life team in carrying out data collection through interviews and questionnaires.

**Katarina Eckerberg** is Professor in Public Administration at the Department of Political Science, Umeå University, Sweden, and was previously Deputy Director at Stockholm Environment Institute in 2006–2010. Her research comprises the field of policy analysis and implementation studies of environment and sustainable development policy from international to local level, often with a European comparative perspective, including studies of land use systems, environmental policy instruments and Local Agenda 21. Currently she is engaged in research on the local politics of nature and water protection and in interdisciplinary research on ecological restoration in policy and practice.

Her publications to date comprise 120 books and book chapters, journal articles, research reports and popular science articles, such as Baker and Eckerberg, eds, *In Pursuit of Sustainable Development: New governance practices at the sub-national level in Europe* (Routledge, London 2008), Nilsson and Eckerberg, eds, *Environmental Policy Integration in Practice: Shaping Institutions for Learning* (Earthscan, London, 2007) and Lafferty and Eckerberg, eds, *From the Earth Summit to Local Agenda 21: Working Towards Sustainable Development* (Earthscan, London, 1998).

**Christian Fertner** is assistant professor at the Department of Geosciences, Natural Resources and Planning at the University of Copenhagen. Christian's research interests include urban and regional development, growth management and spatial planning. He recently finished his Ph.D. dealing with land use change, urban–rural dynamics and planning approaches. In the near future he will work with development challenges of small towns and energy efficient cities. You can follow some of his work on landblend.wordpress.com.

**Jean Paul Gambier**, Communauté d'Agglomération de Montpellier, Direction du Foncier et de l'Aménagement Opérationnel; Montpellier.

**Miroslaw Grochowski**, geographer, lecturer at the University of Warsaw and member of the research team in the Institute of Geography and Spatial Organization, Polish Academy of Sciences. He specializes in studies on urban development, especially metropolitan areas development, including spatial planning, development management and urban renewal. He has participated in several international projects on urban development under EU framework programs and ESPON. Within the PLUREL project he was coordinator responsible for study of Warsaw Metropolitan Area.

**Dagmar Haase** is a landscape ecologist and land use modeller specialised in spatially explicit modelling of land use change and urban ecosystem services. Another subject of her current research refers to socio-environmental effects of urban land use change such as flood risk assessment and climate change impact assessment. Since 2009, Dagmar Haase holds the professorship of Landscape Ecology at the Department of Geography at the Humboldt University of Berlin and has a Lab there for Urban Ecosystem Services. Since than, she works as a guest scientist at the UFZ. Dagmar Haase was engaged in the EU IP PLUREL as module leader for the sustainability assessment modelling of land use changes at regional level. She was futher working on urban flood risk assessment within the FLOODsite EU IP as well as coordinated the Tisza River Basin Case Study in the NeWater EU IP which focuses on conceptual and system dynamics modelling and qualitative mapping to mitigate urban flood issues. Dagmar is the author of a wide range of research paperson land use change modelling and urban ecosystems. She is scientific board member of the Society for Urban Ecology (SURE) and member of the International Association for Landscape Ecology (IALE).

**He Youjun** is an associate Professor of forest ecosystem management, strategy and planning at the Chinese Academy of Forestry. He specialises in multi-purpose forest management and planning, including timber production, carbon sequestration and biodiversity etc., he is also a specialist in forest policy, strategy and climate change. He has written or contributed to several important books on environment management, forest policy and planning and his broad field of research interests are reflected in these. He was co-coordinator in the national project for development model of multi-purpose forestry and its monitoring & evaluation system funded by China Ministry of Science and Technology and within the PLUREL project he was co-coordinator for Hangzhou case in Chinese side.

**Majda Černič Istenič** is associate professor of sociology at Biotechnical Faculty, University of Ljubljana and senior research fellow at the Scientific Research Centre at the Slovenian Academy of Sciences and Arts. Her research activities extend into the domain of various sociological sub-disciplines such as sociology of family, link of sociology and demographic topics (fertility and aging) and rural sociology. In her research work she focuses on gender and intergenerational relations, social aspects of environment, different aspects of life by comparing farm, rural and urban populations, and various actors in the food chain (farmers, extension advisors, consumers, agricultural policy makers). Within the PLUREL project she took the part as a member of social science team.

**Françoise Jarrige** is an economist of public policies and innovation processes in the field of agriculture and territories. She specialises in the issues of farming and farmland in urban policies and projects. She teaches social sciences, with a focus on land and natural resources management in major education programmes at Montpellier Supagro. She contributed to several French and European research projects on periurban agriculture. Within PLUREL project, she coordinated Montpellier Agglomération's case study.

**Mario Köstl** holds a Master degree in meteorology from the Vienna University and a Diploma in technical environmental protection from the University of Technology in Vienna and the Vienna University of Life Sciences. There he specialized in remote sensing and GIS analysis, what is—together with 3D-visalisation, meteorology and climatology—still his research focus. He is working with the Austrian Institute of Technology since 15 years engaged in operative work for various GIS- and spatial analysis-related projects.

**Piotr Korcelli** is an economic and urban geographer, Professor at the Institute of Geography and Spatial Organization, P.Ac.Sc. in Warsaw. He is Ordinary Member of the Polish Academy of Sciences, and of Akademie fuer Raumforschung und Landesplanung (Hannover). His research and major publications have focused on theory of intra-urban structure, national urban systems, interurban as well as international migrations, national urban policies. He lectured at the UIniversity of Warsaw, Technische Universitaet Wien, Univeristy of Minnesota, University of Canterbury. He participated in several projects of the ESPON Programme and served as Coordinator of Poland's Spatial Development Concept, 2030.

**Elżbieta Kozubek** is a geographer, cartographer, and a GIS specialist; she is an assistant professor in the Institute of Geodesy and Cartography, collaborating with the Institute of Geography and Spatial Organization of the Polish Academy of Sciences. Her main research interests include the study of changes in land use, spatial analyses of social infrastructure, the information society and the use of models in anticipating spatial consequences of development processes. She is the co-author of several expertise opinions for the Ministry of Regional Development and the Ministry of Transport, Construction and Maritime Economy. She is a lecturer at the College of Real Estate Management in Warsaw.

**Carlo Lavalle** after graduating in Physics for the University of Bologna, and achieving the specialisation in applied Geophysics, spent the initial part of his professional career as researcher in the Italian CNR in the fields of Atmospheric Dynamics and environmental impact of global changes. He worked for the European Space Agency and in 1990 he joined the Joint Research Centre of the European Commission.

He is with the Land Management Unit of the Institute for Environment and Sustainability, head of the Urban and Regional Development sector.

**Wolfgang Loibl** is a geographer and regional planner with Master and Ph.D. degrees from the Vienna University. After some years working as a spatial planner and researcher he is working since more than 25 years with the Austrian Institute of Technology (AIT). His research is focusing on spatial analysis and modeling with emphasis on regional und urban development, on landscape transition simulation and further on climate change impact analysis at the regional and urban scale. He was head of the Regional and Infrastructure Policy Unit in AIT's Foresight and Policy Development Department for 5 years and is now again concentrating on operative research as a senior scientist, engaged in various national and international projects. For 10 years he was additionally teaching GIS-based Spatial Modelling as lecturer at the Vienna University.

**Vesna Miličić** is a postgraduate student of the Environmental protection at the University of Ljubljana. After finishing her undergraduate studies of Agronomy, she started to work on different research projects related to environmental and resource management and sustainable use of natural resources at the Agronomy department of Biotechnical faculty in Ljubljana. After one year working as a project researcher she applied for the young researcher position and went to Cranfield University where she gained an MSc in Natural Resource Management. At the moment she is finishing her Ph.D. that directly addresses the issues of spatial planning, agricultural land use change projections and agent based modelling with the focus on Koper case study. During PLUREL project she has also completed a four months research exchange at the Centre for the Study of Environmental Change and Sustainability based in the School of GeoSciences at University of Edinburgh.

**Andrej Mlakar** is an architect, graduated at the Faculty of Architecture, University of Ljubljana in 1996. In 2000 he found biro for architecture and urban planning Studio mediterana d.o.o. and he is it chairman since then. He is an expert in urban and spatial planning and has great experience working in all regions in Slovenia especially for coastal region. He is urban planner for municipalities in coastal region. His work extends into urban planning of city centres as well as rural landscape. He also works in field of architecture and worked on several buildings such as Hotel resort on Slovenian coast and residential buildings for different investors. Within the PLUREL project he took the part as a member of case study of city Koper group as a SME.

**Felix Müller** has studied Biology and Geography at the Universities of Kiel and Regensburg. His Ph.D. thesis about soil-geographical investigations on the fate of pesticides and nutrients in ecosystems was published in 1987. Since that time he has been working at the Ecology Centre of the University of Kiel. He was the scientific coordinator of the long-term R&D project "Ecosystem Research in the Bornhöved Lakes District" and has since that time participated in 15 national and international research projects. Since 2010 Felix Müller is affiliated as leader of the Department Ecosystem Management at the Institute of Resource and Nature Conservation of Kiel University.

The main recent research interests are ecosystem analysis, ecosystem modeling, ecosystem services and ecosystem theories, applications of ecosystem approaches at the landscape scale and the derivation of holistic indicator sets for the management of human-environmental systems.

**Felix Müller** has been editor or co-editor of 21 books and special issues and has published more than 130 scientific papers. He is editor-in chief of the journal "Ecological Indicators" and board member of 5 journals, e.g. "Ecological Complexity" and "Ecological Modelling". In 2010 he received the Prigogine Medal of the Wessex Insitute of Technology and the University of Siena.

**Thomas Alexander Sick Nielsen** is Senior researcher at the Technical University of Denmark, Department of Transport. His field of research include transportation, spatial behaviour, regional development, and urban form. Specific topics include preconditions for cycling, developments in regional systems dues to changing interaction patterns, health effects, commuting patterns, and effects of infrastructure. Until 2011 he worked as Senior researcher at the University of Copenhagen where he had the role of project manager of the PLUREL project.

**Stephan Pauleit**, Ph.D., is professor in strategic landscape planning and management at Munich Technical University. Prior, he held positions at Munich Technical University, Wye College, Manchester University and he was professor in landscape planning at the Centre for Forest & Landscape, University of Copenhagen.

Stephan has specialised in urban landscape planning and urban ecology where he undertakes research in strategies for urban green infrastructure, urban forestry and planting design. He developed a special interest in exploring the role of urban green infrastructure for adaptation of the urban environment to climate change. He has broad experience in international research and consultancy, including urban planning and urban greening projects in China and recently, he has been involved in a research project on vulnerability of African cities to climate change.

**Anton Perpar** is an agronomist working as a researcher at Biotechnical Faculty of the University of Ljubljana. He specialises in the field of agricultural economics, with special emphasis on rural development. His research activities are focused on various aspects of rural and agricultural development, economics of agricultural production, environment and nature resources protection as well as some aspects of spatial planning. Within the PLUREL project he was a member of Slovenian research team worked on Koper case study.

**Marina Pintar** is an agronomist and Associate professor of environmental planning at Biotechnical Faculty, University of Ljubljana. Her main field of research is integrated agricultural management, and integrated soil water management. She is a leading expert in natural resource management in agricultural and environmental policy and she has written or contributed to several important books on this field. In the PLUREL project, she was involved in Module 3 as a researcher from the Koper case study. In addition to doing her own research, she coordinateed the work of the University of Ljubljana research group involved in PLUREL on the one hand, and the activities of local stakeholders from Koper municipality and local small and medium-sized enterprise on the other.

**Annette Piorr** holds a Ph.D. degree in agricultural science from the University of Bonn. After a period of teaching and a research fellowship at Bonn University, she was managing director of the research consultancy Ecoscape dealing with sustainable land and water management issues for four years. In 1998, she joined the Leibniz-Centre for Agricultural Landscape Research (ZALF) in Müncheberg as senior scientist. There she is head of the department for 'Land Use Change and Rural Development Policies' at the Institute of Socio-economics. With her scientific expertise covering policy oriented research, multifunctional agriculture, land use change, indicator development, impact assessment and evaluation, participatory research, urban–rural relationships, urban agriculture and metropolitan food chain analysis, she participates in and coordinates several national and EU projects and is active as an evaluator for the European Commission and as a reviewer for several scientific journals.

**Joe Ravetz** is Co-Director of the Centre for Urban & Regional Ecology at the University of Manchester. He is a researcher and designer of sustainable cities and regions, with a wide range of interests in environment policy, spatial development, new economics and governance, innovation and futures studies, systems and complexity science. With a background as an architect, planner and development manager, he is also a creative graphic facilitator and foresight trainer. His main publications include '*City-Region 2020*', '*Environment and City*' (co-authored), and the forthcoming '*Urban 3.0: pathways to synergistic intelligence*' (Routledge). He is a board member of Foresight Journal; coordinator of the Greater Manchester Policy Exchange; Principal at SAMI Consulting; and delivers training courses, seminars, keynotes and review services around Europe.

**Sophie Rickebusch**, Ph.D., is a lecturer Lecturer, Environmental Systems Analysis group, Wageningen University since 2011. She has been a Research Fellow at the Centre for the study of Environmental Change and Sustainability (CECS),

University of Edinburgh from 2008 to 2011. She is interested in various aspects of ecological modelling, such as population dynamics (Ph.D. thesis on tree-line forest dynamics), species distribution and also land-use change. An ecologist by training, she has since then focussed her research on forests and landscape ecology. Sophie's interest in modelling is supported by a background in statistics and an interest in programming and GIS. Sophie was involved in the RUG modelling in PLUREL.

**Dietmar Röhl** was awarded a doctoral degree in Natural Science (Geography) by the Freie Universität Berlin in 1987. He previously worked in market and property research (Prisma Institut Hamburg and GESA Institut Hamburg) and for the Department of Applied Geography, Freie Universität Berlin.

**Mark Rounsevell** is Professor of Rural Economy, Head of the Institute of Geography & the Lived Environment and Director of the Global Environment & Society Academy at the University of Edinburgh. His research focuses on the effects of environmental change on rural and urban landscapes with an emphasis on land-use change modelling and socio-ecological systems analysis. These methods are used to explore the effect of alternative futures of climate and other environmental change drivers on society and land use systems. He was a lead author to the 2nd, 3rd, 4th and 5th Assessment Reports of Working Group II of the IPCC and is a member of the Expert Panel of the UK National Ecosystem Assessment. He has authored or co-authored over 100 international, peer-reviewed publications.

**Alessandro Sarretta** has a Degree in Environmental Sciences from Ca' Foscari University, Venice, a Masters in "Geographical Information Systems and Remote Sensing" from IUAV University of Venice, and a Ph.D. in Environmental Sciences. He has conducted research at the Institute of Marine Sciences (Italian National Research Centre) mainly in sedimentological and morphological studies in the Lagoon of Venice. His current research is on the interoperability of software systems supporting European environmental policies, focusing on application of INSPIRE Implementing Rules to geospatial systems and services developed by JRC to support such policies.

**Nina Schwarz** is an environmental scientist working in urban research. Since 2007, she is working at the Helmholtz-Centre for Environmental Research—UFZ, in Leipzig, Germany. She is the head of the working group on urban form in the department of Computational Landscape Ecology of the UFZ. Her main interests are centered on urban form, thus the spatial configuration of the urban landscape. She is interested in the underlying processes that drive urban form, i.e. urban land use change, as well as in the impacts of urban form on urban ecosystem services. Regarding the latter, she focuses on local climate regulation and the urban heat island. Nina Schwarz was involved as a scientist in the project, working on the case study-related simulations. She studied Environmental Sciences at the University of Lüneburg and afterwards worked at the Centre for Environmental Systems Research at the University of Kassel, Germany, where she received her Ph.D. in 2007.

**Dr. Tomasz Sławiński** is an architect, urban and regional planner specialized in plans for metropolitan areas. He is a deputy director of Mazovian Office for Regional Planning in Warsaw, responsible for development strategy and regional spatial plan for the Mazovia Voivodeship. Tomasz Slawinski is an author of numerous spatial plans and studies, including Warsaw Metropolitan Area study plan and Wilanów Zachodni (new district of Warsaw) spatial plan and several theoretical articles and papers, as well. He is a lecturer in post-diploma spatial planning studies in the Faculty of Architecture of Warsaw Technical University. Dr. Sławiński is a member of Regional Chamber of Town Planners in Warsaw, Society of Polish Architects - SARP, Society of Polish Town Planners—TUP (President of Warsaw Unit till 2012), Architecture and Urbanism Commission in City of Warsaw, Council of Architecture and City Development in Warsaw, Council of Historical Heritage Protection in Warsaw.

**Martin Spiekermann** is a landscape planner and senior project leader at Irene Burkhardt Landschaftsarchitekten und Stadtplaner in Munich, Germany. He is also a lecturer at Hochschule Weihenstephan-Triesdorf and was shortlisted as professor for landscape technology at the same university. He has specialised in ecological, environmental and urban land use planning.

**Dr. Klaus Steinnocher** is a senior scientist at the AIT—Austrian Institute of Technology in Vienna, Austria. His background is in geodetic science and he holds a doctoral degree in remote sensing from the Vienna University of Technology. Dr. Steinnocher has more than 20 years of experience in remote sensing, spatial data analysis and spatio-temporal modeling for different fields of application, including regional and urban planning, environmental analyses, vulnerability assessment, risk and crisis management. His recent work focuses on the integration of socio-economic information and geospatial data for population modeling and disaggregation of multi-level information in both spatial and temporal domains.

**Dr. Iván Tosics** is one of the principals of Metropolitan Research Institute (MRI) in Budapest and holds a doctoral degree in mathematics and sociology. He is research fellow with long experience in urban sociology, strategic development, housing policy and EU regional policy issues. Dr. Tosics is Vice-Chair of the European Network for Housing Research (ENHR) and Coordination Committee member of the European Urban Research Association (EURA).

**Andrej Udovč** is Associate Professor of Agricultural economics and Rural development at Biotechnical Faculty, University of Ljubljana. His main research interests pertain to Farm management, Coexistence of rural economies and high-value natural landscapes and Common pool resources management. He has written or contributed to several articles and books on sustainable nature resource management and rural development. He is member of International Committee for the Research of the Alps and The Center for Trans-disciplinary Study of Institutions, Evolutions and Policies. Within the PLUREL project he was member of the Slovenian team dealing with the agro-economical and institutional questions of the Koper case study.

**Hedwig van Delden** is the scientific and managing director of the Research Institute for Knowledge Systems (RIKS) in Maastricht, the Netherlands and an adjunct associate professor of the University of Adelaide, Australia. Her main fields of research include land use modelling (using approaches such as cellular automata, activity-based modelling and agent-based modelling), model integration (integrating socio-economic and biophysical processes, integrating models with different resolutions and/or built on different paradigms, ...), bridging the science-policy gap (mainly through the design, development, implementation and use of Decision Support Systems) and scenario development (participatory and in-house). She has led the EC FP 7 research project LUMOCAP and was the product leader for the development of the Integrated Assessment Model in the FP6 DeSurvey project. Her works has been published in several books and journals and presented at several conferences. She is currently acting as an expert evaluator for the European Green Capital Award, organised by the European Commission.

**Ms Pam Warhurst** is currently national Chair of the Forestry Commission, a Fellow of the Royal Society of Arts and Manufacturing, and Chair of a social enterprise in the North of England. She also Chairs Pennine Prospects, a regeneration company for the South Pennines. Her other previous posts include Leader of Calderdale Council and Chair of Calderdale NHS Trust, and a member of an RDA. In 2005 Pam was awarded the CBE in recognition of her services to the environment. Pam was Deputy Chair of the Countryside Agency before taking up the appointment of Board Member for Natural England.

**Armin Werner** (m) started his career as scientist of crop production (agronomy) at the University of Bonn. He is head of the Institute of Land Use Systems at the ZALF in Müncheberg, Germany. His work includes the development of new crop production systems for agricultural land use (e.g. precision farming, high efficient crop production, GMO) and studying the economic and environmental impacts of new land use technologies and land use changes. He is an expert on integrating new crop production technologies into concepts for sustainable land use systems as well as developing rule-based decision support systems in crop production and in land use planning. With his group he prepares simulation models for agricultural crops to be used as tools for decision making in crop production or land use planning. In collaboration with scientists from socio-economy and from landscape system analysis he promotes the development of regional models that are applied to assess the ecologic and economic impacts caused by land use changes due to altered political, economic or technologic conditions. Within the PLUREL project he was Land science coordinator on methods and analysis of use relationships in rural–urban regions.

**Piotr A. Werner**, geographer, specialist in geoinformation, Professor of the University of Warsaw. Member of Polish Geographical Society and Polish Association for Spatial Information. Research interests cover application of geographic information systems, modeling of complex spatial systems and geosimulations, including the field of development of information society, urban and regional analysis, land cover and changes of land use and also the potential and used sources of energy. He has participated in national and international interdisciplinary research: Conditions and Measures of Information Society Development in Poland, Information Society Development Level in Carpathian Region, National and International Social Infrastructure in Poland—state and prognosis, Method of research and prognosis of land use changes using map algebra and cellular automata.

**Judith Westerink** is researcher at Alterra, part of Wageningen University and Research Centre. She studies governance and planning of peri-urban areas with special interest in stimulating farming systems and nested markets that can accommodate biodiversity and can contribute to attractive and multifunctional landscapes. From that perspective, she advises the Netherlands Ministry of Economic Affairs, Agriculture and Innovation on Common Agricultural Policy and agri-environment, but she also studies how groups of farmers and citizens influence the development of their own area and take part in policy making. She is project leader of the long-term and trans-disciplinary Farming for Nature project. She is currently involved in the European research project FOODMETRES about food planning in metropolitan regions. In PLUREL, she co-ordinated the case study of The Hague Region.

**Dr. Xin Yi** is an urban planner and architect. He holds a position as lecturer in the Department of Urban Planning at the School of Architecture, Southeast University, Nanjing, and he is Post-doc Researcher at the Chair for Strategic Landscape Planning and Management, Technical University of Munich. In 2011, he completed his dissertation about Mega-Events in Chinese and European cities with Olympic Games as an example. Research is focused on integrated urban planning strategies and urban design theory, which are based on a systematic institutional, political and cultural analysis. He has worked as senior project manager in a German planning office.

**Fei Yang** is a Ph.D student at University of Florida, majoring in Urban and Regional Planning. He has served as the Chair of Student Representative of American Planning Association, Florida chapter, and also as the board of directors of International Association for China Planning. His research interests include land use planning and policy, transportation planning, and the application of Geographic Information Systems in planning. He is currently working on several research projects on climate change issues, with a special attention paid to sea level rise adaptive planning. Within the PLUREL project, he focused on the Hangzhou case study.

**Jianjun Yang** is a professor of urban planning and design at the Department of City and Regional Planning, Zhejiang University, China, and he is a national registered urban planner. From 1998 to 2009 he served as the head of the Department. He has specialized in strategies for urbanization and urban development, city and regional spatial structure planning. He has undertaken research for the National Social Science Foundation of China, the Natural Science Foundation of Zhejiang province and a number of research and planning projects, e.g. to develop urban system plans, urban agglomeration plans, and urban master plans.

**Dr. Ingo Zasada** is post-doctoral scientist at the Leibniz Centre for Agricultural Landscape Research (ZALF) in Müncheberg, Germany. He graduated in urban and regional planning and holds a Ph.D. in agricultural science from the Technical University Munich. In his dissertation thesis he studied multifunctional development perspectives of peri-urban agriculture. Involved in several European research projects, Ingo addresses the topics of rural development policy, urban–rural relationships, land-use and landscape change. In PLUREL, he contributed to the environmental and landscape modelling for impact assessment of urban growth and to the tool development for policy support.

**Affonso Zuin** is senior lecturer in landscape design and ornamental horticulture at at the Federal University of Viçosa, Brazil. He was involved in the PLUREL project as research associate at OPENspace Research Centre at Edinburgh College of Art. With his background in agronomy/horticulture and landscape architecture, his special interest is in research tools for people-environment studies, access to the outdoors, and organic production techniques for tropical flower crops. He has been involved in design projects for public, institutional and private spaces in Brazil and India.

**Marko Zupan** is an agronomist and senior lecturer of soil science at Biotechnical Faculty of the University of Ljubljana. His main research activities are soil identification and mapping, soil/land degradation and contamination, and land evaluation. He is responsible for soil information data base on Centre for Soil and Environmental Science (GIS), including soil maps of Slovenia in scale 1:25.000 and 1:250.000, layer of soil pollution data, and other soil properties. Within the PLUREL project he was a member of Slovenian research team preparing data for modelling and scenarios of Koper case study area.

# Index



K. Nilsson et al. (eds.), *Peri-urban futures: Scenarios and models for land use change in Europe*, DOI 10.1007/978-3-642-30529-0,

**H**

**N**



**P**

**Q**



**R**

MIX
Papier aus verantwortungsvollen Quellen
Paper from responsible sources
FSC® C105338

If you have any concerns about our products,
you can contact us on
**ProductSafety@springernature.com**

In case Publisher is established outside the EU,
the EU authorized representative is:
**Springer Nature Customer Service Center GmbH**
**Europaplatz 3, 69115 Heidelberg, Germany**

Printed by Libri Plureos GmbH
in Hamburg, Germany